Wildlife Research Monographs

Volume 5

The book series on "Wildlife Research Monographs" focuses on all aspects of wildlife biology to fill a vital niche at the intersection of several disciplines. The aim of the series is to summarize knowledge about interaction between wildlife species, their habitats, as well as human activities. Main areas are: wildlife ecology, diseases, toxicology, conservation, management, and the sustainable use of natural resources. The series addresses graduate students, university teachers, scientists and professionals.

More information about this series at http://www.springer.com/series/8826

Vincent Bretagnolle • Juan Traba •
Manuel B. Morales
Editors

Little Bustard: Ecology and Conservation

Springer

Editors
Vincent Bretagnolle
Centre d'Etudes Biologiques de Chizé,
CNRS & La Rochelle Université
Beauvoir sur Niort, France

Juan Traba
Department of Ecology
Universidad Autónoma de Madrid
Madrid, Spain

Manuel B. Morales
Department of Ecology and Research
Centre for Biodiversity and Global Change
Universidad Autónoma de Madrid
Madrid, Spain

ISSN 2366-8733 ISSN 2366-8741 (electronic)
Wildlife Research Monographs
ISBN 978-3-030-84901-6 ISBN 978-3-030-84902-3 (eBook)
https://doi.org/10.1007/978-3-030-84902-3

This Springer imprint is published by the registered company Springer Nature Switzerland AG.
The registered company address is: Gewerbestrasse 11, 6330 Cham, Switzerland

Author of the picture: Dr. Vincent Bretagnolle

Foreword

Paul F. Donald

Few bird families illustrate the world's conservation problems better than the bustards. Of the 26 species recognized by BirdLife International, more than half are globally threatened or near threatened with extinction. They owe this parlous conservation status to the full panoply of threats facing wildlife today: habitat loss, degradation and fragmentation, hunting, infrastructure (particularly power lines), disturbance, agricultural intensification, road traffic, predation and climate change have all played a part in driving many species to the edge of extinction. Bustards have few natural defences against predators and therefore rely on flat, open undisturbed expanses of grassland or other low habitats that allow them to see, and keep, danger at a distance. Unfortunately, these are exactly the sorts of habitats that are most easily converted to other purposes, and many of the natural grasslands of Europe and Asia have been ploughed up for arable cultivation. Some African species remain relatively common and thrive in large national parks and undisturbed desert and semi-desert areas, but all the European and Asian species are now listed as globally threatened or near threatened.

Despite their often large size (the family contains the world's heaviest flying birds), an evolutionary trajectory that has placed a great deal of emphasis on crypsis means that bustards are often hard to observe and to study, and our state of knowledge for many species is poor. It has only been in the last few years, for example, that satellite tracking has revealed that, outside the breeding season, Bengal floricans leave the few protected areas they still nest in and move considerable distances into quite different habitats. The little bustard, being as its name suggests one of the smaller members of the family, is particularly hard to study, but this has not prevented a great deal of research effort being invested in the species, particularly in the European parts of its range. Here, its population has collapsed due to changes in agriculture, and little bustard has become a model species in research assessing the environmental impacts of farming, as the skylark has become elsewhere in Europe. It is therefore both a species of high conservation concern in its own right and an important indicator of the health of the wider landscape. The last two decades have seen a surge in research work on this species, particularly in Spain, Portugal, France and Italy, where it is now restricted to an isolated and threatened population on Sardinia. The recent loss of the species from mainland Italy marks the most recent extinction in a long-term decline that has seen the little bustard being lost as a breeding bird from Germany, Greece, Poland, Slovakia and a number of other European countries. It is no wonder that the little bustard has attracted so much attention from the continent's ornithologists. The results of all their work have thus far been published largely in scientific articles, spread across many national and international journals. What has been lacking, until now, is a comprehensive synthesis of this great body of research.

The western European population is extremely important, but the bulk of the little bustard's global population lies far to the east, especially in the vastness of the Pontic-Caspian steppes. These cover an area of nearly a million square kilometres between eastern Europe and western China. Here, the species is even more difficult to study; not only are these areas remote and often inaccessible but many populations are migratory, adding a further complication to assessing their conservation status and exposing them to many more threats. Massive changes in steppe management since the collapse of the Soviet Union have led to both gains and losses for little bustards in this region, but much of this information has until now been buried in scientific papers (including in the Russian journal *Strepet*, the local name for little bustard).

This remarkable publication, 6 years in the making, represents a landmark in the study of little bustards, for it collects and summarises this mass of disparate information between the covers of a single book, and then uses this synthesis to identify research and conservation priorities to guide future work. The book's chapters are authored by the world's leading authorities on the species and between them they cover a very wide range of subjects, from sexual dimorphism to sex ratios, from phylogeography to phenology, and from national populations to national action plans. What makes the book even more authoritative is that all the chapters have been independently reviewed by other experts in the ecology and conservation of farmland and steppe birds, thus bringing in an even wider range of expertise. This book will serve as the primary reference on the little bustard for many years to come, and I am delighted to have been asked to introduce it.

Contents

Why a Monograph on Little Bustard?

Juan Traba, Manuel B. Morales, and Vincent Bretagnolle

> In its full nuptial plumage the male of this species is to my mind the most beautiful of game-birds found in Europe. It is extremely abundant on the low undulating hills and grass-grown plains of south-west Andalucia, but owing to its peculiar habits of flight is very rarely shot by the sportsman

This is how the British soldier, writer and naturalist Willoughby Verner began his chapter on the little bustard *Tetrax tetrax* in his memoir "My Life among the Wild Birds of Spain", published in 1909. Verner summarizes in a short paragraph all the little bustards' traits that have attracted the attention of ornithologists and naturalists for years, both professional and amateur: their beauty, despite being a bird of flat and treeless landscapes, their surprising behaviour, and their once incredible abundance. We would dare to say that these are, in short, the main motivations to write and edit this book. As ornithologists, we are ceaselessly amazed by the male's clownlike postures and the astonishing camouflage of the females. As ecologists, we have always been fascinated by the species' morphology and spectacular displays, the products of the evolutionary forces at play in its peculiar mating system. As conservation biologists, we see in its current population crash a perfect though pitiful example of the "abundant species paradigm": it is not the scarcity but the speed of decline of a species that should warn us, as mere numbers can be misleading when considering species still abundant in at least some areas, as they can suggest

J. Traba (✉) · M. B. Morales
Terrestrial Ecology Group, Departamento de Ecología, Universidad Autónoma de Madrid (TEG-UAM), Madrid, Spain

Centro de Investigación en Biodiversidad y Cambio Global (CIBC-UAM), Universidad Autónoma de Madrid, Madrid, Spain
e-mail: juan.traba@uam.es; manuel.morales@uam.es

V. Bretagnolle
Centre d'Etudes Biologiques de Chizé, CNRS & La Rochelle Université, Beauvoir sur Niort, France
e-mail: vincent.bretagnolle@cebc.cnrs.fr

V. Bretagnolle et al. (eds.), *Little Bustard: Ecology and Conservation*, Wildlife Research Monographs 5, https://doi.org/10.1007/978-3-030-84902-3_1

only a minor conservation emergency. There is perhaps, a fourth motivation. For us editors, and for sure for all contributors to this monograph, the little bustard is an icon of the vanishing traditional agricultural landscapes of Europe and other parts of the world, once biologically so rich and diverse, and currently so sadly replaced by industrial farmland. This book is a plea for urgent action to recover a biologically friendly agriculture compatible with the existence of the little bustard and so many other wonderful farmland and steppe species.

The little bustard was so common in France and Iberia until the late 1960s and 1980s, respectively, that nobody would have ever believed that the situation could become so critical in only about 15–20 years. However, this is what happened: in Italy, France, Spain, Portugal and Morocco, at slightly different times but during periods that never exceeded two decades, the little bustard populations of these countries just crashed, and in some cases vanished. The red flag was first raised in France, with warning signals by ornithologists in the late 1980s, when the little bustard became scarce or even extinct in former strongholds in the country. This and other bustard species were already the focus of a small network of scientists and conservationists concerned about the alarming status shown by bustards worldwide. They edited a dedicated journal, *Bustard Studies*, published irregularly when funds were available, which contributed to an increased awareness of the fragile conservation status of several bustard species and helped creating an IUCN specialist group. In 1990, the situation was seriously alarming for the little bustard in France, although admittedly, at that time, trend data or long-term surveys were still not available. Thanks to pioneering conservationists a first LIFE programme was set in 1996 in France and dedicated solely to the conservation of the little bustard, a project that included a significant part of research on basic breeding biology, virtually unknown at that time.

The dramatic situation of the little bustard was however not restricted to France. In 1997, a technical international workshop was held in Trujillo, Spain, to discuss the global status of the little bustard in Europe, and to establish conservation priorities. In 2001, as an output of this workshop, a preliminary draft of the European Species Action Plan was elaborated (De Juana 2001). This document highlighted the threatened conservation status of the species in Europe, even though at that time the Iberian population was considered healthy and abundant, and encouraged public administrations to take a series of urgent measures to avoid local and regional extinctions (De Juana 2001). Such actions targeted little bustard habitat, considering the Common Agricultural Policy (CAP) as a potential key tool to promote environmentally friendly practices as well as, at the same time, banning harmful subsidies such as those for afforestation or sunflower crops in steppe or pseudosteppe areas in Spain (De Juana 2001).

In 2010, the European Species Action Plan (SAP) was finally officially released (Íñigo and Barov 2010), trying to put forward specific actions and measures in order to reverse the now dramatic trend undergone by the species during the previous 20 years in most of its range. In 2011, a rapid assessment of the implementation of such measures revealed that almost none of the targets of the draft SAP from 2001 had been achieved, and recommended revising the conservation status category of the species, as well improving implementation of measures throughout the whole range of the species (Barov and Derhé 2011).

Twenty-three years after the 1997 workshop, the situation of the little bustard in Europe has worsened, mainly as a result of continuing destruction of favourable habitats (particularly after the suppression of CAP mandatory fallows after 2008, Traba and Morales 2019), and increasing agricultural intensification at field and landscape scales (to be described in this volume). Results of the second Spanish census of the species, carried out during 2016, revealed a dramatic decline in its main stronghold in Western Europe, with a reduction of 50% in just 10 years (García de la Morena et al. 2018; see also chapter "The little bustard around the world: distribution, global conservation status, threats and population trends"). Wintering censuses showed very similar trends (García de la Morena et al. 2018), as well as an extremely worrying bias in sex-ratio. Indeed, all factors already mentioned in the first draft of the SAP (De Juana 2001) threatening little bustard populations are still active (Morales et al. 2015). Main measures aimed at improving the conservation status of the little bustard were related to land-use policy and modifications of the CAP (Barov and Derhé 2011). However, as necessary as they are, their implementation in a multinational context has proved difficult, due both to the complexity of European legislation and to national particularities. As a consequence, land management measures addressing little bustard conservation have been of limited success, especially outside protected areas. And currently all countries in the Western Palearctic have seen either sharp declines or extinction of the little bustard over the last 75 years, some of them quite recent, like extinctions in north Africa and continental Italy (Morales and Bretagnolle 2021). However, some successful conservation projects were implemented during that period, which provides some hope.

When a species is declining rapidly, there is in general an increase in research effort, both on the causes of decline and on basic breeding biology (to better target conservation effort). The little bustard is no exception. Almost nothing was known of its breeding biology, either in France or in Spain, in the 1980s. Then, following some pioneering studies (e.g. Schulz 1985; De Juana and Martínez 1996; Salamolard and Moreau 1999), the interest in the species has significantly grown, as shown by the papers published in scientific journals (Fig. 1).

Beside conservation issues, the little bustard is an interesting study model for its rather unusual mating system, its suitability to study aspects of sexual selection given its features intermediate between classical leks and resource-based polygyny, and its rather complex habitat requirements and migration patterns (Collar et al. 2016). The biological interest of the little bustard has stimulated a good deal of basic and applied research, as summarized below and displayed throughout this book. The interest is regionally biased, however, as studies carried out in Spain account for nearly 44% of the total number of published papers, France 21% and Portugal 20% (Table 1). This geographic bias is even higher when considering that 89% of the literature comes from the western range of the species, while only 11% comes from countries in the eastern range, which highlights the knowledge gap regarding eastern populations and the need to fill it. In a literature survey aimed at identifying priorities

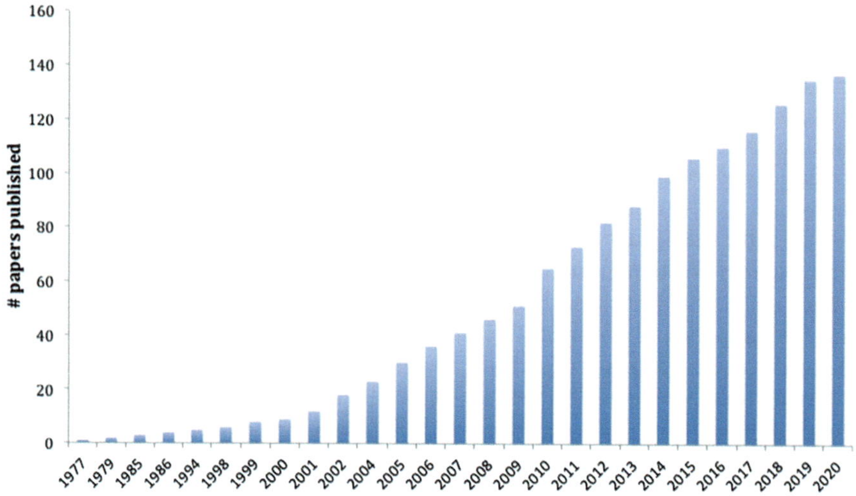

Fig. 1 Cumulative number of published papers per year (1977–2019) about little bustard *Tetrax tetrax* in scientific journals until 2019. Source search: Web Of Science (WOS) with keyword: "little bustard" (updated April 2020)

Table 1 Number of papers published per country between 1978 and 2019

	N studies	%	Distribution range
Azerbaijan	1	0.7	East
Europe	2	1.5	West
France	29	21.2	West
Iran	2	1.5	East
Italy	2	1.5	West
Morocco	1	0.7	West
Portugal	28	20.4	West
Romania	1	0.7	East
Russia	11	8.0	East
Spain	60	43.8	West

Same source as Fig. 1

in conservation-oriented research about steppe birds in Spain, Morales and Traba (2016) considered that little bustard had received a level of scientific attention consistent with its conservation status as Vulnerable in Spain. Research on little bustard has covered all major topics, including population sizes, habitat selection, habitat management, agri-enviromental schemes associated with the CAP and its effects on the species, movements, food requirements, breeding biology and behaviour. In addition, a rather high number of PhD theses have been defended in recent years (Table 2). As a consequence of this collective effort, our knowledge about the basics of the species and its requirements is high enough to propose and implement science-based conservation measures.

Table 2 Number and dates of PhD thesis dedicated to little bustard

Country	Author	Year
France	Frédéric Jiguet	2001
France	Axel Wolf	2001
France	Alex Villers	2010
France	Pierrick Devoucoux	2014
Spain	Paula Delgado	2009
Spain	Anna Ponjoan	2012
Spain	Rocío Tarjuelo	2015
Spain	Eladio L. García de La Morena	2016
Spain	Francesc Cuscó	2019
Portugal (though defended in Germany)	Holger Schulz	1985
Portugal	Joao Paulo Silva	2010
Portugal (though defended in Spain)	Nuno Faria	2015

It was therefore time to synthesize all this breadth of information. The idea of writing this book first came in 2006, when an international workshop was organized in Chizé at the end of the second French LIFE project. Then, Vincent Bretagnolle and Manuel B. Morales thought about writing a monograph on Western Palearctic bustards. However, the idea did not develop, but was never entirely dropped. So, when Beatriz Arroyo and Jesús T. García, as editors of the Wildlife Research Monographs Series, suggested that we could perhaps think of such a book, but focused on little bustard, a very rapid decision was made between all of us: go ahead. It took however 2 further years to gather all authors, write and review the 12 chapters forming the core of this volume.

This book collects, reports and updates all the available information on the species, putting everything together in a comprehensive and compact volume. It starts with a necessary full review of the Otididae family, including origins, phylogeny and systematics (chapter "The little bustard and its family: an overview of relationships") and continues with two general chapters about life history traits (chapter "Natural history of the little bustard: morphology, biometry, diet, sexual dimorphism, and social and breeding behaviour") and the most updated data about distribution, global conservation status, threats and population trends (chapter "The little bustard around the world: distribution, global conservation status, threats and population trends"). Chapter "Breeding biology and demographic traits: population parameters, reproduction and survival" deepens our knowledge of breeding biology, providing recent and novel results about nesting phenology, clutch size, mating system, sex-ratio, productivity and survival rates. Some of these demographic parameters, which are key for population viability, seem to be severely skewed (sex-ratio) or show poor values (productivity), critically affecting recruitment and long-term persistence (see also chapter "Little bustard population dynamics"). The low availability and quality of preferred habitats is behind this poor demographic performance, as the little bustard is a specialist of natural and agrarian steppes and grasslands. Chapter "Habitat selection and space use" makes a thorough revision of habitat selection at large and local scales. Post-breeding movements and migration, which show differences among individuals and populations, are analysed in chapter "Migration, movements and non-breeding ecology". All other aspects relating to behavioural ecology, such as sexual display, mate choice, territoriality and personality traits, are included in chapter "Behavioural ecology of the little bustard: sexual selection and mating systems". Interspecific relationships, including predation, competition with coexisting birds and the effects of parasites, are treated in chapter "Interspecific relationships: predation, competition or coexistence, parasites". Chapter "Captive breeding, handling and care, and the impact of releases on wild populations" evaluates the experience of captive breeding and population reinforcement carried out in France. This chapter also provides useful information about individual handling (capture and tagging). Chapter "Little bustard population dynamics" aims to evaluate population viability, highlighting the effect of key demographic parameters (see above) as well the effects of density dependence, and to explore the causes of adult and chick mortality and dispersal (natal and adult). The relationships of the species with humans are analysed in the next two chapters: chapter "Threats affecting little bustards: human impacts" is centred on threats, reviewing the general causes of the species' decline, as well as recent results about anthropogenic disturbance and stress responses. Chapter "Little bustard and humans: conservation and management" makes a thorough review of conservation initiatives specifically addressing the species, including protected areas, agri-environmental schemes, action plans and LIFE projects. It provides some insights for the European CAP regulations after 2020 to improve the conservation status of the little bustard: mainly, agrarian extensification and the maintenance of fallows, set-asides and grazed pastures, which means a drastic reduction in the application of agrochemicals. Chapter "Conclusions and perspective" summarizes the main conclusions of the book and provides a general prospectus of the species' future.

At this very moment, in winter 2021, there is no doubt about the critical status of the species and the urgent need to undertake major conservation measures. At the time of writing these lines, the little bustard has been included in Appendix I of the Convention on the Conservation of Migratory Species of Wild Animals (CMS), which requires parties to the convention to provide it with the strictest protection. In Spain, an academic initiative has drafted the National Conservation Strategy of the species, in the hope of facilitating public administrations to evaluate periodically the status and viability of the population, its distribution changes, habitat extent and quality and risk factors, and successfully requested the re-listing of the species as "Endangered" in Spain. So far, too few national and international initiatives have been implemented, perhaps with the exception of some LIFE projects in France and Portugal. We hope this book, by providing the most up-to-date review of knowledge of the biology, threats and conservation initiatives for the little bustard all over the world, will boost new projects and management programmes. The decline of little bustard is an allegory of the dramatic decay of natural and agrarian steppes in Europe, and probably in the rest of the world. Little bustard is a recipient not only of the usual threats to agricultural systems, such as agricultural intensification, but also of new ones, such as the expansion of large solar photovoltaic and wind energy installations. The little bustard is an ecological umbrella species; but we confidently feel it is also a conservation umbrella species, representing many others showing the same or even worse trends, perhaps even better than other famous species (Tarjuelo et al. 2014), as its specific habitat requirements match those of other threatened steppe birds such as pin-tailed and black-bellied sandgrouse (*Pterocles alchata* and *P. orientalis*), Eurasian stone-curlew (*Burhinus oedicnemus*), great bustard (*Otis tarda*), Montagu's harrier (*Circus pygargus*) and several passerines (Morales and Bretagnolle 2021). The conservation of the little bustard means the conservation of the steppe bird community and their habitats. It has to be now.

To end, we the editors wish to thank the numerous researchers, from many countries and institutions, who have contributed selflessly to this book. Similarly, every chapter has been revised by external reviewers, to which we want to extend our thanks. Nigel Collar, besides being the main author of chapter "The little bustard and its family: an overview of relationships", made a thorough English revision of the book, and made many important comments. Nigel has also been keen on bustard conservation worldwide, a leader in bird conservation and BirdLife International, and has inspired all of us for the many years we have spent in bustard conservation. We owe him a particular thank you.

References

Barov B, Derhé M (2011) Review of the implementation of species action plans of threatened birds in the European Union (2004–2010). BirdLife International

Collar NJ, Garcia EFJ, De Juana E (2016) Little bustard (*Tetrax tetrax*). In: del Hoyo J, Elliott A, Sargatal J, Christie DA, De Juana E (eds) Handbook of the birds of the world alive. Lynx Edicions, Barcelona

de Juana E (2001) European species action plan. Little Bustard *Tetrax tetrax*: draft action plan. BirdLife International, Cambridge

de Juana E, Martínez C (1996) Distribution and conservation status of the little bustard *Tetrax tetrax* in the Iberian Peninsula. Ardeola 43:157–167

García de la Morena EL, Bota G, Mañosa S, Morales MB (2018) El sisón común en España. II Censo Nacional (2016). SEO/BirdLife, Madrid

Iñigo A, Barov B (2010) Action plan for the Little Bustard Tetrax tetrax in the European Union. SEO/BirdLife and BirdLife International for the European Commission

Morales MB, Bretagnolle V (2021) An update on the conservation status of the Little Bustard Tetrax tetrax: global and local population estimates, trends, and threats. Bird Conserv Int:1–23. https://doi.org/10.1017/S0959270921000423

Morales MB, Traba J (2016) Prioritizing research in steppe bird conservation: a literature survey. Ardeola 63:5–18

Morales MB, Traba J, Arroyo B (2015) El declive del sisón en el centro de España. Quercus 356: 36–43

Salamolard M, Moreau C (1999) Habitat selection by Little Bustard *Tetrax tetrax* in a cultivated area of France. Bird Study 46:25–33

Schulz H (1985) A review of the world status and breeding distribution of the Little Bustard. Bustard Stud 2:131–151

Tarjuelo R, Morales MB, Traba J, Delgado MP (2014) Are species coexistence areas a good option for conservation management? Applications from fine scale modeling in two steppe birds. PLoS One 9:e87847

Traba J, Morales MB (2019) The decline of farmland birds in Spain is strongly associated to the loss of fallow-land. Sci Rep 9:9473

The Little Bustard and Its Family: An Overview of Relationships

Nigel J. Collar and Manuel B. Morales

The Bustard Family: General Traits

The little bustard belongs to the Otididae, a distinctive Old World family of non-passerine birds known in English as bustards, floricans and korhaans, and comprising (by the most recent taxonomic arrangement) 26 species in 12 genera (del Hoyo and Collar 2014; see Table 1 for some of their basic morphological, biological and ecological traits). The earliest presumed bustard fossil dates from the Middle Eocene (48–38 million years ago) (Olson 1985), although now-discredited DNA–DNA hybridisation techniques suggested the family arose as many as 77 million years ago (Sibley and Ahlquist 1985).

Members of the bustard family share a range of features which, in combination, are fully diagnostic, involving moderate to very large body sizes, rather long necks and legs, relatively short bill (never more than the length of the head), powder-down and photoreactive porphyrins at the base of feathers, no uropygial gland, no hind toe (hence no ability to perch in trees) and nidifugous young. They occupy habitats from dry open woodland and high scrub through savanna and steppe to semi-desert and the fringes of desert, based mainly in Africa: the genera *Lissotis* (two species), *Neotis* (four species), *Lophotis* (three species), *Heterotetrax* (three species), *Afrotis* (two species) and *Eupodotis* (two species) are confined to the Afrotropical region, while

N. J. Collar (✉)
BirdLife International, Cambridge, UK

Bird Group, Natural History Museum, Tring, UK
e-mail: Nigel.Collar@birdlife.org

M. B. Morales
Terrestrial Ecology Group, Departamento de Ecología and Centro de Investigación en Biodiversidad y Cambio Global (CIBC-UAM), Universidad Autónoma de Madrid, Madrid, Spain
e-mail: manuel.morales@uam.es

V. Bretagnolle et al. (eds.), *Little Bustard: Ecology and Conservation*, Wildlife Research Monographs 5, https://doi.org/10.1007/978-3-030-84902-3_2

Table 1 Some basic morphological, reproductive and ecological traits of bustard species

Common name	Scientific name	Authority	Weight (g) males/females	Height (cm) males/ females	Clutch size	Distribution	Main habitat	Diet	IUCN Red List category
Little bustard	*Tetrax tetrax*	Linnaeus, 1758	761–1012/ 680–945	43/43	2–6	Eurasia and N Africa	Steppe, pasture-land, cereal farmland	Mainly herbivo-rous, insectivorous in breeding	NT (2018)
Great bustard	*Otis tarda*	Linnaeus, 1758	5800–18,000/ 3300–5300	105/75	2–4	Eurasia and N Africa	Steppe, pasture-land, cereal farmland	Mainly herbivo-rous, insectivorous in breeding, also small vertebrates	VU (2017)
African houbara	*Chlamydotis undulata*	Jacquin, 1784	1800–3200/ 1200–1700	65–75/ 55–65	1–3	N Africa and Canary Islands	Semi-desert, shrub-steppe	Mixed insectivo-rous-herbivorous	VU (2016)
Asian houbara	*C. macqueenii*	Gray, 1832	1500–3175/ 1100–1250	65–75/ 55–65	2–5	Middle East and C Asia	Shrub-steppe, semi-desert, desert	Mixed insectivorous-herbivorous, also small vertebrates	VU (2016)
Hartlaub's bustard	*Lissotis hartlaubii*	Heuglin, 1863	1500–1600/–	60/–	–	E Africa	Grassland and savanna	Mixed insectivo-rous-herbivorous	LC (2016)
Black-bel-lied bustard	*L. melanogaster*	Rüppell, 1835	1800–2700/ 1400	60/60	1–2	W, C, E and SW Africa	Grassland, savanna, open woodland, pas-tureland and farmland	Mainly insectivo-rous, complemented with plants	LC (2016)
Ludwig's bustard	*Neotis ludwigii*	Rüppell, 1837	4200–6000/ 2200–2500	85/–	2	SW Africa	Semi-desert, arid grassland	Mixed insectivorous-	EN (2016)

								herbivorous, also small vertebrates	
Denham's bustard	*N. denhami*	Children & Vigors, 1826	9000–10,000/ 3000	100/80	1–2	W, C, E and S Africa	Grassland, savanna and open woodland	Mixed insectivorous-herbivorous, also small vertebrates	NT (2016)
Heuglin's bustard	*N. heuglinii*	Hartlaub, 1859	4000–8000/ 2600–3000	–/–	2	Kenya and Horn of Africa	Desert, semi-desert, arid grassland	Mixed insectivorous-herbivorous, also small vertebrates	LC (2016)
Nubian bustard	*N. nuba*	Cretzschmar, 1826	5400/–	70/50	2	Sahel	Semi-arid scrub-land and savanna	Mixed insectivo-rous-herbivorous	NT (2016)
Arabian bustard	*Ardeotis arabs*	Linnaeus, 1758	5700–10,000/ 4500	100/75	1–2	Sahel and SW Arabia	Semi-desert, arid grasslands, savanna	Mixed insectivorous-herbivorous, also small vertebrates	NT (2018)
Kori bustard	*A. kori*	Burchell, 1822	10,900–19,000/ 5900	120/90	1–2	E and S Africa	Open grassland and savanna	Mixed insectivorous-herbivorous, also small vertebrates	NT (2016)
Great Indian bustard	*A. nigriceps*	Vigors, 1831	8000–14,500/ 3500–6750	120/90	1–2	N and C India	Semi-desert, open and wooded grassland	Mixed insectivorous-herbivorous, also small vertebrates	CR (2018)
Australian bustard	*A. australis*	Gray, 1829	4300–9300/ 2400–4200	120/90	1–2	N, W and C Australia and S New Guinea	Grasslands, semi-desert and savanna	Mixed insectivorous-herbivorous, also small vertebrates	LC (2016)
Bengal florican	*Houbaropsis bengalensis*	Gmelin, 1789	1250–1700/ 1700–2250	64/68	1–2	S Nepal and N India	Moist grassland, pastureland	Mixed insectivorous-	CR (2018)

(continued)

Table 1 (continued)

Common name	Scientific name	Authority	Weight (g) males/females	Height (cm) males/ females	Clutch size	Distribution	Main habitat	Diet	IUCN Red List category
								herbivorous, also small vertebrates	
Lesser florican	*Sypheotides indicus*	Miller, 1782	450/–	46/51	3–5	India	Grassland, scrub-land and farmland	Mixed insectivorous-herbivorous, also small vertebrates	EN (2016)
Savile's bustard	*Lophotis savilei*	Lynes, 1920	–/–	42	–	Sahel	Arid scrubland and savanna	Unknown	LC (2016)
Buff-crested bustard	*L. gindiana*	Oustalet, 1881	675–900	50	1–2	E Africa	Arid scrubland	Mixed insectivo-rous-herbivorous	LC (2016)
Red-crested bustard	*L. ruficrista*	Smith, 1836	680	50	1–2	E Angola to S Mozambique and N South Africa	Dry grassland, scrubland and savanna	Mixed insectivo-rous-herbivorous	LC (2016)
Karoo bustard	*Heterotetrax vigorsii*	Smith, 1831	1600/1350	60	1–2	SW Africa (Cape Province)	Desert and semi-desert	Mainly herbivo-rous, complemented with arthropods	LC (2016)
Rüppell's bustard	*H. rueppelii*	Wahlberg, 1856	–/–	60	1	SW to NW Namibia	Desert and semi-desert	Mixed insectivorous-herbivorous, also small vertebrates	LC (2016)
Little brown bustard	*H. humilis*	Blyth, 1856	700	40	2	Horn of Africa	Semi-desert, arid grassland	Mixed insectivo-rous-herbivorous	NT (2016)
	Afrotis afra		700	50	1–2	Cape region			

Southern black bustard		Linnaeus, 1766					Semi-arid scrubland and farmland	Mixed insectivorous-herbivorous	VU (2016)
Northern black bustard	*A. afraoides*	Smith, 1831	700	50	1–2	Inland S Africa	Desert, semi-desert, dry grassland and savanna	Mainly insectivorous, complemented with plants	LC (2016)
White-bellied bustard	*Eupodotis senegalensis*	Vieillot, 1820	1400–1500/ <1400	50–60/ <50	1–3	Sahel, C, E and S Africa	Grassland and savanna, farmland	Mixed insectivorous-herbivorous	LC (2016)
Blue bustard	*E. caerulescens*	Vieillot, 1820	1120–1612	55	1–3	SE Africa	Grassland, pastureland, farmland	Mixed insectivorous-herbivorous, also small vertebrates	NT (2017)

Data are compiled from Collar et al. (1986), superseded by Ziembicki (2010) for Australian bustard, Cramp and Simmons (1980) for Asian houbara and various sources (including M.B. Morales own data) for little bustard. The most recent classification in IUCN world conservation status categories is also presented, along with year of last assessment in parentheses (IUCN Red List; *CR* Critically Endangered, *EN* Endangered, *VU* Vulnerable, *NT* Near Threatened, *LC* Least Concern)

Ardeotis has two species in the Afrotropics (one penetrating the Arabian peninsula), one in the Indian subcontinent and one in Australia and southernmost New Guinea. The monotypic *Tetrax* and *Otis* plus *Chlamydotis* (two species) are Palearctic, but all have populations in North Africa; only *Sypheotides* and *Houbaropsis* occur entirely outside Africa, both in the Indian subcontinent but the latter also in Cambodia (Table 1).

Phylogenetic Relationships and Life-History Traits

Unsuspected Relatives

Increasingly confident broad-scale studies of the relationships of birds have in the past 10 years redefined the longstanding working assumption in avian taxonomy that bustards are part of or close to the Gruiformes (cranes, crakes and various allies). Hackett et al. (2008) located '*Choriotis*' (= *Ardeotis*) and *Eupodotis* between *Opisthocomus* (Hoatzin) and the Cuculiformes (cuckoos), with 'Gruiformes' (sic, in inverted commas) sister to Cuculiformes and with Musophagiformes (turacos) also distantly indicated. Building on this, McCormack et al. (2013) found 'surprisingly close relationships between phenotypically divergent bird families, such as bustards (Otididae) and turacos (Musophagidae)'. Moreover, Prum et al. (2015) produced a cladogram which inserts *Ardeotis* between the turacos and the cuckoos in what they call an 'Otidimorph' clade, sister to a 'Columbimorph' clade of mesites, sandgrouse and pigeons and grouping with these latter in a wider category named 'Columbaves', this being sister to Gruiformes on the one side and to 'Strisores' (nightjars, hummingbirds, swifts) on the other. These are unsettling findings: the notion that bustards are potentially closest to such ecologically and phenotypically divergent birds as cuckoos and turacos is very hard to credit, but the evidence appears to be consistent (Jetz et al. 2012; Jarvis et al. 2014; Prum et al. 2015).

An Unresolved Family Tree

No satisfactory molecular phylogeny of the Otididae has been published, and the work that has been done to date leaves as many questions as answers over the relationships of the species. Two papers appearing within a year of each other have been the only contributions that address the family as a taxonomic entity. Pitra et al. (2002), using both mitochondrial (cytochrome b, cyt_b, for all taxa, and control region for a subset of 10 taxa) and nuclear (CHD1_Z intron, for 9 taxa) evidence, sampled the most species, only missing two (*Heterotetrax humilis* and *H. vigorsii*). Based on cyt_b evidence, they found that the genera *Neotis* and *Ardeotis* did not separate out into two clades; in their tree *Neotis heuglinii* sits between the two Afrotropical *Ardeotis* and the two from Asia and Australia (the latter two are very similar but structurally very distinct from the Afrotropical species, and very possibly merit their own genus); the two *Lissotis* were sister to the clade with *Neotis* and *Ardeotis*. In a separate clade from all these came *Otis* and *Chlamydotis*, as sisters, along with *Afrotis* and *Eupodotis* (sensu del Hoyo and Collar 2014); *Houbaropsis* and *Sypheotides* were—albeit not closely—related to each other and together were sister to *Lophotis*, while *Tetrax* emerged as an isolated form but sister to the previous three genera. Most unexpectedly and inexplicably, *Eupodotis* (now *Heterotetrax*) *rueppelii* was recovered as basal to all other bustards.

Broders et al. (2003) used only the cytochrome *b* gene to determine the relationships of seven genera, but the sequences were more than twice as large as Pitra et al. (2002), i.e. 1143 bp instead of 444. They also found that *Otis* and *Chlamydotis* cluster together, and again detected a link between *Lophotis* and *Tetrax*. However, in their tree *Afrotis* emerged as basal to the other taxa studied, while *Eupodotis* (now *Heterotetrax*) *rueppelii* was sister to *Ardeotis* (at least the two Afrotropical species); they judged that *rueppelii*'s anomalous position in the Pitra tree was probably attributable to degraded DNA or a mislabelled sample. However, *Eupodotis senegalensis* emerged as the next closest relative of *Afrotis* and far from *E. rueppelii*, and for this reason Broders et al. suggested the adoption of *Heterotetrax* for *rueppelii* and its relatives (a proposal followed by Dickinson and Remsen 2013; del Hoyo and Collar 2014).

It is the inconsistencies rather than the consistencies between these two studies that matter; they undermine the confidence with which any of the findings can be treated. Regrettably this did not trigger a more comprehensive review of the family, but one further piece of work produced another unexpected result. In seeking to add *Heterotetrax humilis* to the Pitra et al. tree, and using the same sequence and bp length from cytochrome b as in that study, Horreo et al. (2014) discovered that much its closest relative was *Tetrax* rather than other *Heterotetrax*. The authors therefore speculated that these two species, very different phenotypically, shared a recent common ancestor and that the distinct sexual plumage dimorphism in *Tetrax* must have evolved over a very short period of time (riskily inferring that 'many of the phenotypic traits used to classify members of the Otididae are not phylogenetically informative'). However, they also wisely concluded that errors in process may have occurred and 'further analyses including additional sequence from unlinked nuclear genome markers will be necessary to confirm phylogenetic relationships in the family Otididae'. Figure 1 compiles the phylogenies, respectively, produced by Pitra et al. (2002), Broders et al. (2003) and Horreo et al. (2014), highlighting the position of the little bustard in each tree.

In fact, advances of this type had already been made, but the results have so far remained unpublished except as a PhD thesis. Cohen (2011) used mitochondrial and nuclear DNA (5341 nucleotide bp; three mitochondrial markers: cyt_b, NADH, Ctr_II; five nuclear markers: Fib5, TGFB,GAPDH, ODC, CHD1Z) as well as a morphological/behavioural dataset of 'all 27 species of Otididae' (he recognised *Eupodotis senegalensis barrowii* as a full species) to arrive at a series of alternative but largely complementary conclusions concerning the relationships of the bustards, synthesised in a consensus phylogenetic tree, reproduced in Fig. 2. In his phylogeny

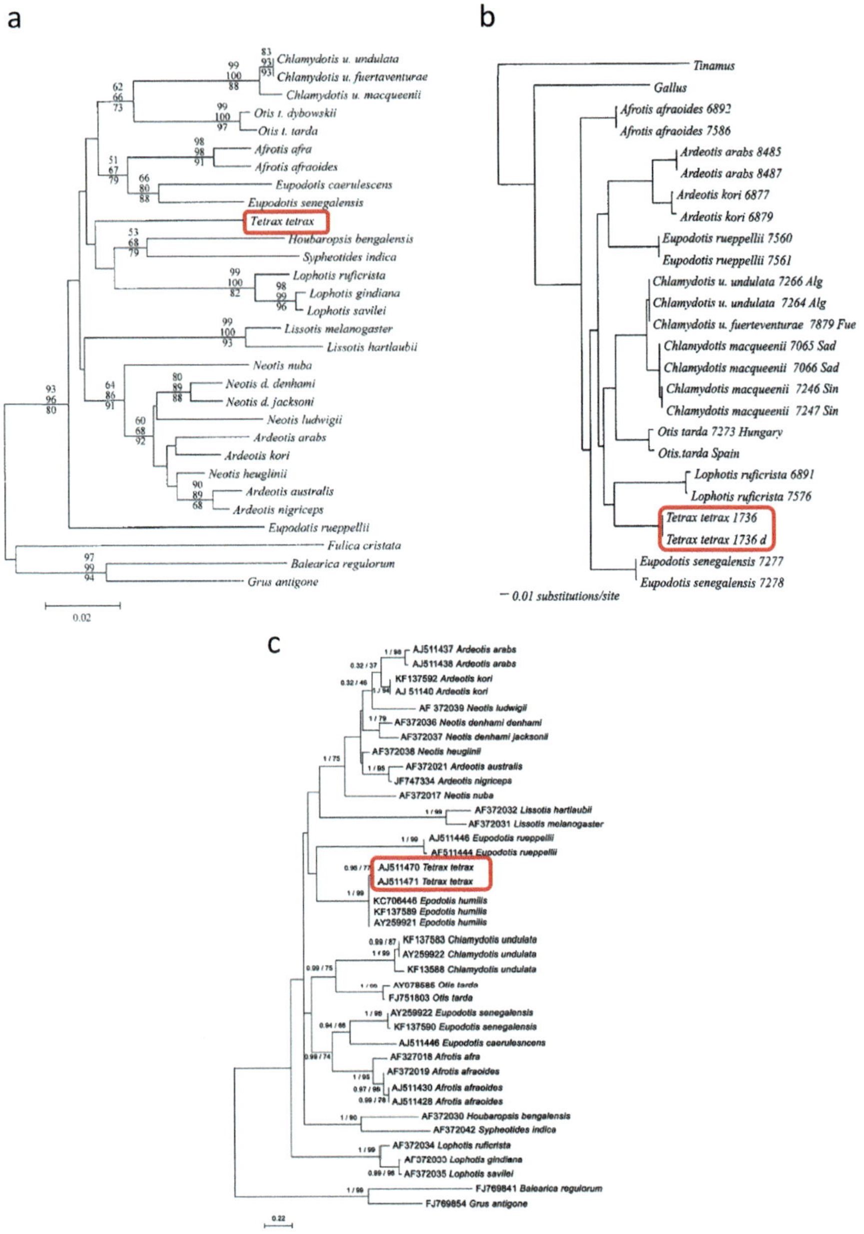

Fig. 1 Compilation of three phylogenetic trees of the Otididae using the mitochondrial cyt_b marker, highlighting the position of the little bustard in each phylogeny. Composed figure based on Pitra et al. (2002) (**a**), Broders et al. (2003) (**b**) and Horreo et al. (2014) (**c**). See text for marker sequence details. In tree (**c**) '*Epodotis*' should be read '*Eupodotis*'

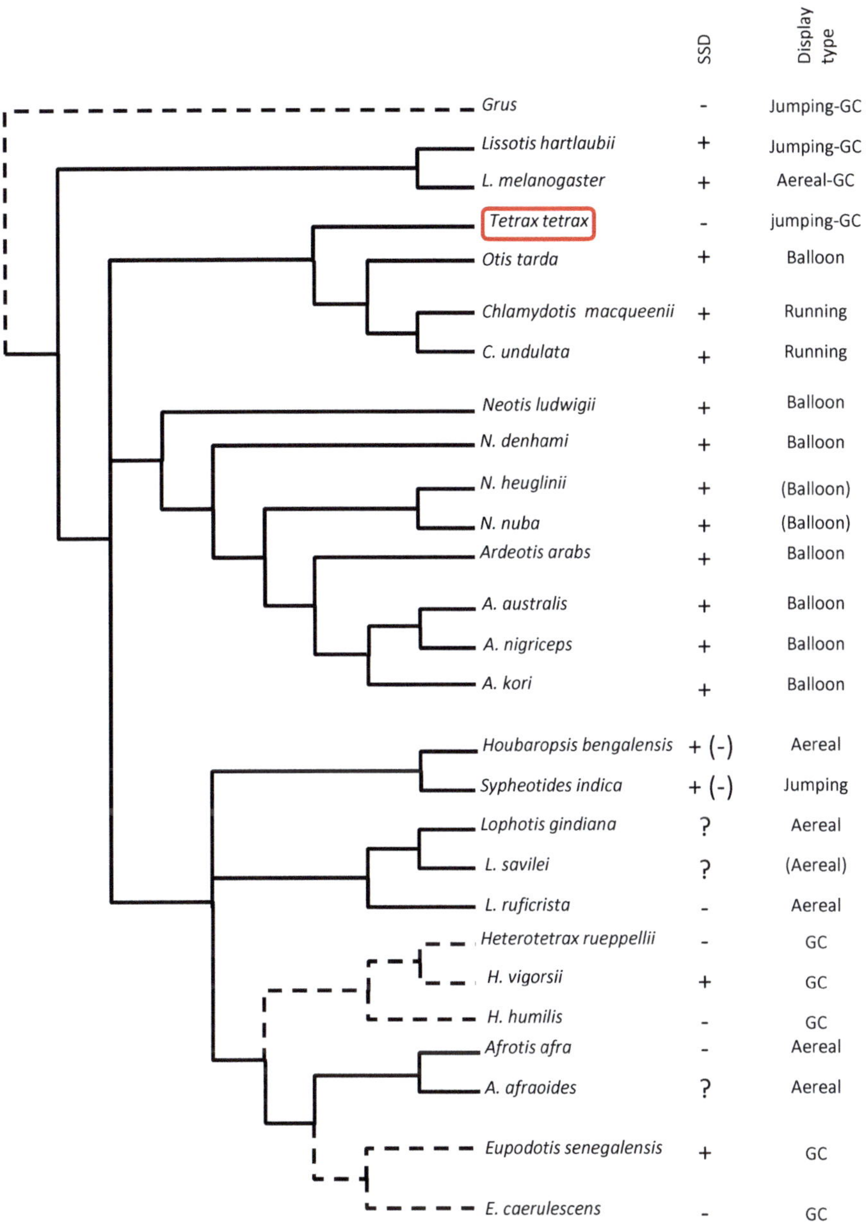

Fig. 2 Variation of sexual size dimorphism (SSD; present: +, absent: –), type of sexual display (GC = ground-calling) and mating strategy across the bustard phylogenetic tree (based on the consensus phylogeny provided by Cohen 2011, see text for details). On the basis of values provided in Table 3, a species is considered dimorphic when the percentage difference is larger than 10% for body weight and/ or larger than 5% for wing length (see Höglund 1989). Cases of inverse SSD (females larger than males) are indicated as (–), while interrogation marks denote lack of information. Solid branches indicate polygynous (i.e. exploded lek) lineages, while dashed branches denote monogamous lineages. The position of the little bustard in the tree is highlighted

Table 2 Position of genus *Tetrax* in bustard phylogeny according to existing studies

Source	Position in phylogenetic tree	Closest genera
Pitra et al. (2002)	Derived isolated single-genus clade	*Houbaropsis*, *Sypheotides*, *Lophotis* forming sister clade
Broders et al. (2003)	In small clade sister to large one formed by *Ardeotis*, *Heterotetrax rueppelii*, *Chlamydotis* and *Otis*	*Lophotis* as sister genus in clade
Cohen (2011)	In small basal clade	*Otis* and *Chlamydotis* in sister sub-clade
Horreo et al. (2014)	In small derived clade sister to *H. rueppelii*	*Heterotetrax humilis* as sister in clade

Sources are ordered chronologically

Lissotis tended to emerge as basal, *Tetrax* (highlighted) grouped with *Otis* and *Chlamydotis*, *Neotis* and *Ardeotis* again failed to emerge as two separate lineages, and the split between *Eupodotis* and *Heterotetrax* was upheld, albeit with *H. humilis* proving close to both *H. rueppelii* and *H. vigorsii* and having nothing to do with *Tetrax*. Table 2 summarises the different positions of *Tetrax*, the genus of our subject species, assigned in the four phylogenetic studies reviewed here.

Yang et al. (2010) sequenced the whole mitochondrial genome (16,849 bp) for *Otis tarda*, and used 12S, 16S and VAL sequences on four other bustard species to produce a tree, in which *Tetrax tetrax* was basal. The sequence of species generated by Pitra et al. (2002) formed the basis for the treatment of the family in Dickinson and Remsen (2013), whereas Cohen's (2011) multiple arrangements were used to decide the sequence in del Hoyo and Collar (2014). This latter places the little bustard first, implying a basal position. Next to it is the great bustard *Otis tarda*, reaffirming a relationship that was expressed for many decades by the frequent placement of little bustard in *Otis* (an assuredly mistaken habit which persisted into the 1970s; see, e.g., Harrison 1978). These two species share several characteristics: a relatively short bill with distinctly decurved upper mandible, a proportionately rather less elongate neck and slightly shorter legs than in other species, stockier, less tapered build, marked seasonal sexual plumage dimorphism, and largely overlapping herb-rich steppeland/farmland habitats and distributions. They diverge sharply in their self-advertisement behaviour (Table 3 and Fig. 2), the little bustard performing a striking but little-known display that involves accelerating food-stamping, a little leap, wing-flash and simultaneous call from a small prepared display-point in low light conditions (see chapter 'Natural history of the little bustard: morphology, biometry, diet, sexual dimorphism, and social and breeding behaviour'), the great bustard undertaking a celebrated 'balloon-display' with extraordinary contortions of wings and tail at various apparently random points within a lekking ground (see, for example, Cramp and Simmons 1980).

The African Houbara *Chlamydotis undulata* and Asian Houbara *C. macqueenii*, clearly very closely related to each other, replace the great and little bustard in the more arid areas of the Palearctic. There is some potential for temporary sympatry

Table 3 Degree of body weight and wing length sexual dimorphism, dominant mating system and type of courtship display in bustard species

Species		Weight dimorphism (%)	Wing length[a] dimorphism (%)	Mating system	Type of display[b]
Common name	Scientific name				
Little bustard	*Tetrax tetrax*	9.11[c, d]	0.99	Exploded lek[e]	Jumping and ground-calling
Great bustard	*Otis tarda*	176.74[c]	21.26	Exploded lek[f]	Balloon
African houbara	*Chlamydotis undulata*	72.41[c]	8.7	Exploded lek[e, g]	Running
Asian houbara	*Chlamydotis macqueenii*	72.41[c]	8.7	Exploded lek[h]	Running
Hartlaub's bustard	*Lissotis hartlaubii*	–	7.4	(Exploded lek)[e]	Aerial and ground-calling
Black-bellied bustard	*L. melanogaster*	60.71[c]	8.26	Exploded lek[e]	Aerial and ground-calling
Ludwig's bustard	*Neotis ludwigii*	56.92[c]	15.97	Exploded lek[e]	Balloon
Denham's bustard	*N. denhami*	216.67[c]	17.35	Exploded lek[e]	Balloon
Heuglin's bustard	*N. heuglinii*	114.29[c]	14.53	Exploded lek[e]	(Balloon)
Nubian bustard	*N. nuba*	–	15.25	(Exploded lek)[e]	(Balloon)
Arabian bustard	*Ardeotis arabs*	74.44[c]	13.04	(Exploded lek)[e)]	Balloon
Kori bustard	*A. kori*	77.12[i]	16.12[i]	Exploded lek[e]	Balloon
Great Indian bustard	*A. nigriceps*	119.51[c]	27.00	Exploded lek[e]	Balloon
Australian bustard	*A. australis*	106.06[c]	11.5	Exploded lek[j]	Balloon
Bengal florican	*Houbaropsis bengalensis*	−25.32[c]	−19.04	Exploded lek[e]	Aerial
Lesser florican	*Sypheotides indica*	–	−6.38	Exploded lek[e]	Jumping
Savile's bustard	*Lophotis savilei*	–	–	(Exploded lek)[e]	(Aerial)
Buff-crested bustard	*L. gindiana*	–	–	(Exploded lek)[e]	Aerial
Red-crested bustard	*L. ruficrista*	–	2.91	Exploded lek[e]	Aerial

(continued)

Table 3 (continued)

Species		Weight dimorphism (%)	Wing length[a] dimorphism (%)	Mating system	Type of display[b]
Common name	Scientific name				
Karoo bustard	*Heterotetrax vigorsii*	18.52[c]	6.5	Monogamy[e]	Ground-calling
Rüppell's bustard	*H. rueppelii*	–	4.76	Monogamy[e]	Ground-calling
Little brown bustard	*H. humilis*	–	1.96	Monogamy[e]	Ground-calling
Southern black bustard	*Afrotis afra*	–	3.03	Exploded lek[e]	Aerial
Northern black bustard	*A. afraoides*	–	–	(Exploded lek)[e]	Aerial
White-bellied bustard	*Eupodotis senegalensis*	–	8.26	Monogamy[e]	Ground-calling
Blue bustard	*E. caerulescens*	–	0.99	Monogamy[e]	Ground-calling

Degree of dimorphism is calculated as the percent increase of male relative to female body weight. Data used for weight calculations are the median of male and female weight ranges provided in Table 1, while those for wing length are based on Johnsgard (1994), except for the case of Kori bustard, which are provided by Osborne and Osborne (1998). Dashes denote insufficient information for dimorphism calculation. Parentheses in mating system and display type categories indicate only (educated) guessed information

[a]Johnsgard (1994)
[b]Osborne et al. (1984)
[c]Collar et al. (1986)
[d]M. B. Morales unpublished data
[e]Corresponding reference in Morales et al. (2001)
[f]Morales and Martin (2002)
[g]Hingrat and Jalme (2005)
[h]Riou and Combreau (2014)
[i]Osborne and Osborne (1998)
[j]Ziembicki (2010)

between members of these four species in places, but the ecological niche of *Chlamydotis* is distinct; the ability of bustards to survive without water is particularly marked in these two species and in their Sahelian equivalent the Nubian bustard *Neotis nuba* Collar et al. (1986). Male *Chlamydotis* perform a remarkable self-advertising display in which they erect their black-and-white neck-side ruffs so that these billow over the retracted head while the bird runs at speed, seemingly unable to see where it is going but in fact highly alert to the environment, in lines and circles across a considerable area of terrain (for differences in the displays of the two species, see Collar and Combreau 2017).

Despite the conspicuousness of this breeding-season display, it was virtually unobserved and unrecorded by any ornithologist anywhere across the vast range of the two species until performances by captive birds in Israel led to its first serious description as little as 40 years ago (Mendelssohn et al. 1979). The facts that such a distinctive behaviour could go undocumented for so long in this genus, and that the full male display repertoire of the little bustard was only first reported even later, in Cramp and Simmons (1980), may help explain why it is that the displays of the African bustards remain so poorly documented, as reflected in Table 3. Bustard displays take place at times of day when—or in parts of the landscape where—human observers are not usually present. The displays of all *Neotis* and African *Ardeotis* bustards are very poorly described—an astonishing thing in the case of the well-known and conspicuous kori bustard *A. kori*—while the repertoires of *Lissotis*, *Afrotis* and *Lophotis* are no fully known.

This dearth of information is a serious drawback to our appreciation of the species, and it denies us the opportunity to make comparisons (see section "Relationships between mating strategies, courtship display and sexual-size dimorphism") which might, in the absence of molecular analyses, provide phylogenetic insights. Twenty years have passed since Collar et al. (1986) expressed the hope that the taxonomic position of the seemingly anomalous *Chlamydotis* might be resolved as more information emerged on the displays of other bustard species, notably those in *Neotis*, but the record has remained depressingly blank. Collar et al. (1986) speculated that the Nubian's replacement of the African Houbara in the southern fringes of the Sahara might indicate a closer relationship than appearances would allow, but the genetic evidence, although weak, tends to reject the speculated close relationship between *Chlamydotis* and *Neotis*.

Little bustard and lesser florican *Sypheotides indicus*, the two smallest and lightest members of the family, are the only two species of bustard that perform brief display-leaps on a fixed spot, in both cases in fairly rapid succession (unlike in *Afrotis*, *Lophotis* and *Houbaropsis*, in which the aerial displays are much longer in both duration and periodicity), and this evidently led Collar et al. (1986) to assume a potential phylogenetic link between them. Such a link seemed also to be indicated by the molecular analyses of Pitra et al. (2002). However, no such relationship emerged in Cohen's (2011) phylogeny, and it is instead perhaps just as likely that the two species evolved on-the-spot display-leaps independently. In the case of the florican, the leaps are made from within the concealment of vegetation; in the case of the little bustard, they are made from within the semi-concealment of half-darkness (see chapter 'Behavioural ecology of the little bustard: sexual selection and mating systems').

These leaps are evidently used by females in the area to assess male quality. Only the pale-bellied and sometimes group-living *Eupodotis* and *Heterotetrax* species, which defend territories by calling but are not known to perform any other kind of self-advertisement display, aerial or terrestrial, are believed to form monogamous pair-bonds and to tend offspring jointly, although male kori bustards will offer food items to females and male Denham's bustards *Neotis denhami* have been observed apparently attending nests, as if some pair-bond might sometimes exist Collar et al.

(1986). The males of all other bustard species appear or are known to play no part in the rearing of their young, and are, therefore, subject to scrutiny and evaluation by the females they seek to attract and fertilise, resulting in the striking variety of male plumage pattern and size increment (Table 3). In the non-passerines only the families Phasianidae (pheasants, partridges, turkeys and grouse), Anatidae (ducks, geese and swans) and Trochilidae (hummingbirds) reflect similar levels of plumage and/or size dimorphism based on sexual selection (see, for example, Johnsgard 1994). The breeding male little bustard, with its modified seventh primary for sound production, longer neck feathering and black-and-white gular and pectoral pattern unique in the plumages of birds, certainly represents one of the most divergent and distinctive evolutionary entities within the remarkable radiation of the Otididae.

Relationships Between Mating Strategies, Courtship Display and Sexual Size Dimorphism

A remarkable behavioural trait of the bustard family is the dominance of socially polygynous species, particularly those that gather in exploded leks (Morales et al. 2001, see also Table 3 and chapter 'Behavioural ecology of the little bustard: sexual selection and mating systems'). In lek mating systems, displaying males closely aggregate to attract females, which select mates based only on male phenotypic quality (thus, seeking good genes), while in exploded leks male clustering is much looser and females may also use the resources defended by the displaying males as a means of assessing their quality (Bradbury 1981; Ligon 1999). Although bustards are among the few avian families in which lek-like mating tends to be the rule (Johnsgard 1994), as mentioned above some species have been cited as monogamous: those formerly composing the genus *Eupodotis* (Collar et al. 1986), that is, *E. senegalensis* and *E. caerulescens*, plus the three species currently classified in *Heterotetrax* (*H. vigorsii, H. rueppelii* and *H. humilis*).

However, the degree to which mating system may reflect phylogenetic proximity remains unclear. In the first place, the phylogenetic position of genera, including *Tetrax*, differs between studies. Second, both monogamous and lekking genera have been considered basal according to different authors (*Heterotetrax* in the former case, *Afrotis, Lissotis, Tetrax*, among the latter). Finally, the supposedly socially monogamous mating strategy (i.e. with shared parental care) assigned to *Eupodotis* and *Heterotetrax* still needs to be confirmed through field studies, as does the exploded lek behaviour attributed to some scarcely known species of polygynous genera. In this regard, it is important to bear in mind that mating systems can be highly plastic within species and populations, varying with local factors such as population density and resource distribution (Emlen and Oring 1977; Ligon 1999). Therefore, the information provided for the least known species (e.g. most African bustards) in Table 3 must be considered preliminary. By contrast, the mating strategies of the four Palearctic species, namely great bustard, little bustard, African houbara and Asian houbara, are fairly well known and have been studied under different population and environmental conditions (Jiguet et al. 2000, 2002; Morales et al. 2003, 2014; Hingrat and Jalme 2005; Alonso et al. 2010; Riou and Combreau 2014).

As is the case in other avian families where polygyny is the prevailing mating system, bustard species tend to be sexually dimorphic (e.g. Johnsgard 1994). In addition to plumage dimorphism, moderate to strong male-biased body size dimorphism seems to be the rule. Nevertheless, among species where information is available (see Table 3 and Fig. 2), the percent increase in male in relation to female size ranges from strongly pronounced (e.g. Denham's and great bustard) to negative, the latter indicating inversed size dimorphism (e.g. the Bengal florican *Houbaropsis bengalensis*). However, it is difficult to infer a clear relationship between mating strategy and degree of sexual size dimorphism in the family from existing information on species' morphology and breeding behaviour. For instance, at least one supposedly monogamous species, karoo bustard *Heterotetrax vigorsii*, is moderately dimorphic in weight (18.52%) and slightly in wing length (6.5%, see Table 3), while the polygynous (exploded-lekking) little bustard is practically size-monomorphic (9.11% and 0.99% for body weight and wing length, respectively), according to the criteria for size dimorphism in birds proposed by Höglund (1989). Nevertheless, in the extremely size-dimorphic great bustard, mating success is usually highly skewed among lekking males, and body mass clearly plays a role in this pattern (Alonso et al. 2010).

Sexual size dimorphism has also been related to the type of sexual display, both in some bustards and in other avian groups (Johnsgard 1994; Székely et al. 2000), so that males of species that perform acrobatic courtship behaviour tend to be smaller in relation to females than species with ground display. However, the information compiled in Table 3 does not allow any clear conclusion in this regard. Although ground displays (balloon, running and ground-calling) are dominant in the family, in some species males perform leaps and acrobatic flights. The degree of sexual size dimorphism among ground-displaying species ranges from moderate to strongly marked. However, among jumping and aerially courting bustards, we find all patterns of size dimorphism, from monomorphic and inversely dimorphic species (little bustard and floricans, respectively) to the black-bellied bustard *Lissotis melanogaster*, a species showing clear male-biased size dimorphism. The latter species probably shares male-biased weight dimorphism with its congeneric Hartlaub's bustard *L. hartlaubii*. The variation in sexual size dimorphism and type of sexual display across the family's consensus phylogeny produced by Cohen (2011) is shown in Fig. 2. Again, future field and comparative research on bustard breeding ecology and behaviour will hopefully help clarify the evolutionary relationships between mating strategies and sexual size dimorphism in the family.

Little Bustard Phylogeography

What is the phylogenetic diversity within the subject species of this book, the little bustard? As described in the next chapter, the little bustard was historically split into two subspecies (*T. t. tetrax* and *T. t. orientalis*), based on slight morphological differences, a distinction that seems no longer supported (Cramp and Simmons 1980; Collar et al. (1986); Dickinson and Remsen 2013; del Hoyo and Collar 2014). However, the large range and the variety of bioclimatic regions occupied by the little bustard suggest that the species should exhibit a relatively high level of genetic diversity.

The use of molecular markers to unravel the degree of genetic relatedness between the species's populations across geographical space, that is, its phylogeography, has been so far restricted to a sector of the western sub-range (see chapter 'Natural history of the little bustard: morphology, biometry, diet, sexual dimorphism, and social and breeding behaviour')—more precisely, to a subset of populations across Spain and France, from Andalusia (southern Iberia) to Poitou-Charentes (central-western France), including intermediate sites such as Galicia (north-western Iberia), Catalonia (north-eastern Iberia) and Mediterranean France. García et al. (2011), using a mitochondrial DNA marker, found very low genetic diversity (only four haplotypes) and no strong genetic structure, that is, differentiation, across these populations, which form the core of the species's western range. One haplotype was common to, and dominant in, all populations, but no haplotype was exclusive to a single population. Nevertheless, some geographical differences appeared: one haplotype was only present in Iberian populations and another was particularly frequent in Mediterranean France. Populations from Galicia and Catalonia were the most genetically diverse because they had both exclusively Iberian haplotypes and haplotypes found there and in French populations. Moreover, Catalonia shared a haplotype exclusively with Mediterranean France.

Based on these genetic diversity patterns and making use of the molecular clock, García et al. (2011) also estimated the historical periods of the species's population expansion and contraction during the Pleistocene (see also chapter 'Migration, movements and non-breeding ecology'). Expansion corresponded to cold periods of Pleistocene glacial stadials during which cold steppe ecosystems reached their maximum extent south of the European ice sheet (Pons and Reille 1988; Allen et al. 1999). Conversely, during warmer interstadials forest ecosystems expanded northwards from Mediterranean refugia, leading to the population contraction of the little bustard and other steppe fauna (Geroudet 1961; Santos and Suárez 2005). The periods of little bustard population expansion identified by García et al. (2011) occurred at different Pleistocene temperature minima, ranging from around 228,000 years ago to approximately 108,000 years ago. This scenario prompts the consideration that, had the anthropogenic expansion of open agricultural habitats not occurred, Palearctic bustards and other steppe fauna would currently be confined to scattered relic steppes associated with local microclimate and particular edaphic conditions (Bennett and Provan 2008).

Genetic diversity in newly colonised areas is usually a subset of the diversity found in refugia (Hewitt 1996). Based on this principle, but also on their finding that northern Iberian populations harbour higher diversity than others, García et al. (2011) outlined a scenario in which the little bustard populations of Galicia and Catalonia represent admixture zones where birds confined to southern refugia during interstadials came into contact with individuals from outside Iberia during periods of range expansion. Furthermore, the north-eastwards increase in the frequency of haplotypes shared between northern Iberia and France suggests the existence of a refugium in Central or Eastern Europe.

Even so, the phylogeographic picture of the little bustard remains largely incomplete, lacking information from south-western Iberia (Extremadura and Portugal) and Sardinia, which still hold viable populations. In addition, the now virtually extinct populations in the western sub-range, in Morocco and continental Italy (see chapter 'The little bustard around the world: distribution, global conservation status, threats and population trends'), are important missing pieces. Finally, the genetic structure of the eastern sub-range is still entirely unexplored. An unpublished study by García (2017) filled some of these gaps. He gathered 212 *Tetrax tetrax* specimens collected in 15 different countries across the species range, both from historic (museum) and contemporary (field-collected) specimens. Only eight haplotypes were found concatenating ND2 and CO1 genes (for a total of 1722 bp), highlighting a low level of variation. One haplotype was shared by 101 of 117 individuals (García 2017). These results were further supported by microsatellite (13 loci) data and analyses. Population genetic analyses confirmed overall a slight, but significant, geographic structuring among populations from Spain, France and Kazakhstan (García 2017). Future genetic analysis of these eastern populations should address at least three questions. First, it should be possible to refute (or confirm) the existence of a distinct eastern race or subspecies. Second, the phylogeographic position of geographically intermediate but now extinct populations (Panonian region, the Balkans) could be assessed with the use of museum specimens. And, third, the phylogeography of the eastern range should be resolved using samples from southern Russia, Kazakhstan and western China. Such a complete picture would provide the criteria for determining and prioritising genetically defined management units (i.e. evolutionarily significant units [ESUs]: Moritz 1994) which could be integrated in a global conservation strategy for the species.

Acknowledgements We are indebted to Jesús T. García, Carlos A. MartÚn and V. Bretagnolle for their valuable comments on a previous version of this chapter.

References

Allen JRM, Brandt U, Brauer A, Hubberten HW, Huntley B, Keller J, Kraml M, Mackensen A, Mingram J, Negendank JFW, Nowaczyk NR, Oberhänsli H, Watts WA, Wulf S, Zolitschka B (1999) Rapid environmental changes in southern Europe during the last glacial period. Nature 400:740–743

Alonso JC, Magaña M, Palacín C, Martín C (2010) Correlates of male mating success in great bustard leks: the effects of age, weight, and display effort. Behav Ecol Sociobiol 64:1589–1600
Bennett KD, Provan J (2008) What do we mean by 'refugia'? Quaternary Sci Rev 27:2449–2455
Bradbury JW (1981) The evolution of leks. In: Alexander RD, Tinkle DW (eds) Natural selection and social behaviour. Chiron Press, New York, pp 138–169
Broders O, Osborne T, Wink M (2003) A mtDNA phylogeny of bustards (family Otididae) based on nucleotide sequences of the cytochrome b-gene. J Orn 144:176–185
Cohen C (2011) The phylogenetics, taxonomy and biogeography of African arid zone terrestrial birds: the bustards (Otididae), sandgrouse (Pteroclidae), coursers (Glareolidae) and Stone Partridge (*Ptilopachus*). PhD thesis, Percy FitzPatrick Institute of African Ornithology, University of Cape Town
Collar NJ, Combreau O (2017) Only one species of houbara? A response to Cowan. Br Birds 110: 760–763
Collar NJ, Goriup PD, Osborne PE (1986) Bustard family. In: Urban EK, Fry CH, Keith S (eds) The birds of Africa, vol 2. Academic Press, London, pp 148–179
Cramp S, Simmons KEL (eds) (1980) The birds of the western Palearctic, vol 2. Oxford University Press, Oxford
del Hoyo J, Collar NJ (2014) The HBW–BirdLife international illustrated checklist of the birds of the world, 1: non-passerines. Lynx Edicions, Barcelona
Dickinson EC, Remsen JV Jr (2013) The Howard and Moore complete checklist of the birds of the world, Non-passerines, vol 1. Aves Press, Eastbourne
Emlen ST, Oring IW (1977) Ecology, sexual selection and the evolution of mating systems. Science 197:215–223
García JT (2017) Genetic characterization and affinity of Italian and other populations of Tetrax tetrax. Unpublished report. Instituto de Investigación en Recursos Cinegéticos (IREC-CSIC-UCLM), Ciudad Real, Spain
García JT, Mañosa S, Morales MB, Ponjoan A, García de la Morena EL, Bota G, Bretagnolle V, Dávila JA (2011) Genetic consequences of interglacial isolation in a steppe bird. Mol Phyl Evol 61:671–676
Geroudet P (1961) Les Passeraux, vol I. Delachaux et Niestlé, París
Hackett SJ, Kimball RT, Reddy S, Bowie RCK, Braun EL, Braun MJ, Chojnowski JL, Cox WA, Han KL, Harshman J, Huddleston CJ, Marks BD, Miglia KJ, Moore WS, Sheldon FH, Steadman DW, Witt CC, Yuri T (2008) A phylogenomic study of birds reveals their evolutionary history. Science 320:1763–1767
Harrison CJO (1978) Bird families of the world. Peerage Books (Elsevier Phaidon), London
Hewitt GM (1996) Some genetic consequences of ice ages, and their role in divergence and speciation. Biol J Linn Soc 58:247–276
Hingrat Y, Jalme S (2005) Mating system of the houbara bustard Chlamydotis undulata undulata in eastern Morocco. Ardeola 52:91–102
Höglund J (1989) Size and plumage dimorphism in lek-breeding birds: a comparative analysis. Am Nat 134:72–87
Horreo JL, Alonso JC, Milá B (2014) DNA sequences from the Little Brown Bustard *Eupodotis humilis* suggest its close phylogenetic relationship to the Little Bustard *Tetrax tetrax*. Ostrich 85: 97–101
Jarvis ED et al (2014) Whole-genome analyses resolve early branches in the tree of life of modern birds. Science 346:1320–1331
Jetz W, Thomas GH, Joy JB, Hartmann K, Mooers AO (2012) The global diversity of birds in space and time. Nature 491:444–448
Jiguet F, Arroyo B, Bretagnolle V (2000) Lek mating systems: a case study in the Little Bustard *Tetrax tetrax*. Behav Process 51:63–82
Jiguet F, Jaulin S, Arroyo B (2002) Resource defence on exploded leks: do male little bustards, *Tetrax tetrax*, control resources for females? Anim Behav 63:899–905

Johnsgard PA (1994) Arena birds: sexual selection and behavior. Smithsonian Institution Press, Washington DC
Ligon JD (1999) The evolution of avian breeding systems. Oxford University Press, Oxford
McCormack JE, Harvey MG, Faircloth BC, Crawford NG, Glenn TC, Brumfield RT (2013) A phylogeny of birds based on over 1,500 loci collected by target enrichment and high-throughput sequencing. PLoS One 8(1):e54848
Mendelssohn H, Marder U, Stavy M (1979) Captive breeding of the Houbara (*Chlamydotis undulata macqueenii*) and a description of its display. Bull Internatn. Council Bird Pres 13: 134–149
Morales MB, Martin CA (2002) Great bustard *Otis tarda*. BWP Update 4(3):217–232
Morales MB, Jiguet F, Arroyo B (2001) Exploded leks: what bustards can teach us. Ardeola 48:85–98
Morales MB, Alonso JC, Martín C, Martín E, Alonso JA (2003) Male sexual display and attractiveness in the great bustard *Otis tarda*: the role of body condition. J Ethol 21:51–56
Morales MB, Casas F, García de la Morena EL, Ponjoan A, Calabuig G, Martínez-Padilla J, García JT, Mañosa S, Viñuela J, Bota (2014) Density dependence and habitat quality modulate the intensity of display territory defence in an exploded lekking species. Behav Ecol Sociobiol 68: 1493–1504
Moritz C (1994) Defining 'evolutionarily significant units' for conservation. Trends Ecol Evol 9: 373–375
Olson SL (1985) The fossil record of birds. In: Farner DS, King JR, Parkes KC (eds) Avian biology, vol 8. Academic Press, New York, pp 79–252
Osborne T, Osborne L (1998) Ecology of the Kori bustard in Namibia. First Annual Report for the Ministry of Environment and Tourism Permit Office, Windhoek
Osborne P, Collar N, Goriup P (1984) Bustards. Dubai Wildlife Research Centre, Dubai
Pitra C, Lieckfeldt D, Frahnert S, Fickel J (2002) Phylogenetic relationships and ancestral areas of the bustards (Gruiformes: Otididae), inferred from mitochondrial DNA and nuclear intron sequences. Molec Phylogen Evol 23:63–74
Pons A, Reille M (1988) The Holocene and upper Pleistocene pollen record from Padul (Granada, Spain): a new study. Palaeogeogr Palaeoclimatol Palaeoecol 66:243–263
Prum RO, Berv JS, Dornburg A, Field DJ, Townsend JP, Lemmon EM, Lemmon AR (2015) A comprehensive phylogeny of birds (Aves) using targeted next-generation DNA sequencing. Nature 526:569–573
Riou S, Combreau O (2014) Male territories and the lek-like mating system of MacQueen's bustard Chlamydotis macqueenii. J Ornithol 155:959–967
Santos T, Suárez F (2005) Biogeography, population trends of Iberian steppe birds. In: Bota G, Morales MB, Mañosa S, Camprodon J (eds) Ecology and conservation of steppe-land birds. Lynx Edicions and Centre Tecnològic Forestal de Catalunya, Barcelona, pp 69–102
Sibley CG, Ahlquist J (1985) The relationships of some groups of African birds, based on comparisons of the genetic material. In: Schuchmann KL (ed) Proceedings of the symposium on African vertebrates. Museum A. Koenig, Bonn, pp 115–161
Székely T, Reynolds JD, Figuerola J (2000) Sexual size dimorphism in shorebirds, gulls, and alcids: the influence of sexual and natural selection. Evolution 54:1404–1413
Rong Yang, Xiaobing Wu, Peng Yan, Xia Su, Banghe Yang (2010) Complete mitochondrial genome of Otis tarda (Gruiformes: Otididae) and phylogeny of Gruiformes inferred from mitochondrial DNA sequences. Mol Biol Rep (2010) 37:3057–3066
Ziembicki M (2010) Australian bustard. CSIRO Publishing, Sidney

Natural History of the Little Bustard: Morphology, Biometry, Diet, Sexual Dimorphism, and Social and Breeding Behaviour

Vincent Bretagnolle, Santi Mañosa, and Manuel B. Morales

Introduction

The little bustard *Tetrax tetrax* belongs to the family Otididae, a family that used to be placed within the order Gruiformes. However, more recent avian phylogenies have quite dramatically reshaped sister relationships between bustards and other birds, placing them in their own order Otidiformes (see details in chapter "The little bustard and its family: an overview of relationships"). Indeed, bustards have many unique life history traits, which are reflected in their anatomy, biology, and ecology. We present here some main characters and life history traits of the species, detailing in particular those that are not covered in other chapters; but since mating system and mate choice, habitat selection, breeding biology, fecundity, survival and movements are all covered in other chapters they will not be dealt with here. This chapter should not either be considered as a detailed natural history account of the species (despite the fact that none is unfortunately yet available). Rather it describes some general life history traits that have not been covered so far in great detail. Other general accounts of little bustard general habits, breeding biology and behaviour can be found in

V. Bretagnolle (✉)
Centre d'Etudes Biologiques de Chizé, CNRS & La Rochelle Université, Beauvoir sur Niort, France
e-mail: vincent.bretagnolle@cebc.cnrs.fr

S. Mañosa
Departament de Biologia Evolutiva, Ecologia i Ciències Ambientals, Institut de Recerca de la Biodiversitat (IRBio), Universitat de Barcelona, Facultat de Biologia, Barcelona, Catalonia, Spain
e-mail: smanosa@ub.edu

M. B. Morales
Terrestrial Ecology Group, Departamento de Ecología & Centro de Investigación en Biodiversidad y Cambio Global (CIBC), Universidad Autónoma de Madrid, Madrid, Spain
e-mail: manuel.morales@uam.es

V. Bretagnolle et al. (eds.), *Little Bustard: Ecology and Conservation*, Wildlife Research Monographs 5, https://doi.org/10.1007/978-3-030-84902-3_3

Cramp and Simmons (1980), Schulz (1985), Boutin and Métais (1995), Glutz and Bauer (1994), and Collar in del Hoyo et al. (1996).

Morphology and Anatomy

The little bustard is one of the smallest species of the Otididae family (25–27 species, see chapter "The little bustard and its family: an overview of relationships"). Its body size is about 40–45 cm, for a wingspan of 110 cm (see section below on biometrics) and an average weight of about 850–900 g (Cramp and Simmons 1980; see also below for details). It is the smallest of the three species of bustards found in the western Palearctic (there were formerly four, but Arabian bustard *Ardeotis arabs* has been extirpated: Cramp and Simmons 1980). Like most bustard species, the little bustard is highly dimorphic, and more specifically, sexually dichromatic (Plate 1a, b).

General morphology in bustards, and the little bustard in particular, is characterized by small head, long neck, and long legs. The little bustard is a relatively heavy bird for its size. Like all bustards, little bustards are terrestrial birds that do not use trees, and thus do not perch. They lack a hind toe, have broad claws and long toes. They have very slender necks, especially females. They have a powerful flight, with very little gliding: although little bustards spend most of their time on the ground, they are not reluctant to fly and can actually cover very large distance without

Plate 1 Little bustard adults and young; (**a**) (i) adult male in breeding plumage (Photo: Vincent Bretagnolle), (ii) adult breeding male showing inflated neck while singing (Photo: Carlos Palacín); (**b**) (i) adult female and (ii) adult female and few-days old chicks (Photos: Carlos Palacín); (**c**) detail of male wing showing emarginated seventh primary feather (Photo: Eladio García); (**d**) adult female (i) and male (ii) in winter plumage (Photos: Vincent Bretagnolle); (**e**): juvenile during first autumn (Photo: Carlos Palacín)

stopping (see chapter "Migration, movements and non-breeding ecology" on migration). A unique trait of male little bustards is the short fourth primary that produces a whistling sound when they are flying, regardless of season (Plate 1c).

Bustards in general have several peculiar anatomical characters, which were noted by early anatomists and have actually obscured their phylogenetic position; they were once placed close to cranes—see Yang et al. (2010)—although they are now classified closer to turacos (Prum et al. 2015, chapter "Migration, movements and non-breeding ecology"). Bustards are almost unique among birds for their hexagonal scales and scutellations on the legs (Collar in del Hoyo et al. 1996). Little bustards (like all bustards) lack an oil-preen gland, and have a pink powder-like down at the base of their feathers. Although the mechanism behind the coloration of this down was long known (it is based on porphyrin pigments), the function of pink-powder down has remained unclear. A recent study on great bustard *Otis tarda*, in which porphyrin pigments were demonstrated to disappear very rapidly (no more than 12 min), suggested that males showing females such pigments around cloaca were thus virgin, and may be preferred by the females (Galván et al. 2016). However, since other bustard species have also porphyrin pigments but do not perform such displays, limits the generality of this explanation. Little bustards also show a particular digestive anatomy and tract (Bailey et al. 1997). Bustards do not possess a crop (Bailey et al. 1997), and in general have two, relatively long, intestinal ceca (review in Clench and Mathias 1995, see Plate 2). For example, the length of ceca from an adult female little bustard found dead in Malpartida de Cáceres (Extremadura, Spain) was 15 and 19 cm, which represents 19% and 24%, respectively, of the animal's intestine length, while ceca from an adult male little bustard recovered dead near Toledo (central Spain) were 34 and 30 cm long, representing 33% and 29% of its total intestine length (M.B. Morales unpubl. data). These figures, even if isolated, suggest that the species' size dimorphism (see below) may have correlates at the physiological level and point to the greater importance of herbivory in males in summer. The function of long ceca is not very well known, although it is most likely involved in the digestion of cellulose and fibre, and possibly also in water retention. Long ceca were suggested in great bustard to lower seed germination capacity in *Solanum nigrum* (Bravo et al. 2016). Sexual anatomy has not been studied in detail in little bustard, nor its physiology such as endocrine and hormones, although studies are available for other species of bustards, like Kori *Ardeotis kori* (Penfold et al. 2013) and great (Bautista et al. 2013) bustards.

Little bustard identification is easy, at least in spring. Males are black and white on the neck and head, brown on the dorsal parts, and white on the ventral, while females are brown and cream with many dots and vermiculated feathers (Plate 1a, b). Within the species, age and sex identification is however more complicated, especially in winter plumages, as detailed below.

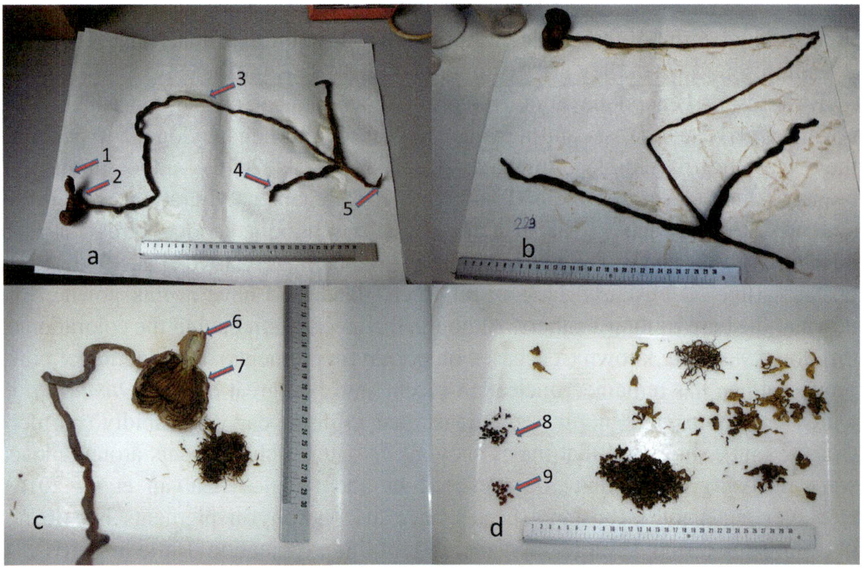

Plate 2 Little bustard stomachs, intestines, and stomach contents; (**a**) female stomach and intestine (1: proventriculus; 2: stomach; 3: intestine; 4: intestinal ceca; 5: cloaca); (**b**) male stomach and intestine (the difference in relative size of ceca in relation to female is remarkable); (**c**) Dissected proventriculus (6) and stomach (7); (**d**) example of stomach contents (winter), dominated by plant shoot and leaf material but with non-negligible presence of coleopterans (8) and seeds (9). Photos by Carolina Bravo

Biometry and Variation in Size

Taxonomy and Subspecies

There were formerly two recognized subspecies of little bustard (*T. t. tetrax* from France, Iberia, Italy, and North Africa) and *T. t. orientalis* for the rest of its current distribution (see chapter "The little bustard around the world: distribution, global conservation status, threats and population trends" for distribution maps). Where exactly the two subspecies were supposed to split is difficult to assess because bustard populations have gone extinct from all countries where they were recorded between France and Ukraine (see chapter "The little bustard around the world: distribution, global conservation status, threats and population trends"). Since the geographical range of the Little Bustard was less fragmented than it is today (it was more or less continuous in the Mediterranean), the location of the historical split is not quite obvious. The taxon *orientalis* was said to be slightly larger, paler on the back (greyer), with more or coarser marks. The two subspecies are no longer recognized in recent checklists, owing to large overlap in biometrics and individual variation in colouration (see Cramp and Simmons 1980).

Specimens were measured in Berlin, Tring (UK), Paris, Vienna, Coimbra, New York, and Madrid museums of natural history, yielding a grand total of 125 adult specimens, by far the largest sample size gathered so far for such an analysis. The assignation of specimens to one or other subspecies was not easy in some cases (even though museum labels often contained an indication of subspecies). The extreme case was represented by two birds shot in the UK in winter and tentatively assigned to *orientalis* (as were birds seen and shot in winter in the Netherlands or Germany; see Cramp and Simmons 1980). We therefore conducted a descriptive analysis, with no a priori assignment to any form, using either country or subcontinental area. Previous analyses suggest that there is clinal variation in size, with south-western birds (Iberia, Italy) being overall smaller than eastern birds (Russia, Central Europe). Table 1 provides average values of measurements from the literature (Cramp and Simmons 1980) and our own. Although *orientalis* is said to be larger, this is actually only true for wing and tail lengths; tarsus length does not differ between former subspecies, and actually bill (culmen and bill depth) is larger in *tetrax* than in *orientalis*. Therefore the magnitude and trend in the cline varies with biometric character, since culmen length is actually larger in south-western birds (Fig. 1). Although bill measurements from Cramp and Simmons (1980) differ from our own measurements, *tetrax* birds have larger bills, in both sexes and whatever the data set used (see Table 1). Using wing length and culmen size, birds from the two putative subspecies were plotted in a scatter plot. Substantial variation occurs, but on a bivariate plot the two forms can be actually separated (Fig. 1). There are, however, some misclassified individuals. This does not mean that the subspecific designation should be kept, but further investigation in regard to morphometrics and genetic markers is clearly required, especially because these two populations (Western Europe and Asia) currently face strongly diverging and opposed fates (chapter "The little bustard around the world: distribution, global conservation status, threats and population trends").

Geographical and Seasonal Variation in Size Within Iberia and France

Table 2 provides average biometric measurements for several populations studied in France and Spain. There is little variation among populations. In particular the three French populations, despite the fact that one is migratory while the other two are sedentary, show almost identical measurements, except for body weight where the La Crau population is on average 100 g lighter, but these birds were all captured and measured in winter (see below). In addition, as found in museum specimens, Spanish birds are smaller than French ones.

Seasonal variation in body weight is known for some species (and may be up to 50% in Australian bustard *Ardeotis australis*: Collar in del Hoyo et al. 1996). We found slight evidence of such pattern in the Little bustard (Fig. 2), since in two study

Table 1 Measurements of Little bustard, by sex and former subspecies, as obtained from Cramp and Simmons (1980), noted as "C&S", and our own measurements from museums (more than 80% of measurements made by VB)

Sex Subspecies	*N*	Wing	Tail	Tarsus	Culmen	Bill height	Bill width	Body weight
Males								
Tetrax	C&S	250 (238–259)	101 (95–107)	68.8 (64–73)	17.4 (16–19)			940–975
Orientalis	C&S	254 (244–264)	106 (97–114)	66.6 (62–73)	16.7 (15–18)			794–907
Tetrax	41	243.5 ± 5.9 (231–256)	101.2 ± 5.0 (91–115)	68.0 ± 2.9 (59.3–73.0)	21.7 ± 1.4 (18.5–24.7)	9.7 ± 0.7 (8.5–12.0)	13.5 ± 1.6 (9.8–16.8)	
Orientalis	30	249.8 ± 5.3 (235–260)	107.9 ± 6.4 (99–124)	65.8 ± 3.9 (53.1–73.0)	20.1 ± 1.9 (16.7–25.5)	9.0 ± 0.9 (7.7–12.8)	13.2 ± 1.6 (10.3–16.7)	
Females								
Tetrax	C&S	245 (240–250)	95.4 (86–105)	65.9 (62–71)	17.5 (16–19)			
Orientalis	C&S	255 (247–269)	102 (96–112)	66.3 (62–70)	15.6 (14–17)			740–910
Tetrax	25	241.1 ± 7.3 (228–257)	93.7 ± 9.9 (65–109)	66.6 ± 5.0 (60.0–84.0)	21.6 ± 1.7 (19.2–25.8)	9.0 ± 1.0 (6.0–10.3)	13.2 ± 1.5 (10.9–17.8)	
Orientalis	27	250.9 ± 7.2 (239–263)	104.5 ± 5.5 (89–117)	65.0 ± 2.7 (60.2–73.0)	19.5 ± 1.4 (16.7–21.6)	8.3 ± 0.7 (7.2–10.0)	12.5 ± 1.1 (10.6–14.7)	

Data are provided as mean and (range) for those obtained from C&S and as mean ± SD for museum measurements

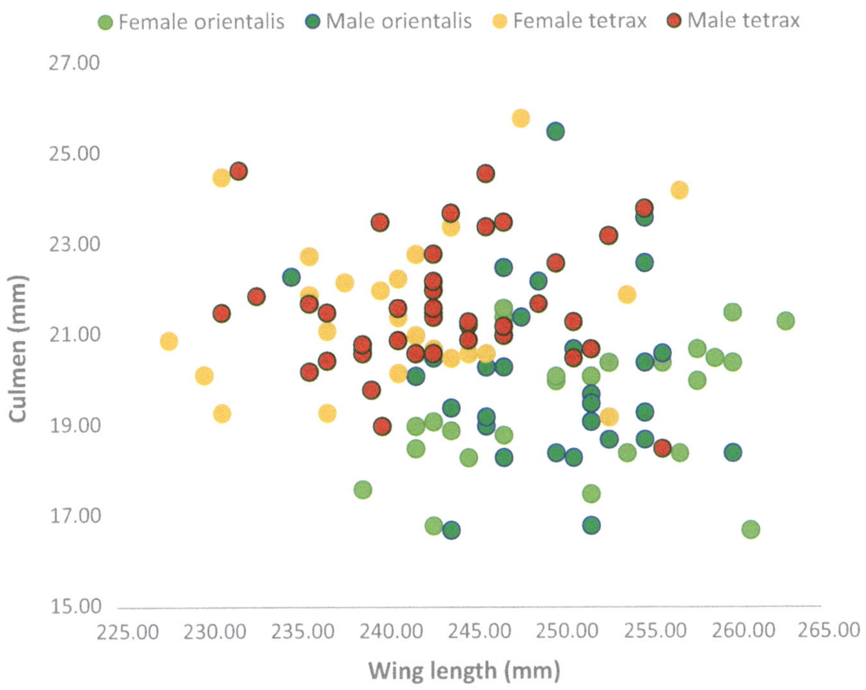

Fig. 1 Scatter plot showing variation in culmen versus wing length among little bustard males and females belonging to western (*tetrax* subspecies) and eastern (*orientalis* subspecies) populations. While few sexual differences are apparent, *tetrax* individuals of both sexes present overall larger culmens and shorter wings than *orientalis* birds. All measurements from museum specimens (see text)

sites in France males and females were slightly heavier in April and May, respectively. This corresponds to the peak activity of display and mate choice, with 1 month lag between these two study sites. In winter, birds lose weight. Interestingly, there is some support from the graph that birds tend to lose weight just prior to migration, since even the sedentary birds from southern France make some movements in March and late August (P. Devoucoux unpubl. Data; see chapter "Migration, movements and non-breeding ecology"). Female weight measurements from Lleida (Catalonia) are substantially smaller than elsewhere, in both absolute and relative terms (Table 2). A possible explanation for this lower mass of Catalonian females is that they were captured and weighed in early summer (late June–early July), when rearing chicks. Their lighter weight might be the result of mass loss during incubation and chick rearing, suggesting the existence of potential breeding costs.

Table 2 Measurements of Little bustard, by sex, obtained from several studies carried out in different regions of France and Spain

Population	Culmen	Wing	Tarsus	Body weight
France				
Deux Sevres[a]				
Males	25.0 ± 1.4 (18)	249.4 ± 4.9 (18)	65.8 ± 3.1 (18)	908.0 ± 72.2 (18)
Females	24.7 ± 1.6 (12)	246.8 ± 5.0 (12)	65.7 ± 3.9 (12)	883.4 ± 86.7 (12)
Crau (winter)[a]				
Males	23.7 ± 1.7 (31)	250.5 ± 6.0 (31)	68.8 ± 3.0 (31)	802.8 ± 89.3 (31)
Females	24.3 ± 1.9 (38)	248.2 ± 5.3 (38)	66.1 ± 3.0 (38)	783.3 ± 68.9 (38)
Gard[a]				
Males	25.9 ± 1.7 (36)	249.2 ± 5.6 (36)	66.6 ± 2.5 (36)	892.9 ± 80.7 (36)
Females	26.0 ± 1.5 (7)	245.9 ± 3.3 (7)	64.6 ± 3.3 (7)	843.6 ± 66.3 (7)
Spain				
Galicia[b]				
Males	–	241.5 ± 3.5 (2)	68 ± 0.7 (2)	851 ± 7.1 (2)
Castilla y León[b]				
Males	–	–	71.7 ± 1.4 (2)	909 ± 63.6 (4)
Navarra[b]				
Males	–	245.0 (1)	74.5 (1)	912 ± 11.3 (2)
Cataluña[c]				
Males		249 ± 18 (31)	74 ± 4 (33)	923 ± 74 (34)
Females		243 ± 8 (25)	68 ± 3 (26)	684 ± 46 (28)
Madrid[b]				
Males	21.3 ± 3.2 (3)	245.3 ± 6.9 (17)	69 ± 3.6 (16)	911.6 ± 70.1 (13)
Females	22 ± 0.2 (2)	241.0 (1)	63 ± 0.3 (2)	705 ± 145.7 (2)
Toledo (winter)[b]				
Males	22.3 ± 2.2 (4)	246.4 ± 5.0 (10)	69.3 ± 3.5 (10)	836.8 ± 46.3 (10)
Females	24.3 (1)	240.6 ± 5.5 (5)	66.3 ± 2.9 (5)	786.0 ± 80.0 (5)
Ciudad Real (winter)[b]				
Males	22.8 (1)	240 ± 11.7 (8)	68.6 ± 6.3.2 (8)	795.8 ± 52.7 (9)
Females	–	238.9 ± 6.1 (9)	73.2 ± 8.7 (8)	726.4 ± 36.0 (8)

Data are provided as mean ± SD (sample size)
[a]V. Bretagnolle, A. Wolff and P. Devoucoux unpublished data
[b]M.B. Morales and E. García de la Morena, unpublished data
[c]S. Mañosa and G. Bota, unpublished data

Sexual Dimorphism, Seasonal Variation in Plumage, and Moult

The major seasonal change in social behaviour is accompanied by changes in male plumage, which shifts from highly conspicuous (for the purpose of display) during the breeding period to camouflaged in winter (Plate 1e). In contrast, female and juvenile birds remain in camouflaged plumage in both winter and summer (Plate 1d

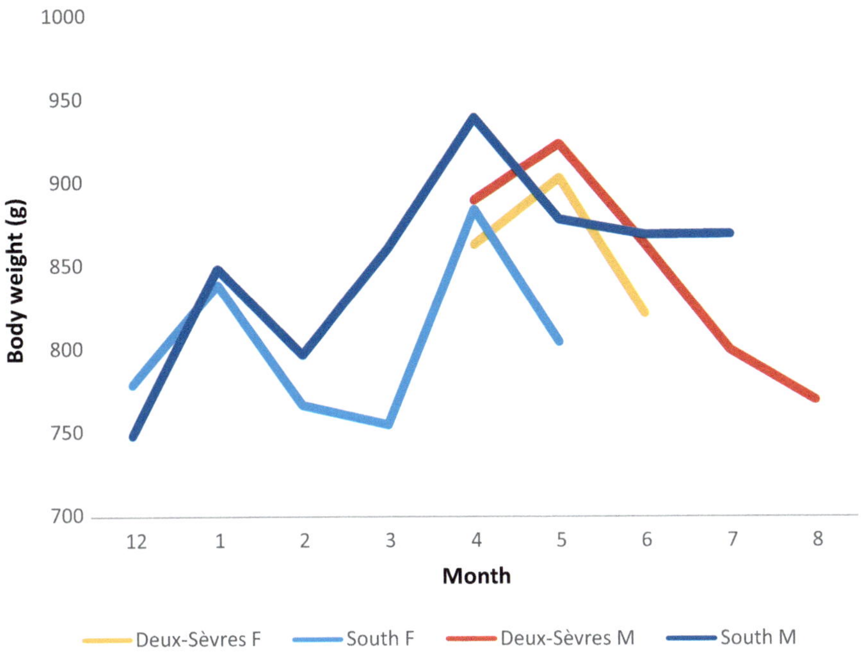

Fig. 2 Seasonal (months indicated by their number) variation in average body weight of adult little bustards (M: males; F: females) from two different French populations, one migratory (Deux-Sèvres) and one resident (South; two localities: La Crau and Nimes). In total, 124 birds of either sex and populations were captured and used here. Data provided in part by Axel Wolff, Pierrick Devoucoux, and Alex Villers

(i), e). Perhaps one of the most striking features of the Little Bustard is the emarginated fourth primary (when counting from wing tip, seventh primary otherwise). Fledgling and 1-year-old males lack this feature; it comes after the 1-year-old completes its moult in late summer (Géroudet 1978; Jiguet and Wolff 2000), so is only found in mature males, i.e. from 16 to 18 months onwards, in both summer and winter (Duquet 2018; Plate 1c). Winter and summer plumages are usually alike in bustards, and little bustards show perhaps the most extreme case of seasonal change in plumage, possibly in relation to the fact that more northerly populations are migratory, while the other species are mostly or entirely sedentary. In the field, fledglings (up to first-winter), first-year males in spring, females, and winter-plumage adult males can all be distinguished by plumage. The most difficult task is to separate fledglings from adult females in autumn post-breeding flocks (September–November). Making this distinction is important, since in many populations and field surveys it is the only way to estimate productivity at the population level.

Jiguet and Wolff (2000) have described the autumn–winter male, female, and juvenile plumages in detail and include drawings and photos of such plumages. In winter adult males there is a sharp change from the brown chest to the white belly, and their dorsal plumage, as in spring, shows a fine vermiculated pattern; the breast shows two triangular brown patches, sometimes separated from the neck by a cream band. Frequently, adult males exhibit patches of nuptial feathers on the lower neck, either as remnants of previous nuptial plumage or, in late winter, first signs of newly developing ones (Plate 1e). Finally, the male silhouette is more robust than that of females, showing broader necks and heads. In adult females, there is a gradual plumage transition from neck to chest, which shows black dots scattered over the whitish or greyish ventral parts (Plate 1b, d). The contrast between neck and ventral parts is a fairly reliable criterion to distinguish wintering males from females and juveniles in the field.

In lateral view, females show a few black-dot alignments and their dorsal parts and underwing-coverts show much thicker vermiculation, with black dots mixed with paler areas. The overall female colouration is darker than in males, but more reddish than in young birds, which present an ochre tone in early autumn. Juveniles younger than 60 days can be visually distinguished from adult females due to their smaller size and more gracile silhouette (see appendix in Bretagnolle et al. 2018). Since their plumage is completely new, they present a single generation of feathers. Their underwing-coverts are beige or ochre. However, this trait becomes useless when covert growth is completed. The middle coverts show an x-pattern close to their tip, which contrasts with the beige background (Plate 1e). After their first autumn moult, these traits become undetectable and differentiation between young and adult females becomes impossible.

The species' moult pattern is described in some details in Cramp and Simmons (1980), thus we will not extensively deal with it here. To summarize, there is a complete moult after breeding (i.e. July–September), except perhaps for some primaries (the last three or four). These are not moulted in winter, since moult stops completely between October and spring; thus they are moulted the following year. Body feathers are moulted again in early spring, just after arrival from migration in earliest birds, when they acquire their breeding plumage. Those arriving later from migration have already moulted their body feathers. Fledgling birds moult in winter, and usually complete their moult by December or January.

One of the most striking features of feather growth in chicks is that primaries are the first feathers to grow, and by the age of 15–20 days chicks are able to fly over short distances (20–30 m maximum) to escape predators. Their body is still covered with down at this age. Then, with age, body and tail feathers start to grow.

Natural History: Habitats, and Social and Breeding Behaviour

Habitats

Little bustards live in open habitat, including steppe, pseudosteppe, low and open scrubland, and extensive as well as intensive agricultural habitats. They avoid mountainous areas (avoid slopes higher than about 10°), although they can be found up to 2000 m asl. They also avoid marshlands (although they can winter, for instance, in Spain, in wetlands and very close to water), and forested areas. They therefore select large open and flat or nearly flat areas. A peculiar characteristic of little bustard ecology is that habitat selection depends on sex, particularly at fine scale (e.g. Morales et al. 2008; review in Devoucoux et al. 2019), and season (Suárez-Seoane et al. 2008). More details on habitat selection are provided in chapter "Habitat selection and space use".

Despite the fact that little bustards are large birds that live in fairly open habitats they can be surprisingly well hidden in the vegetation—not only females and young with their highly cryptic plumages, but even breeding plumage males. In the latter, although head and neck are basically black and white, birds can hide in vegetation by lowering their head, thus concealing the black and white colouration.

Most studies of Little Bustard ecology have focused on the breeding and post-breeding periods (Martinez 1994; Salamolard and Moreau 1999; Wolff et al. 2001, 2002; Silva et al. 2007; Traba et al. 2008; Morales et al. 2008, 2013; Delgado et al. 2010). Far less information is available on winter ecology (but see Silva et al. 2004; García de la Morena 2015; Cuscó et al. 2018), when bustards congregate in large flocks. Breeding and winter ecology are indeed different, in terms of social behaviour (i.e. gregariousness), food preferences and foraging (Bravo et al. 2016, 2017), home range size and daily activity patterns (Wolff 2001). There is a strong seasonal variation between sexes (Tarjuelo et al. 2013; Morales et al. 2008). In spring, males and females differ not only in plumage but also in relation to their habitats (Traba et al. 2015; see chapter "Habitat selection and space use"). Sex differences in spring and summer result from the species' mating system, which is an exploded lek system (Jiguet et al. 2000; Jiguet and Bretagnolle 2006). Indeed, females visit leks solely for the purpose of mating, and they will remain hidden in the vegetation most of the time, for nesting and rearing the chicks. Conversely, males remain mainly in open habitats, in order to be visible to females, and to attract them. Females and males therefore respond to rather opposite habitat characteristics. This results especially in strong habitat differences between the sexes in arable habitat, where the mosaic of crops results, at almost any time in the season, in a variety of crop heights. Conversely, in pastureland, which is more homogeneous at a large scale, vegetation growth is more or less simultaneous, and differences between the sexes at a macro-habitat scale are less pronounced (Devoucoux et al. 2019; see also chapter "Habitat selection and space use").

General Social Behaviour

Basically, birds of both sexes are more or less solitary in spring and summer (Morales et al. 2005), and become gregarious in winter. Flock size can be huge in winter, and although the average is usually around 100–200 birds (García de la Morena 2015) there are repeated records of flocks of thousands in south-west Asia (e.g. Gauger 2007), but also in Iberia (García de la Morena 2015) and southern France (Wolff

2001; Devoucoux 2014). Social bonds between breeding males and females are minimal. Both sexes tend to aggregate by sex, males in exploded leks (Ponjoan et al. 2012; see below), and females visiting leks in small groups, often 3–7 individuals. When males and females do encounter each other in spring, males very often harass females for copulations, so females try to avoid them. Males, despite their territorial behaviour, congregate in loose leks (Jiguet et al. 2000; see section "Spacing behaviour" and chapter "Behavioural ecology of the little bustard: sexual selection and mating systems"). Their densities were regularly monitored in Portugal over 7 years and shown to vary between years according to rainfall as well as percent grassland cover (Delgado and Moreira 2000). Weather influence on annual density was also recorded in central Spain (Delgado et al. 2009).

Other social bonds include family behaviour. Females alone incubate the clutch and rear their chicks, so family bonds are restricted to relationships between the mother and her chicks (Lett et al. 2000; Silva et al. 2014). These relationships are strong, since females take care of the brood for about 55–65 days. Females do not seem to feed chicks (except at the very first stage, until the age of about 7 days, as observed in captivity: V. Bretagnolle, pers. obs.), but show them insects that they will eventually catch. Females also guard their brood, and do not hesitate to take risks in face of predators (raptors, mammals) including humans. For instance, several females have been seen using the broken-wing display of many waders to draw predators or danger away from their chicks. Females also use particular alarm calls to warn the chicks, and chicks have also special contact calls (a whistle) to keep contact between themselves and with their mother. Scientists actually use these various behaviours by breeding females, particularly with chicks, to locate nests or at least the fields within which a nest or a brood is located (see Bretagnolle et al. 2018).

In winter, birds of either sex or age aggregate in flocks. Apart from some agonistic inter-individual interactions, there are no particular bonds or social links between birds in wintering flocks, either between sexes or between adults and first-year birds. Although wintering birds often commute between foraging areas (fields) and roosts, anecdotal evidence gathered from La Crau and Lleida with radio-tagged birds showed that groups are not constant in composition, so that from day to day individual birds change roosts and foraging fields, and thus the individual composition of flocks changes every day (Wolff 2001; S. Mañosa, G. Bota and F. Cuscó, unpublished data). In any case, little bustard winter habitat use seems to be largely constrained by the species' characteristic behavioural traits, such as conspecific attraction and site fidelity; wintering females exhibit inter-individual consistency in habitat selection, in regard to terrain slope, alfalfa, corn stubble, and irrigated cereal stubble availability, but these interact strongly with social factors (Cuscó et al. 2018).

Spacing Behaviour

During the breeding season, displaying male little bustards concentrate in more or less loose aggregations (Schulz 1985; Jiguet et al. 2000). The nature of these aggregations, and how variable this pattern might be, are still a matter of study and debate (see, e.g., Morales et al. 2001, 2014; Jiguet and Bretagnolle 2006, 2014). The causes of such clumping (present in all lekking species) have long intrigued behavioural ecologists, and some research on this subject has been conducted with the little bustard as well. One possibility is that male aggregation is simply the result of some kind of habitat clumping (*landmark aggregation* hypothesis: see Höglund and Alatalo 1995), which means that one of the conditions of lekking—namely, that male aggregation must be independent from habitat features—would not be fulfilled. Nevertheless, although habitat may obviously cause some degree of aggregation at larger scales, several studies have revealed that male little bustards aggregate even within patches of homogeneous habitat where there is no apparent habitat clumping (Jiguet et al. 2000), which discounts the possibility of landmark aggregations in these circumstances. However, it is not always easy to separate the effect of habitat clumping in the aggregation pattern. In highly intensified agricultural habitats, optimal patches for display or feeding might be highly clumped, promoting a higher aggregation pattern than would be expected otherwise. An experimental manipulation of lek size using male decoys in various habitat landscapes (Jiguet and Bretagnolle 2006) seemed however to rule out this possibility (see chapter "Behavioural ecology of the little bustard: sexual selection and mating systems" for more results).

Conversely, females seem to be territorial in the breeding season, particularly when incubating. Although they do not defend territories with fixed boundaries like many other birds such as passerines or raptors, there are many examples and instances of females chasing other females, especially at the laying period, presumably to avoid or reduce intra-specific competition for food during chick rearing. However, there is also repeated evidence of females nesting in the same field (e.g. Devoucoux et al. 2019; Bretagnolle et al. 2018). M.B. Morales and E.L. García found two active nests in the same small fallow field in 2002, and up to three active nests were found in the same small alfalfa field in 1998 by F. Jiguet and V. Bretagnolle. Interestingly, those nests were found because the field was mown, i.e. females were very much hidden in the field at the time of laying. Such observations therefore raise the possibility of egg dumping (a female laying in another female's nest, i.e. a kind of parasitism), a phenomenon observed in captive breeding little bustard programmes in a single large aviary in Chizé (France) between 1997 and 2003, when we found eggs from different females (recognized by colour pattern, size, and shape) in the same nest (they were not incubated by a female as they were removed to be artificially incubated).

Overall therefore, in spring and summer (breeding season), males aggregate in loose concentrations where they defend small territories, while females do not aggregate (except when visiting leks) and repel each other, although not necessarily defending a fixed territory but simply excluding other females at close range. Both sexes become highly gregarious in winter, and gather in medium to large flocks.

Breeding Behaviour and Life History Traits

Extremely little is known about copulations in Little Bustard, and indeed bustards in general. Copulation has been either described or filmed only in Great bustard (Hellmich 1991) and Kori (Hellmich 1988), as well as both species of houbara (mostly in captivity). However, in the little bustard there is no single published description of copulation behaviour, and we know only a few instances when observations of this behaviour have been reported in the wild. This is strange, since copulations in lekking species are usually numerous (Höglund and Alatalo 1995), and should be observable. Therefore, copulation probably takes place at a time where it is very difficult to watch, probably just before night in the evenings, or even at night. Interestingly, although there are nowadays several facilities where little bustards are kept in captivity for conservation purposes, sometimes in fairly large numbers, copulations have not been seen either. Alternatively, copulations in this species may not be so common, which would be inconsistent with the observations that suggest a relatively important sperm competition (see below and chapter "Behavioural ecology of the little bustard: sexual selection and mating systems").

Once copulation has been achieved, females search for a place for nest location. Anecdotal evidence (Jiguet and Bretagnolle, unpubl. data) based on radio-tracked females suggest that there may be up to 2 weeks between copulation time and egg laying, leaving enough time for sperm competition to arise (see also chapter "Behavioural ecology of the little bustard: sexual selection and mating systems") and suggesting that females can store male sperm for some time. Nests are very simple, only built by females (which take care of all breeding tasks), and consist usually of a scrape in the ground, sometimes with a few pieces of vegetation added (see chapter "Breeding biology and demographic traits").

Clutches are comprised of large eggs, oval and green (presumably coloured mostly if not only with biliverdin). There is, however, a wide range of shading, even within a clutch, from dark olive eggs to very pale ones, perhaps especially so in captive-bred females. There is some evidence that late eggs within a clutch are paler than first-laid eggs, suggesting that females, who synthesize pigment, may become short of it. Although egg colouration in little bustard may not appear as cryptic as in other species (e.g. stone curlew *Burhinus oedicnemus*), eggs are very difficult to see in the nest, which is most often in the shadow of vegetation.

Little bustards are rather long-lived birds. Males usually do not breed in their first year (i.e. they do not display nor copulate), although there are rare exceptions. In the captive breeding facilities at Chizé (Deux Sèvres, France), some 1-year-old males have displayed in their first spring, despite being in first-year plumage. Some of them are even able to perform calls. There is even evidence (also from captive breeding stock in Chizé) that these first-year males are actually able to mate with females and fertilize eggs. Finally, there is at least one observation of a wild male juvenile performing the jump display (see below) during the breeding season of 2017 in Lleida (north-eastern Spain; G. Bota personal communication). Females, on the other hand, are able to breed in their first year of life, since radio-tagged birds from France were shown to nest successfully (Jiguet and Bretagnolle unpubl. data).

Survival rates are dealt with in detail in chapter "Breeding biology and demographic traits". Longevity is known for captive birds: the oldest male in a captive breeding stock in France is at least 17 years old, and several males in the same stock have reached the age of 14. In the wild, a male (incidentally, a captive-raised bird) survived until 15 years old, but the current oldest-recorded birds are two GPS-tagged males, one in Deux Sèvres (France), which was 18 years old in 2018, and a male in Lleida (Mañosa et al. 2018), which was re-observed in spring 2019 (Roger Guillem, pers. comm.) as a 15-year-old. Causes of mortality have been studied in detail in birds equipped with VHF, satellite, and GPS transmitters (Marcelino et al. 2017). Of 151 birds tracked over 12 years, collision with powerlines was found to be the cause of death of many birds (3.4–3.8%/year), followed by illegal killing (2.4–3%/year). Poaching was however the main cause.

Parental Care and Lekking

There is general agreement about the fact that little bustard males do not perform any sort of parental care (Schulz 1985; Jiguet et al. 2000). This is the dominant pattern among bustard species, in most of which males take no part in protecting and rearing their offspring (see chapter "The little bustard and its family: an overview of relationships"), and it is also found in some other avian groups such as the galliforms, waders, and ducks (Johnsgard 1994). Why is this so? In fact, this is an exception among birds where, in most species, males contribute to some extent to nest building, incubation, food provisioning and/or protection of the female and young, or other sorts of parental care. However, little bustard chicks are precocial, and can feed by themselves early in life, so it is possible that male assistance is of little help during the brooding phase, although males of certain other precocial species, such as partridges, do give parental care. Other factors are thus probably involved. The more or less strict herbivorous diet of adult little bustards may play some role, as grasses and weeds are relatively abundant and easy to find around the nests, so that incubating females can feed by themselves relatively easily, and thus the help of a provisioning or an incubation-sharing male would be of little value. In these circumstances, a little bustard male may obtain much greater fitness benefits by trying to mate with as many females as possible (hence a positive selection for the lek mating strategy) and leaving them in charge of egg and chick caring, than by getting involved in parental care of a single or several broods. Theoretical, empirical, and even experimental studies have been performed in order to better understand why little bustards show an exploded lek mating system (Morales et al. 2001; Jiguet and Bretagnolle 2006, 2014).

Lekking is a peculiar type of polygynous mating system in which males aggregate and defend small territories that females visit only with the purpose of mating (Höglund and Alatalo 1995). Lekking is thus a non-resource-based mating system in which females select mates among males (Bradbury and Gibson 1983). In such systems, males have no, or very little, contribution to their offspring fitness, apart from their genes. But males have evolved a suite of characters (morphological, behavioural) that signal their "good" genes in order to attract females (Andersson 1994; Höglund and Alatalo 1995). In the Little Bustard, males have evolved particular distinctive plumages, call and display activities (see below). But another striking peculiarity is that lekking in the Little Bustard is associated with territorial defence of rather large territories, in a particular form of lekking, the exploded lek, that is best characterized by male spacing behaviour. Exploded leks are intermediate between classical leks, in which males are highly clumped, and resource-based polygyny, in which females benefit from resources controlled by territorial males (Höglund and Alatalo 1995). The variation between these three types of mating systems (that should rather be seen as a gradient) is explored in more details in chapter "Behavioural ecology of the little bustard: sexual selection and mating systems".

Foraging, Diet, and Feeding Behaviour

Diet

Almost all bustards are omnivorous and, considering the whole life cycle, this is also the case for the little bustard. Quantitative studies on little bustard diet have been conducted in France and Spain (Jiguet 2002; Bravo et al. 2017), but in general most information is based on casual data and qualitative assessments, obtained from field observations or stomach content analysis (Table 3). The scarcity of diet studies is surprising given the relatively large number of works on little bustard ecology (Morales and Traba 2016) as well as the species' own ecological flexibility. Little bustard diet consists mainly in green plant material (leaves, stems, and shoots), seeds and arthropods (Cramp and Simmons 1980), with very rare use of vertebrates. However, this general description hides seasonal, sexual, and age-related variation in diet.

Diet during the breeding season has been quantitatively studied in western France (Jiguet 2002). The proportion of plant material was >97% in adult faeces, the rest being arthropod remains. No difference was found in this respect between males and females, but the proportion of arthropods was significantly higher in summer than in spring. In the Transvolga region, Shlyakhtin et al. (2004) also found a preponderance of plant material (>90%) in adult faeces early in the breeding season, which decreased to 66% in the post-breeding season. However, there are reports of little bustards consuming large amounts of locusts (Devishev 1958), in Shlyakhtin et al. 2004), indicating that the little bustard can adapt its feeding habits to the availability

Table 3 Little bustard diet across the species' range. Data are obtained from different studies in France and Spain, based on either stomach or faeces contents. The average percentage of abundance of each diet component is provided (± SD when available) at the family level

	France		Spain	
Taxon	Stomach[a] (Breeding, $N = 3$)[a]	Faeces[a] (March–October, $N = 388$)[a]	Stomach[b] (Winter, $N = 3$)[b]	Faeces[c] (Winter, $N = 357$)[c]
Plants				
Asteraceae	–	–	27.0 ± 29.6	9.1 ± 12.8
Brassicaceae	–	–	11.9 ± 11.8	23.4 ± 20.3
Caryophyllaceae	–	–	1.3 ± 2.33	0
Convolvulaceae	–	–	0	1.8 ± 5.4
Fabaceae (cultivated)	–	–	53.1 ± 31.6	46.7 ± 34.5
Fabaceae (wild)	–	–	0	1.0 ± 3.0
Labiatae	–	–	0	4.1 ± 7.3
Papaveraceae	–	–	1.7 ± 1.5	4.7 ± 13.7
Plantaginaceae	–	–	1.0 ± 1.7	
Poaceae (cultivated)	–	–	0	5.8 ± 9.0
Poaceae (wild)	–	–	3.5 ± 3.14	1.9 ± 2.1
Scrophulariaceae	–	–	0.5 ± 0.9	0
Others	–	–	0	1.5 ± 1.5
Invertebrates				
Coleoptera	67.3 ± 34.1	55.9	47.6 ± 44.2	–
Carabidae	–	–	12.5 ± 16.6	–
Crisomelidae	–	–	14.9 ± 25.8	–
Curculionidae	–	–	20.3 ± 31.2	–
Dermaptera	4.4 ± 4.9	20.6	0	–
Orthoptera	21.4 ± 34.1	11.0	0	–
Hymenoptera	0	8.1	19.0 ± 23.0	–
Formicidae	0	–	19.0 ± 23.0	–
Lepidoptera	1.2 ± 2.1	3.1	0	–
Heteroptera	5.6 ± 5.0	0.4	0	–
Diptera	0	0.4	0	–
Opilionae	0	0.6	0	–
Araneidae	0	0.2	0	–

[a]Based on Jiguet (2002)
[b]M.B. Morales and C. Bravo, unpublished data
[c]From Bravo et al. (2017)

of food resources and/or its dietary needs. The plant component of the diet of breeding adults varies according to the availability of plants in any particular region, but is dominated by annual dicotyledonous plants (basically legumes) and grasses (Gramineae), but also brassicaceae, asteraceae, chenopodiaceae, and papaveraceae (Cramp and Simmons 1980; Glutz von Blotzheim and Bauer 1994; Boutin and Métais 1995; Shlyakhtin et al. 2004).

Chicks and fledglings are almost exclusively insectivorous during the 2–3 first weeks of life. Jiguet (2002) found a predominance of coleopterans, followed by orthopterans, dermapterans, and himenopterans (mainly ants). The relative proportion of each group changed over the season, with orthopterans increasing in relative importance during the chick rearing stage.

In autumn and winter all birds, regardless of sex or age, become fully folivorous, although on warm days they can eat some insects such as ants. Diet during winter in the Iberian Peninsula is almost entirely based on herbs and annual plants of different families (Fabaceae, Brassicaceae, Asteraceae, and Amarantaceae, subfamily Chenopodioideae), but with a preference for legumes (Fabaceae) (Bravo et al. 2017). As during the breeding season, the taxonomic composition varies at a regional level and depends on the availability of plants, which is also influenced by the farming regime. In some steppe or semi-desert areas of Azerbaijan, Little Bustards have been observed feeding on *Artemisia* during the winter (Gauger 2007), and the main plant eaten in the Transvolga region in autumn was the asteracean *Lactuca tatarica* (Shlyakhtin et al. 2004). The stomach examination of two adult males and one female found dead in winter in central and western Spain (M.B. Morales and C. Bravo, own unpublished data) showed that 89% of contents was plant material (plant leaves and shoots: 84.7%; seeds: 2.9%; flowers: 1.3%), while only 8% consisted of insects and 3.1% of small pebbles (presumably to grind vegetation to help with digestion). Moreover, analysis of winter diet has shown a marginal presence of seeds in little bustard droppings (seeds present in 0.8% of 357 droppings, with mean dry weight proportion of 0.001%: Bravo et al. 2017).

Foraging

Foraging behaviour in the little bustard has not received strong attention. One main driver of foraging behaviour relies obviously on prey type, as foraging on highly mobile and agile grasshoppers requires different behaviour compared to rapeseed or alfalfa leaves. Indeed, associated with variability of diet, foraging behaviour is also variable. Little bustards capture grasshoppers by sight (no evidence of hunting by ear, as in other birds), often running after them, depending mainly on grasshopper age since very young grasshoppers are easier to catch. They prey upon coleopterans opportunistically, pecking animals as they see them or foraging actively by walking. The relatively high percentage of ants in the total of arthropods, revealed by the examination of some stomachs' contents (Table 3), suggests that birds may in some

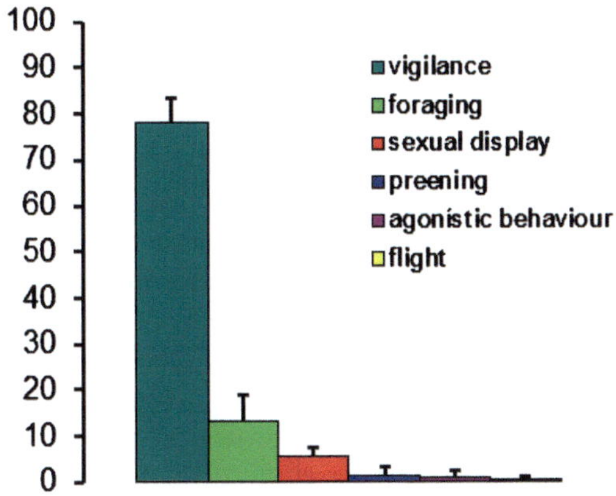

Fig. 3 Mean ± SD ($N = 13$) percentage of time dedicated to different activities by little bustard males during daily activity peaks of the mating season. Overall differences were statistically significant (ANOVA $F_{5,72} = 499.72$, $p < 0.0001$). Only the percentage of time devoted to sexual display differed significantly between individuals (see text)

instances peck for long periods at the entrance to ant nests, as observed in the great bustard (Lane et al. 1999).

Foraging times and diurnal rhythm have been little studied (but see Jiguet and Bretagnolle 2001). M.B. Morales made a complete account of breeding males' time budget during the peaks of diurnal activity (early mornings and late afternoons) in April and May 2001, based on focal observations (30 min of continuous recording) of 13 males identified by phenotype (Arroyo and Bretagnolle 1999). As shown in Fig. 3, males devoted 13% of their time to foraging activities. Casas et al. (2009) studied the trade-off between vigilance and other activities, including feeding, during the hunting season in France (early October to early December) and found that the frequency of flights and vigilance time increased on hunting days to the detriment of resting time, although no significant decrease in time devoted to foraging was detected (see also chapter "Migration, movements and non-breeding ecology").

Metabolism and Food Requirements

We know almost nothing about Little Bustard energy requirements. There has been no measure so far of either Resting Metabolic Rate or Basal Metabolic Rate (BMR). Some data are available on energy requirements on a daily basis, at least for chicks. Whatever their age, chicks need on average 200 grasshoppers per day, the size of the grasshoppers being adjusted to chick age, that is, they feed on juvenile grasshoppers (c. 1 cm body length or even less) in their first days, medium-sized individuals at 1 month of age (1–2 cm body length), and larger adult grasshoppers afterwards. For instance, adult or large chicks often eat large *Calliptamus italicus* females, often longer that 3 cm (Jiguet 2002). Daily foraging time has been little investigated (but see Casas et al. 2009).

In winter, presumably, little bustards require more food. However, evidence gathered from non-migrating birds that winter in Deux Sèvres (France) suggests rather that birds try to minimize their energy expenditure. For instance, in 1997, a small group of three birds spent the entire winter in a single oilseed rape field, and within that field mostly remained within an area not larger than 1000 m^2. They were only eating rapeseed leaves. Sedentary behaviour is a common feature now in south-western France, and little bustards can rather easily tolerate negative temperatures (V. Bretagnolle pers. obs.). However, they strictly avoid snow; in snowy episodes, they move south. Evidence obtained with GPS-tagged birds has shown that on such occasions birds only move 100 or 200 km south, and come back to their usual wintering field as soon as snow has melted. In many areas of the Spanish northern plateau, where snowfalls are not uncommon, the species is absent in winter (García de la Morena et al. 2015, see also chapter "Migration, movements and non-breeding ecology").

Daily Activity Patterns

Based on M.B. Morales' 2001 study, males devote most of their time to vigilance (78%), followed distantly by foraging (13%) and sexual display (5.4%), while the time devoted to other activities was marginal (agonistic interactions: 1.1%; preening: 1.6%; flight: 0.4%; Fig. 3). Most of the time between the morning and evening activity periods is devoted to resting, though there is very little quantitative information on little bustard activity during the day, particularly in the breeding season. Moreover, female activity, either during or between display peaks is largely unexplored. General description of daily activity patterns are found in Jiguet and Bretagnolle (2001), Silva et al. (2015), and Gudka et al. (2019). The most studied activity has been calling rate (snort call, see below) in males. In a detailed study on the effects of ambient air temperature of male general activity, including breeding and non-breeding season, Silva et al. (2015) found that temperature was a main driver and trigger of several behaviours in males. Calling rate varies throughout the season, with a gradual increase at the start of the breeding season (Gudka et al. 2019); it peaks at sunrise and sunset, as in other bustard species, but unlike other bustards calling continues all day long (at least under mild climate, since temperatures above 25 °C were shown to limit, or even stop, any activity: see Silva et al. 2015; Gudka et al. 2019). M.B. Morales found marginally significant differences between the 13 individuals sampled in snort call rate and the total percentage of time devoted to display (Kruskal–Wallis test, $H\ (12, 34) = 21.09$, $p = 0.05$ and $H\ (12, 34) = 20.98$, $p = 0.05$, respectively). However, no differences were found between individuals in the percentages of total observation time of either vigilance, foraging, preening, agonistic interactions, or flight. High inter-individual variability in snort call rate was also found in another study population of central Spain, where this rate and acoustic snort call traits were analysed in 17 males in relation to anthropogenic noise (Barrero et al. 2021). In a recent study using GPS data and accelerometer (Gudka et al. 2019), snort-call display was found to decrease with increasing temperature, especially above 30 °C; birds snorted more in the early morning, while daily snort call activity (proportion of display events in each day) also decreased with increasing daily mean daytime temperature.

Visual Displays and Calls

The first very detailed description of little Bustard behaviour was published by Schulz (1986). Male visual displays of bustard are quite spectacular, and this is also the case for the little bustard. There are however surprisingly few descriptions and accounts of little bustard male displays (but see Cramp and Simmons 1980; Schulz 1986; Jiguet and Bretagnolle 2001), which appears to be both variable and complex. Indeed, there is a gradient of male sexual displays, with not all males using the same display types. There is also within-male variation. The function of each display is not well known (see also chapter "Behavioural ecology of the little bustard: sexual selection and mating systems"). Only two studies attempted to describe displays in detail and provide their function (Jiguet and Bretagnolle 2001, 2014). Visual displays during breeding can be split into five categories:

Neck Display Males have the ability to erect some feathers of their neck, to a varying extent (see Plate 3a for a range of such displays). When resting, all collar feathers lie along the neck (left of Plate 3a). In slight erection only some feathers are erected. A paroxysmal version of this display can occur when males chase other males out of their territory, or when they pursue females for copulation, and also in flight (see extreme right on Plate 3a). When chasing females, males adopt a slightly different version of the neck display in which the neck is not elongated at all and feathers are erected at their maximum.

Plate 3 Little Bustard visual and acoustic displays. (**a**) various postures and variation in neck display, from normal (vigilance) posture to flight escort or female pursuit (Photos: Vincent Bretagnolle). (**b**) snort call display (Photo: Carlos Palacín). (**c**) wing-flash display (Photo: Didier Buysse). (**d**) jump display (Photo: Louis-Marie Preau). (**e**) males fighting (Photo: Vincent Bretagnolle)

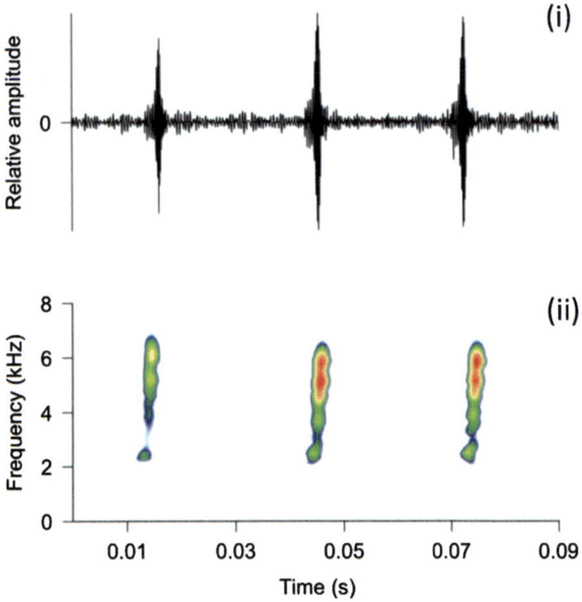

Fig. 4 Acoustic display (snort call) of a little bustard male recorded in the field: (i) Oscillogram and (ii) spectrogram. Modified from Barrero et al. (2021)

Snort call is the basic display, consisting of a brief vocalization uttered with a sharp toss of the head (Plate 3b). Males stand upright with slightly flattened wings, usually (but not always) with the neck display, and then at frequent intervals jerk back their head and utter a monosyllabic snort "*prrtt*", which although not very loud can be heard at quite long distances. The sound is to some extent "ventriloquial", because it sounds alike whether at 25 or 500 m, and can give the impression that a bird calling may come from different locations. The snort call consists of three notes whose sound frequency ranges from ca. 1500 to ca. 7000 Hz and are separated by intervals of ca. 0.025 s, so that the entire call duration is ca 0.06 s. Maximum acoustic energy is concentrated in the range of ca. 4000–6000 Hz. Figure 4 shows the oscillogram and spectrogram of the snort call by a particular male recorded in the field (see also Barrero et al. 2021). The snort call can be preceded by 1–4 stereotyped foot steps (*stamped snort call*). Actually, because males usually use the same well-defined display sites, grass tends to get beaten down and the area covered becomes bare ground (see photos in Boutin and Métais 1995). Snort call rates apparently do not differ whether a female is close to the male or not (Jiguet and Bretagnolle 2001). The snort call display is mainly a territorial behaviour, as suggested by playback tests of snort call on territorial males that always respond with a snort call; in addition, males perform the snort call display almost all day long, indicating that the snort call is mainly involved in male–male interactions (Jiguet and Bretagnolle 2001, 2014). Whether the snort call also has an intersexual function is not known, although it cannot yet be discarded (Morales et al. 2014).

Wing-flash display (called wing-beat display in Schulz 1985), in which the bird stamps its feet on the ground, utters the snort call and then beats its wings with feet still touching ground, producing a characteristic whistle from the seventh primary (Plate 3c). The wing-flash is always preceded by foot stamping, of variable duration. It results in a sudden white flash which is highly striking, especially in twilight. Wing-flash display, conversely to snort call, has mainly a sexual function and is involved in intersexual communication. First, it is most likely performed when a female visits the display site (Jiguet and Bretagnolle 2001), although it is not known whether female presence is the cause or a consequence of this behaviour. Vegetation height was also found to affect the occurrence of wing-flash display, and it is likely that in habitats that preclude the propagation of visual cues (e.g. high vegetation) males will not invest in wing-flash display and may be more inclined to use the jump display: indeed there is anecdotal evidence that the wing-flash display is comparatively rarer in southern France than in Deux-Sèvres, contrary to the jump display (pers. obs.). Wing-flash display is more costly to produce than the snort call, and actually some males were never seen to use this display; it also reduces snort call rate when used (Jiguet and Bretagnolle 2001). Schulz (1986) proposed that the wing-flash display had also a territorial (thus intra-sexual) function. Jiguet and Bretagnolle (2001) also found that the probability that a given male used wing-flash display increased with group size (see also Petretti 1993), thus supporting a territorial function for this display, depending on local population density (i.e. lek size).

Jump consists of foot stamping, snort calling and finally a jump with the wings beating to enhance both the whistle and the visual signal of the white wings (Plate 3d). The height of the jump varies from 20 to 100 cm, and usually the bird lands at the very same position. The jump display has an obvious intersexual function, as it is only performed in the presence of a female, whatever the time of the day (conversely to wing-flash which is restricted to early morning and late evening), although it is observed more often in the morning or in the evening (Schulz 1986; Petretti 1993; Jiguet and Bretagnolle 2001). The jump display can be seen in direct response to the arrival of a female on the display site, although it is not systematically performed in such circumstances. It is the most extravagant display, highly variable between but not within individuals, and is the best candidate of all behaviour traits to be involved in active female attraction (Jiguet and Bretagnolle 2001, 2014). Moreover, M.B. Morales analysed the frequency of jump displays among the focal males whose activity time budget is described above (see Foraging) in relation to the symmetry of their neck design (Arroyo and Bretagnolle 1999), and found that symmetric males jumped significantly more often than asymmetric ones (Fisher's exact probability test, $p = 0.03$, $N = 13$). Symmetry in bilateral phenotypic traits is correlated with the individual's capacity to maintain homeostasis (Møller 1990; Møller and Höglund 1991; Andersson 1994), and symmetric traits have been associated with increased fitness (Badyaev et al. 1998), so that they may function as an honest signal of individual quality. Therefore, neck-symmetric little bustard males would be in better condition to perform this costly intersexual display at higher frequencies. Indeed, Jiguet and Bretagnolle (2006) experimentally showed that leks containing more symmetric male decoys were more attractive to females.

Agonistic Behaviour Males are often seen in agonistic behaviour. It starts with an aggressive posture (Schulz 1986), which differs from the alarm posture (see below) by the angle of the neck and bill. Threat display includes wings opened and running. Fights occur quite often between males (and sometimes, although rarely, between females), and are usually preceded by a side-by-side or face-to-face position, at 1 m from each other (Plate 3e). If no male flees (usually by flight), then a fight occurs (see description in Schulz 1986). In the course of different capture attempts and experiments using male decoys to attract territorial males, the latter have been heard to produce a prolonged agonistic groan accompanying the aggressive posture (M.B. Morales, G. Bota and E. García de la Morena own observations).

Submissive or Fleeing Behaviour Satellite males (or sneakers, see chapter "Behavioural ecology of the little bustard: sexual selection and mating systems"), do not usually display (but see exceptions mentioned above), and are always hiding, or moving slowly within the vegetation cover. The neck is never elongated, and they walk slowly and gently.

Tail Raised Display This display is mostly used by females, but 1-year-old males also use it when chased by adult males, and on some occasions adult males also perform it. It is also often seen in post-breeding groups, by all sexes and ages. The meaning of the display is unclear: females use it clearly against pressing males, and thus it could be interpreted as a threat display against copulation. But both males and females also use it sometimes outside the copulation period.

Ritualized Flight In the breeding season, males often fight on the displaying site. Quite often one of the males, and in many cases two, take flight. They engage in escort flight, in which a territorial male pursues in flight an intruder, escorting him outside his territory. Escort flight is known in many species, but in little bustard it is ritualized as males keep a particular posture. Males also use escort flight toward females, but for exactly the opposite reason, since they try to keep the female or the female group *within* their territories. In such cases, they usually do not use the ritualized flight posture.

Alarm Posture Neck elongated at maximum and head pointing slightly above horizontal, with all feathers of the body and the neck pressed against the body.

Acknowledgements The authors wish to thank the museum curators, Josefina Barreiro, curator of vertebrate collections at the Natural Science Museum of Madrid, Pascal Eckhoff and Sylke Frahnert from the Museum of Berlin, Mark Adams from Tring (UK), Jerôme Fuchs from the Muséum National d'Histoire Naturelle (Paris), the late Anita Gamauf from Vienna, and Paul R. Sweet from the AMNH in New York, who all kindly gave us access to preserved little bustard specimens. The authors wish to thank also Axel Wolff, Pierrick Devoucoux, and Alex Villers for providing unpublished data. Nigel Collar extensively commented on a previous version of this chapter and kindly revised the English.

References

Andersson M (1994) Sexual selection. Princeton University Press, Princeton, NJ

Arroyo B, Bretagnolle V (1999) Field identification of individual Little Bustard Tetrax tetrax males using plumage patterns. Ardeola 46(1):53–60

Badyaev AV, Etjes WJ, Faust JD, Martin TE (1998) Fitness correlates of spur length and spur asymmetry in male wild turkeys. J Anim Ecol 67:854–852

Bailey TA, Mensah-Brown EP, Samour JH, Naldo J, Lawrence P, Garner A (1997) Comparative morphology of the alimentary tract and its glandular derivatives of captive bustards. J Anat 191: 387–398

Barrero A, Llusia D, Traba J, Iglesias-Merchán C, Morales MB (2021) Vocal response to traffic noise in a non-passerine bird: the little bustard *Tetrax tetrax*. Ardeola. https://doi.org/10.13157/arla.68.1.2021.ra8

Bautista LM, Silván G, Cáceres S, Martínez-Fernández L, Bravo C, Illera JC, Alonso JC, Blanco G (2013) Faecal sexual steroids in sex typing and endocrine status of great bustards. Eur J Wildl Res 59:815–822

Boutin JM, Métais M (1995) L'outarde canepetière. Eveil Editeur, Paris

Bradbury JW, Gibson RM (1983) Leks and mate choice. In: Bateson PPG (ed) Mate choice. Cambridge University Press, Cambridge, pp 107–138

Bravo C, Ponce C, Bautista LM, Alonso JC (2016) Dietary divergence in the most sexually size-dimorphic bird. Auk 133:178–197. https://doi.org/10.1642/AUK-15-206.1

Bravo C, Cuscó F, Morales MB, Mañosa S (2017) Diet composition of a declining steppe bird the Little Bustard (*Tetrax tetrax*) in relation to farming practices. Avian Conserv Ecol 12:3. https://doi.org/10.5751/ACE-00938-120103

Bretagnolle V, Denonfoux L, Villers A (2018) Are farming and birds irreconcilable? A 21-year study of bustard nesting ecology in intensive agroecosystems. Biol Conserv 228:27–35

Casas F, Mougeot F, Viñuela J, Bretagnolle V (2009) Effects of hunting on the behaviour and spatial distribution of farmland birds: importance of hunting-free refuges in agricultural areas. Anim Conserv 12:346–354

Clench M, Mathias J (1995) The avian caecum: review. Wilson Bull 107:93–121

Collar NJ (1996) Family Otididae (bustards). In: del Hoyo J, Elliott A, Sargatal J (eds) Handbook of the birds of the world, vol 3. Lynx Edicions, Barcelona, pp 240–273

Cramp S, Simmons KEL (eds) (1980) The birds of the western Palearctic, vol 2. Oxford University Press, Oxford

Cuscó F, Cardador L, Bota G, Morales M, Mañosa S (2018) Inter-individual consistency in hábitat selection patterns and spatial range constraints of female little bustards during the non-breeding season. BMC Ecol 18:56

Delgado A, Moreira F (2000) Bird assemblages of an Iberian cereal steppe. Agric Ecosyst Environ 78:65–76

Delgado MP, Morales MB, Traba J, García de la Morena EL (2009) Determining the effects of habitat management and climate on the population trends of a declining steppe bird. Ibis 151: 440–451

Delgado MP, Traba J, García de la Morena EL, Morales MB (2010) Habitat selection and density-dependent relationships in spatial occupancy by male Little Bustards *Tetrax tetrax*. Ardea 98(2): 185–194

Devishev RA (1958) (translated: Methodological letter to the schools of Saratov oblast devoted to activities aimed at protection and attraction of useful birds). Saratov. Gos. Univ., Saratov

Devoucoux P (2014) Conséquence et impacts prévisibles d'une perte d'habitat majeure sur une espèce menace aux exigences écologiques complexes: Dynamique de la population d'outardes canepetières des Costières de Nîmes et construction de la Ligne à Grand Vitesse CNM. Octubre 2014. PhD thesis, University of Poitiers, Poitiers

Devoucoux P, Besnard A, Bretagnolle V (2019) Sex-dependent habitat selection in a high-density Little Bustard *Tetrax tetrax* population in southern France, and the implications for conservation. The Ibis 16:310–324

Duquet M (2018) A propos du sifflement d'aile de l'Outarde canepetière *Tetrax tetrax* mâle en vol. Ornithos 25-6:370–381

Galván I, Camarero PR, Mateo R, Negro JJ (2016) Porphyrins produce uniquely ephemeral animal colouration: a possible signal of virginity. Sci Rep 6:39210

García de la Morena E (2015) Ecología y movimientos migratorios del sisón del sisón común (*Tetrax tetrax*) fuera del periodo reproductor. Autónoma University of Madrid, Dissertation

García de la Morena EL, Morales MB, Bota G, Silva JP, Ponjoan A, Suárez F, Mañosa S, De Juana E (2015) Migration Patterns of Iberian Little Bustards Tetrax tetrax. Ardeola 62:95–112

Gauger K (2007) Occurrence, ecology and conservation of wintering Little Bustards *Tetrax tetrax* in Azerbaijan. Arch Nat Landsch 46:5–22

Géroudet P (1978) Grands Echassiers, Gallinacees et Râles d'Europe. Delachaux-Niestle, Paris

Glutz von Blotzheim UN, Bauer, KM (1994) Handbuch der Vögel Mitteleuropas. Band 5: Galliformes und Gruiformes. Akademische Verlag, Frankfurt am Main

Gudka M, Santos CD, Dolman PM, Abad-Gomez JM, Silva JP (2019) Feeling the heat: elevated temperature affects male display activity of a lekking grassland bird. PLoS One 14:e0221999. https://doi.org/10.1371/journal.pone.0221999

Hellmich J (1988) Zum Balz-Verhalten der Riesentrappe (Ardeotis kori). Zool Garten 58:345–352

Hellmich J (1991) La Avutarda en Extremadura. Monografías Alytes 2, Adenex, Mérida (Spain)

Höglund J, Alatalo RV (1995) Leks. Princeton University Press, Princeton, NJ

Jiguet F (2002) Arthropods in diet of Little Bustards *Tetrax tetrax* during the breeding season in western France. Bird Study 49:105–109

Jiguet F, Bretagnolle V (2001) Courtship behaviour in a lekking species: individual variations and settlement tactics in male Little Bustard. Behav Process 55:107–118

Jiguet F, Bretagnolle V (2006) Manipulating lek size and composition using decoys: an experimental investigation of lek evolution models. Am Nat 168:758–768

Jiguet F, Bretagnolle V (2014) Sexy males and choosy females on exploded leks: correlates of male attractiveness in the Little Bustard. Behav Process 103:246–255

Jiguet F, Wolff A (2000) Déterminer l'âge et le sexe des Outardes canepetières *Tetrax tetrax* à l'automne. Ornithos 7(1):30–33

Jiguet F, Arroyo B, Bretagnolle V (2000) Lek mating systems: a case study in the Little Bustard *Tetrax tetrax*. Behav Process 51:63–82

Jiguet F, Jaulin S, Arroyo B (2002) Resource defence on exploded leks: do male Little Bustards, *T. tetrax*, control resources for females? Anim Behav 63(5):899–905

Johnsgard, PA (1994) Arena birds. Sexual selection and behaviour. Smithsonian Institution Press, Washington, DC

Lane SJ, Alonso JC, Alonso JA, Naveso MA (1999) Seasonal changes in diet and diet selection of great bustards (Otis t. tarda) in north-west Spain. J Zool Lond 247:201–214

Lett JM, Guyomarch JC, Boutin JM (2000) Occupation de l'espace par les femelles reproductrices chez l'outarde canepetière (*Tetrax tetrax*). Chantiers d'Éthologie 20:1–20

Mañosa S, Guillem R, Bota G (2018) Rècord de longevitat del sisó Tetrax tetrax. L'Abellerol 56:23

Marcelino J, Moreira F, Mañosa S, Cuscó F, Morales MB, García de la Morena EL, Bota G, Palmeirim JM, Silva JP (2017) Tracking data of the Little Bustard Tetrax tetrax in Iberia shows high anthropogenic mortality. Bird Cons Int 28(4):509–520. https://doi.org/10.1017/S095927091700051X

Martinez C (1994) Habitat selection by the Little Bustard *Tetrax tetrax* in cultivated areas of Central Spain. Biol Conserv 67:125–128

Møller AP (1990) Fluctuating asymmetry in male sexual ornaments may reliably reveal male quality. Anim Behav 40:1185–1187

Møller AP, Höglund J (1991) Patterns of fluctuating asymmetry in avian feather ornaments: implications for models of sexual selection. Proc R Soc London 245:1–5

Morales MB, Traba J (2016) Prioritising research in steppe bird conservation: a literature survey. Ardeola 63:137–150

Morales MB, Jiguet F, Arroyo B (2001) Exploded leks: what bustards can teach us. Ardeola 48:85–98

Morales MB, García JT, Arroyo B (2005) Can landscape composition changes predict spatial and annual variation of Little Bustard male abundance? Anim Conserv 8:167–174

Morales MB, Traba J, Carriles E, Delgado MP, García de la Morena EL (2008) Sexual differences in microhabitat selection of breeding Little Bustards *Tetrax tetrax*: ecological segregation based on vegetation structure. Acta Oecol 34:345–353

Morales MB, Traba J, Delgado MP, García de la Morena EL (2013) The use of fallows by nesting Little Bustard *Tetrax tetrax* females: implications for conservation in mosaic cereal farmland. Ardea 60:85–97

Morales MB, Casas F, de la Morena EG, Ponjoan A, Calabuig G, Martínez-Padilla J, Garcia JT, Mañosa S, Viñuela J, Bota G (2014) Density dependence and habitat quality modulate the intensity of display territory defence in an exploded lekking species. Behav Ecol Sociobiol 68: 1493–1504

Penfold LM, Hallager S, Boylan J, de Wit M, Metrione LC, Marcie O (2013) Differences in Fecal Androgen Patterns of Breeding and Nonbreeding Kori Bustards (Ardeotis kori). Zoo Biol 32: 54–62

Petretti F (1993) Notes on the lek behaviour of the Little Bustard in Italy. Avocetta 17:19–22

Ponjoan A, Bota G, Manosa S (2012) Ranging behavior of little bustard males, Tetrax tetrax, in the lekking grounds. Behav Process 91:35–40

Prum RO, Berv JS, Dornburg A, Field DJ, Townsend JP, Lemmon EM, Lemmon AR (2015) A comprehensive phylogeny of birds (Aves) using targeted next-generation DNA sequencing. Nature 526:569–573

Salamolard M, Moreau C (1999) Habitat selection by Little Bustard *Tetrax tetrax* in a cultivated area of France. Bird Study 46:25–23

Schulz H (1985) Grundlagenforschung zur Biologie der Zwergtrappe *Tetrax tetrax*. PhD Dissertation. Staatlichen Naturhistorischen Museum, Braunschweig

Schulz H (1986) Agonistisches Verhalten, Territorialverhalten und Balz der Zwergrtrappe (Tetrax tetrax). Journal fúr Ornithologie 127(2):125–204

Shlyakhtin GV, Tabachishin VG, Khrustov AV, Zav'yalov EV (2004) Ecological segregation of bustards (Otididae) in the north of the lower Volga region: evolutionary and adaptive aspects. Russian J Ecol 35:247–253

Silva JP, Pinto M, Palmeirim JM (2004) Managing landscapes for the little bustard Tetrax tetrax: lessons from the study of winter habitat selection. Biol Conserv 117:521–528

Silva JP, Faria N, Catry T (2007) Summer habitat selection and abundance of the threatened Little Bustard in Iberian agricultural landscapes. Biol Conserv 139(1–2):186–194

Silva JP, Estanque B, Moreira F, Palmeirim JM (2014) Population density and use of grasslands by female Little Bustards during lek attendance, nesting and brood-rearing. J Ornithol 155(1): 53–63

Silva JP, Catry I, Palmeirim JM, Moreira F (2015) Freezing heat: thermally imposed constraints on the daily activity patterns of a free-ranging grassland bird. Ecosphere 6:119. https://doi.org/10.1890/ES14-00454.1

Suárez-Seoane S, García de la Morena EL, Morales Prieto MB, Osborne PE, de Juana E (2008) Maximum entropy niche-based modelling of seasonal changes in Little Bustard (*Tetrax tetrax*) distribution. Ecol Model 219:17–29

Tarjuelo R, Delgado MP, Bota G, Morales MB, Traba J, Ponjoan A, Hervás I, Mañosa S (2013) Not only habitat but also sex: factors affecting spatial distribution of Little Bustard *Tetrax tetrax* families. Acta Ornithol 48:119–112

Traba J, Morales MB, García de la Morena EL, Delgado MP, Krištín A (2008) Selection of breeding territory by Little Bustard (*Tetrax tetrax*) males in Central Spain: the role of arthropod availability. Ecol Res 23:615–662

Traba J, Morales MB, Carmona CP, Delgado MP (2015) Resource partition and niche overlap in a steppe bird assemblage. Community Ecol 16:178–188

Wolff A (2001) Changements agricoles et conservation de la grande avifaune de plaine: étude des relations espèce-habitats à différents échelles chez l'outarde. PhD Thesis University of Montpellier, Montpellier

Wolff A, Paul JP, Martin JL, Bretagnolle V (2001) The benefits of extensive agriculture to birds: the case of the Little Bustard. J Appl Ecol 38:963–975

Wolff A, Dieuleveut T, Martin JL, Bretagnolle V (2002) Landscape context and Little Bustard abundance in a fragmented steppe: implications for reserve management in mosaic landscapes. Biol Conserv 107(2):211–220

Yang R, Wu X, Yan P, Su X, Yang B (2010) Complete mitochondrial genome of Otis tarda (Gruiformes: Otididae) and phylogeny of Gruiformes inferred from mitochondrial DNA sequences. Mol Biol Rep 37:3057–3066

The Little Bustard Around the World: Distribution, Global Conservation Status, Threats and Population Trends

Manuel B. Morales and Vincent Bretagnolle

Current World Range and Large-Scale Distribution Features

The world range of the little bustard encompasses a vast longitudinal belt of the Palearctic biogeographical region, lying between 30° and 55° north and extending from Portugal to north-western China. Although the flat and open landscapes dominated by the steppes, grasslands and cereal farmland preferred by little bustards are widespread throughout this range, the species' current distribution is far from continuous. Two main, markedly disjoint, sub-ranges exist: the western breeding range, whose core is located in the Iberian Peninsula, but which also includes the populations still remaining in France, Italy and Morocco, and the eastern breeding range, which harbours the populations of south Russia, Ukraine, Georgia, Kazakhstan, Kyrgyzstan, north-west China, northern Iran and Turkey (Fig. 1). Formerly, the populations comprising these two nuclei were assigned to different subspecies, respectively, named *T.t. tetrax* and *T.t. orientalis* (see also chapters "The little bustard and its family: a review of relationships" and "Natural history of the little bustard: morphology, biometry, diet, sexual dimorphism and social and breeding behaviour"), based on the larger size and somewhat darker underparts of the latter, but nowadays the species is considered monotypic (Collar 1996). This is, in fact,

M. B. Morales (✉)
Terrestrial Ecology Group, Departamento de Ecología, Universidad Autónoma de Madrid, Madrid, Spain

Centro de Investigación en Biodiversidad y Cambio Global-CIBC, Universidad Autónoma de Madrid, Madrid, Spain
e-mail: manuel.morales@uam.es

V. Bretagnolle
Centre d'Etudes Biologiques de Chizé, CNRS & La Rochelle Université, Beauvoir sur Niort, France
e-mail: vincent.bretagnolle@cebc.cnrs.fr

V. Bretagnolle et al. (eds.), *Little Bustard: Ecology and Conservation*, Wildlife Research Monographs 5, https://doi.org/10.1007/978-3-030-84902-3_4

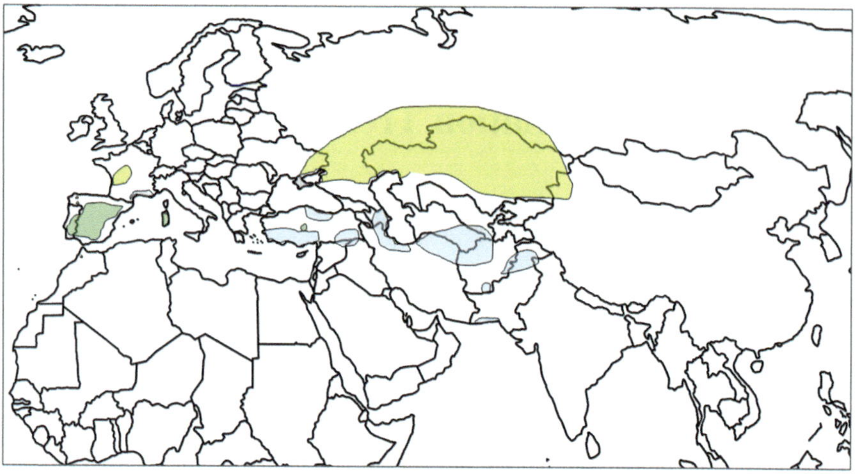

Fig. 1 Approximate world distribution of the little bustard. Green patches represent areas where the species is mainly resident and yellow ones correspond to areas where it is basically a breeding visitor. Blue patches indicate wintering areas. Based on Iñigo and Barov (2010)

consistent with the presence of now extinct little bustard populations in central and eastern European countries such as Germany, Hungary, Serbia and Bulgaria (Cramp and Simmons 1980), once probably connecting the extremes of the species' range. Nevertheless, genetic studies examining the relationships between western and eastern populations may eventually change this view.

The distribution range described includes migratory as well as resident populations (see chapter "Migration, movements and non-breeding ecology"), and both types of strategies are present in the western and eastern sub-ranges. In fact, each sub-range encompasses most of the wintering areas used by its corresponding migratory populations. For example, all migratory individuals breeding in France and northern Spain winter in the southern half of Iberia (Villers et al. 2010; García de la Morena et al. 2015), and the migrants from central Asia most probably winter close to breeding areas south of the Caspian Sea (Gauger 2007; Sehhatisabet et al. 2012), although wintering sites as far south as southern Pakistan have been cited (BirdLife International 2020).

The areas occupied by the little bustard over its breeding range are characterized by their flat or gently undulating relief and low human population density and occur over a broad altitudinal range, from sea level to nearly 2000 m.a.s.l. (Delgado et al. 2011; Estrada et al. 2016). In terms of climate, these areas show relatively low annual precipitation, although little bustards tend to avoid sites with too little or too high soil moisture, which indicates a preference for breeding grounds of moderate primary production (Delgado et al. 2011). They prefer relatively high temperatures during the breeding season, although the relationship between little bustard habitat favourability and breeding season (April to July) temperature is unimodal, so that the species tends to avoid too cool or too warm breeding areas. Indeed, Silva et al. (2015) found that breeding and post-breeding little bustards markedly reduced their activity when summer temperatures rose above 25° C, which in the face of global warming and the expected temperature trends in large parts of the species' range might imply a strong constraint on breeding activities and thus a significant loss of those areas' suitability (see Estrada et al. 2016 for distribution maps under climate scenarios). Little bustards also prefer small annual temperature ranges (Delgado et al. 2011; Estrada et al. 2016). In fact, in those breeding areas where climate is markedly continental, the little bustard behaves as a typically migratory bird, travelling south to mild winter sites (García de la Morena et al. 2015; Heiss 2013). Similarly, in dry Mediterranean areas, little bustards perform summer post-breeding movements to fresher localities at higher latitudes or altitudes, or to irrigated farmland, where primary productivity is higher (García de la Morena et al. 2015).

Global Conservation Threats and Status

Despite the size of its range, the little bustard exhibits worrying and even alarming trends in many areas, mainly in the western sub-range (see below). However, the situation in the eastern range is poorly known. For that reason, the little bustard is globally classified as Near Threatened in the IUCN Red List (BirdLife International 2020). Its decline throughout Europe, however, has led the species to be listed as Vulnerable there, as well as specifically in the European Union (BirdLife International 2015), and it is listed in Annex I of the European Union (EU) Birds Directive as well as in Annex II of the Bern Convention. Finally, it has been included in Annexes I and II of the Bonn Convention on Migratory Species (https://www.cms.int/en/species/appendix-i-ii-cms). As discussed below and in later chapters, the main cause of the species' decline is the transformation of extensively (i.e. traditionally) managed grasslands and cereal farmland into intensive agrarian landscapes that has occurred in most of Europe over the second half of the 20th and the first decades of the present century (see also Collar et al. 2016). Agricultural intensification has brought loss of habitat diversity at both landscape (loss of natural grasslands and fallows, spread of irrigation and woody crops) and small spatial scale (homogenization of vegetation structure, disappearance of wild plants and arthropods through increased herbicide and pesticide application), reducing the availability of key resources such as insect food, cover and adequate nesting sites, and thus of suitable habitats (see chapters "breeding biology and demographic traits" and "Little bustard population dynamics"). Food reduction, particularly of insects, has a direct negative impact on juvenile survival and thus population recruitment, which is a key parameter for population growth (Morales et al. 2005; Inchausti and Bretagnolle 2005). The conversion of extensively farmed landscapes into intensive croplands and irrigated fields reduces nesting habitat suitability (Brotons et al. 2004; Morales et al. 2013), negatively affecting breeding success, while the use of modern harvesting machinery is an important source of female and juvenile mortality, as well as nest failure (Inchausti and Bretagnolle 2005; Faria et al. 2016; Bretagnolle et al. 2018), which could also account, at least partially, for the strongly male-skewed sex ratios observed in many populations (see chapter "Breeding biology and demographic traits"), another important constraint of population viability (Morales et al. 2005).

Urban and infrastructure development have been identified as another threat to little bustard habitats in certain areas where urban sprawl has been important in recent years, for example, before the global economic crisis of 2008 in Spain. The Madrid region, in central Spain, provides a good example of infrastructure development followed by unplanned urban growth causing direct habitat disappearance and loss of suitability due to habitat fragmentation and edge effects, since breeding little bustards avoid the proximity of large linear infrastructures and urban areas (Suárez-Seoane et al. 2002, 2008; Silva et al. 2010). In addition, disturbance caused by human presence, which increases near urbanized areas, has been shown to induce physiological stress responses in little bustards (Tarjuelo et al. 2015), a threat difficult to assess at large scale, but potentially important in the increasingly human-frequented European rural landscape.

Land-use changes also threaten the species in northern Africa (see below) and central Asia. In this latter region, after a period of agricultural abandonment following the fall of the Soviet Union, which appears to have benefited many steppe birds in countries such as Russia and Kazakhstan (Shlyakhtin et al. 2004; Kamp et al. 2011), the recent resurgence of intensive agriculture seems certain to decimate eastern populations of little bustard and other steppe birds (Kamp et al. 2011; Collar et al. 2017).

Another global process threatening the little bustard throughout its range is climate change. As noted, the species' climate niche requirements may not be met under the conditions forecasted for the end of this century in its current range (Gudka et al. 2019). In fact, the species' European range was predicted by Huntley et al. (2007) to shift northwards (see also Estrada et al. 2016), so that it would disappear from many southern areas, but could colonize northern European regions. However, when forecasts incorporate suitable habitat availability and spatial constraints related to philopatry and conspecific attraction, the favourable range remains fairly unchanged (Estrada et al. 2016), which means that the little bustard will be left stranded in regions increasingly mismatching its climate niche requirements.

Changes in land-use and climate are not, however, the only global threat to little bustard populations. Hunting pressure and collision with powerlines, among other factors, have been identified as relevant risks in different European countries (De Juana and Martínez 2001; Iñigo and Barov 2010; Silva et al. 2010). For example, Marcelino et al. (2018) found that 42% of mortality events recorded in a large sample of birds ($N = 139$) satellite and radio-tracked in Spain and Portugal were human-induced, with poaching as the leading cause (32% of total mortality), although collision with powerlines (see also Silva et al. 2010) and vehicles also accounted for a significant share of the casualties (6.45 and 3.23%, respectively). While shooting is probably linked to poaching and opportunistic hunting in Europe (De Juana and Martínez 2001), legal hunting pressure may still be an important threat for eastern populations, particularly on the wintering grounds. For instance, the species is legally shot in Iran, although authorities banned little bustard hunting from 2006 to 2009 due to the perceived risk of an avian influenza outbreak (Sehhatisabet et al. 2012).

Historical Distribution and Population Trends

Until the early twentieth century, the little bustard was probably present throughout the longitudinal belt defined above, breeding in most countries of southern and central Europe, northern Africa, the Middle East and central Asia (Cramp and Simmons 1980). As regards the western Palearctic, and according to Schulz (1985), until 1940 the species still bred in large areas of the Iberian Peninsula, France, the Pannonian region and Danube valley, Ukraine and southern Russia, and maintained more restricted ranges in north-eastern Germany, Italy (on both the continent and the two large islands of Sicily and Sardinia) and northern Greece, as well in north-western Morocco, northern Algeria, Tunisia, western Anatolia, Syria and the Caucasus. However, by 1980, the little bustard breeding range in the western Palearctic had dramatically shrunk (Schulz 1985). Apart from the strongholds of the Iberian Peninsula and southern Russia, the species had disappeared from most of its European domains. The German populations were extinct, while the Pannonian region and Ukraine still presented two relict nuclei in Hungary and Crimea, respectively. The small breeding range of northern Greece had also vanished, and the Italian populations were confined to two remaining nuclei: a relict one with at most a few tens of individuals in Apulia (south-eastern Italy), and the Sardinian population which still seemed to include 1500–2000 individuals (Petretti 2006; Santangeli et al. 2010) though more recent estimates suggest that less than 500 birds may persist (Nissardi and Zucca 2011). The species' range in France had also notably contracted (see the species' historical distribution in Yeatman 1976), being reduced basically to the west central plains of the country, some site close to the Massif Central, and the Mediterranean fringe of the country (Issa and Muller 2015). Basically, this was the European distribution reflected in the first European Atlas of Breeding birds, which compiled data from the previous 25 years up to the mid-1990s (Hagemeijer and Blair 1997). Similarly in northern Africa and the Middle East, by 1980 only three relict nuclei remained in northern Morocco, while the persistence of the species in Turkey and Syria could not be confirmed (Schulz 1985; Palacín and Alonso 2009).

In summary, over the twentieth century the geographical range of the little bustard underwent a dramatic process of fragmentation and contraction leading to the current extremely disjunct distribution formed by the eastern and western breeding sub-ranges. In this process the species' extinction has been verified at least in nine countries of the western Palearctic (Table 1) and has probably also occurred in different eastern Palearctic regions. Below we review trends over recent decades in those countries or regions where the data are sufficient to determine the fate of the populations.

Table 1 Year or decade of breeding little bustard extinction over the twentieth century in different countries of the western Palearctic

Country	Year
Germany	1907
Poland	1909
Austria	1921
Former Czechoslovakia	1945
Hungary	1970s
Moldova	1930
Sicily	1950s
Greece	1950s
Serbia	1948
Rumania	Before 1900
Bulgaria	1950
Tunisia	1930
Algeria	Before 1990
Azerbaijan	1940s

Compiled from Schulz (1985), Snow and Perrins (1998) and Patrikeev (2004) for Azerbaijan

Table 2 Different historical estimates of the Spanish breeding and wintering little bustard populations

	Estimate			
Year	Breeding	Winter	Remarks	Authors
1985	50,000–70,000		Individuals	Schulz (1985)
1996	100,000–200,000		Males	De Juana and Martínez (1996)
2006	71,112–147,763		Individuals Based on an estimated sex ratio of 1.4 males per female	García de la Morena et al. (2006)
2006	43,000–71,700		Individuals Re-estimated with updated habitat availability	García de la Morena et al. (2006)
2006		16,429–35,929	Individuals	García de la Morena et al. (2006)
2016	51,808		Individuals Based on an estimated sex ratio of 1 female every 3 males	García de la Morena et al. (2018)
2016		6668–29,848	Individuals	García de la Morena et al. (2018)

Little Bustard Trends in Spain

Schulz (1985) estimated the Spanish little bustard population at 50,000–70,000 individuals (Table 2). The species was considered widespread in the country's main agricultural plains, i.e. the Northern (Castilla y León) and Southern (Castilla-La Mancha and Madrid) Plateaus, Extremadura, and the Ebro and Guadalquivir valleys. However, on the basis of censuses carried out in 1993 and 1994 in known little bustard sites distributed over those regions, De Juana and Martínez (1996) reached a much higher estimate and extended the species' range in Iberia to the south and east. This geographical distribution was confirmed by the Spanish Atlas of Breeding Birds (García de la Morena et al. 2003). According to De Juana and Martínez (1996), Spain harboured between 100,000 and 200,000 singing males (Table 2). On the other hand, there was no information at that time about what the trend of such large population could be, although De Juana and Martínez (1996) already detected a pattern of population fragmentation and isolation, particularly in peripheral areas that had been previously considered important for the species such as the Northern Plateau, Andalucía and the Ebro Valley.

In the following years, various local studies yielded population size and trend estimates in different regions where little bustards were still present, including those with relict and isolated presence, such as Galicia in the north-west. These studies were reviewed by García de la Morena et al. (2004), confirming the range fragmentation process indicated by De Juana and Martínez (1996). This review highlighted the negative trends of the little bustard in most Spanish regions during the 1990s and early 2000s, both in peripheral populations (Galicia, Andalucía, Navarra, Cataluña) and in more central ones, such those of the Northern Plateau and Extremadura. García de la Morena et al. (2004) provided an accurate idea of this general regression process: in Extremadura, a traditional stronghold of the species in Iberia, the population declined by 40% in 9 years, with a density decrease from 6 to 2.4 individuals/km^2 in Llanos de Cáceres, one of its main nuclei (López Ávila and Hidalgo de Trucios 1998), while in the peripheral populations of Navarra and Cataluña the species decreased by 27% in 4 years and 25–54% in 6 years, respectively. However, no reliable trend could be estimated for the entire country, largely due to the lack of information on the Southern Plateau, which harboured the bulk of the Spanish population.

The first Spanish national breeding census was carried in 2005, resulting in a new but broad estimate of 41,482–86,195 singing males and 71,112–147,763 total individuals (Table 2; see García de la Morena et al. 2006 for estimate calculations). The highest mean density was recorded in provinces of the Southern Plateau (Toledo and Ciudad Real), with 3 males/km^2. In another 8 provinces (including core and peripheral ones), mean density was 1–2 males/km^2, while in the remaining 26 it was below 1 male/km^2. However, these figures were considered to be overestimations resulting from the extrapolation of observed numbers to an outdated surface area of presumed suitable habitat. Indeed, when these extrapolations were made in regions where updated information on suitable habitat availability existed (Northern Plateau and Aragón in the Ebro valley), estimates turned out to be much smaller because the area of suitable habitat in those regions had clearly decreased in recent years. Conservatively assuming an equivalent habitat regression in the rest of the country, García de la Morena et al. (2006) corrected their estimate down to 43,000–71,700 individuals (Table 2).

Although the corrected estimate reached from the national census results cannot be directly compared to that provided by De Juana and Martínez (1996) due to important methodological differences, the much lower values of the former clearly suggested an overall decline of the species in Spain between the mid-1990s and the mid-2000s, which was consistent with the regional decreases that were being reported in that period (García de la Morena et al. 2006). The national census also allowed the detection of other worrying patterns. On the one hand, it confirmed the fragmentation and isolation of populations in peripheral areas like Galicia and the south-east. It also showed an increasing rarity in traditional core areas like Extremadura and the Northern Plateau, many of whose 10×10 km^2, although occupied at the time of the breeding atlas, contained no individuals during the national census (García de la Morena et al. 2006).

A national winter census was also carried out during the winter 2005–2006. The total area covered was determined on the basis of historical records of wintering birds (García de la Morena et al. 2006; García de la Morena 2015). This census yielded a markedly smaller estimated range of 16,429–35,929 individuals (Table 2). Again, significant methodological differences between the breeding and winter censuses may explain the difference between the corresponding total estimates (see details in García de la Morena et al. 2006), raising the obvious question of which total estimate should be considered more reliable. Given that the difference between estimates roughly ranges from 38 to 50%, the answer has clear conservation implications both at national and international levels.

In the years following the Spanish national census, evidence from different regions has continued to indicate an alarming general decline. De Juana (2009) reported a steady significant decline of 65% in the number of males in Extremadura's best areas for the species between 1993 and 2008 (some local densities in these areas had historically reached 24 males/km^2). Morales et al. (2007) detected a mean decline of 61% in the number of breeding males (ranging from 30 to 100%) in eight localities of the best cereal steppe areas of Madrid province, protected as a Special Protection Area (SPA), between 2000 and 2007. A formerly important area for the little bustard, Campo de Calatrava, also protected as a SPA, sited in the Southern Plateau (which hosts the bulk of the Spanish population), witnessed a density decline of roughly 55% (Casas et al. 2019). Finally, Mañosa et al. (2015) reported a 50% regression in the number of singing males but an estimated 75% fall in the number of breeding females between 2002 and 2014 in the cereal steppes of Lleida, in the Catalonian sector of the Ebro valley. This last result is particularly worrying, since strong female shortages can seriously compromise populations' potential for recovery (Morales et al. 2005). Surveys of females at other study sites show the same trend (Morales et al. 2007; Casas et al. 2019; Serrano-Davies et al. submitted, see also chapter "Little bustard population dynamics"). According to the results of the Spanish Common Breeding Bird Monitoring Programme (SACRE), coordinated by SEO/BirdLife and involving surveys since 1998, the little bustard in Spain showed an overall population trend of −76.19%, and a mean yearly population trend of −5.4 (CI: −6.3, −4.5) (SEO/BirdLife 2018, see Fig. 2), which indicates an alarming population decrease.

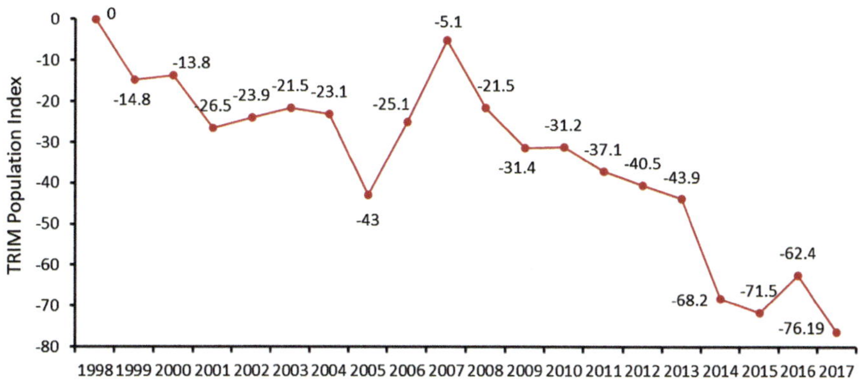

Fig. 2 Little bustard trends in Spain between 1998 and 2017 calculated as TRIM population index from data provided by the Spanish Common Bird Monitoring Programme (SACRE, SEO/Birdlife 2018). Annual change values with respect to the first year of the series are shown

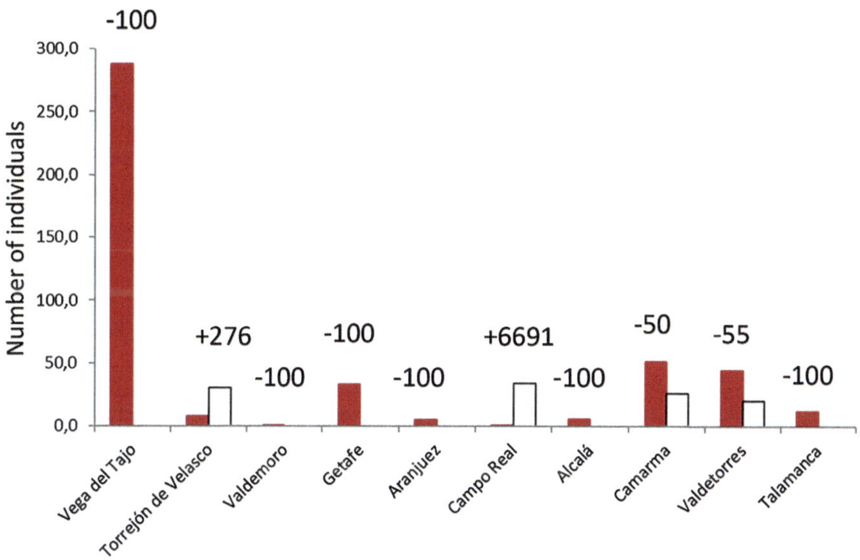

Fig. 3 Evolution of little bustard wintering numbers in different localities of central Spain between winters 2003–2004 (red bars) and 2013–2014 (white bars). Numbers above bars indicate the percentage of increase (+) or decrease (−) in the number of individuals. Based on Morales et al. (2015)

Winter surveys further confirmed these trends. A comparison of winter censuses carried out in 14 10 × 10 km Atlas squares of Madrid and Toledo (Southern Plateau) in 2003–2004 and 2013–2014 indicates an overall decline of 76% even in some of the best wintering areas for the species in Spain (Morales et al. 2015, Fig. 3). Therefore, little bustard population trends in Spain are clearly negative in most parts of its range, both central and peripheral, although distribution shifts due to land-use changes occurred in the last decade may partly account for the species regression or disappearance in some sectors.

A new Spanish national breeding census and another winter survey were carried out in spring 2016 and winter 2016–2017, respectively (see details in García de La Morena et al. 2018), allowing the real magnitude of the little bustard decline in its main historical stronghold to be assessed. This second national census yielded an estimate of 38,856 breeding males (27,037–59,136), which would imply a total population of 51,808 individuals (after applying a mean sex ratio of one female every three males; see García de la Morena et al. 2018 for details). Eighty-seven percent of the population was concentrated in the cereal steppes of the southern half of the country. Castilla-La Mancha was the region with the greatest population, harbouring about 65% of the total census. The population of the northern half was considerably smaller, with most of the birds concentrated in the Ebro valley, followed by some areas in the north-western fringe of the Duero valley. The winter mean total population estimate was 14,643 (6668–29,848) individuals. The estimate of the total population of breeding males in Spain declined by 48% in 11 years, which corresponds to an average yearly decrease of 5.7%. On average, male densities decreased by 38% (25–67%). As regards the wintering population, it declined from 28,775 to 15,022 individuals, which also implies a decrease of roughly 48% in 11 years, and an average yearly decrease of 5.7%, which was particularly strong in the Southern Plateau and the Ebro Valley, traditional wintering headquarters for the species in the Iberian Peninsula. Overall, these figures again point out to a crash in population size.

Little Bustard Trends in Portugal

Portugal holds the second largest little bustard population in the species' western range. In fact, it is largely connected with the Spanish population of Extremadura and could be considered as a part of the little bustard metapopulation of south-western Iberia (Silva 2010). First published estimates of this population's size were between 10,000 and 20,000 individuals (Goriup 1994), although several nineteenth century reports indicate that the little bustard was extraordinarily abundant in central and southern Portugal, and a widely hunted game species (Smith 1868). The first national census, carried out between 2003 and 2006, provided an estimate of 17,515 breeding males, which would imply a total of 35,030 individuals assuming a 1:1 adult sex ratio (Silva and Pinto 2006). However, this latter figure should be treated with caution given the biased sex ratios detected in other Iberian and European populations (see above and Silva 2010). Between 90 and 95% of this population occurs in the Alentejo region, south of the Tagus River, where the little bustard is widespread and presents some of the highest known local breeding populations worldwide, with 9–10 males/km^2 in the Castro Verde area and a mean regional density of 4.8 males/km^2 (some earlier studies report local densities at Castro Verde of 26 males/km^2) although in some sites densities are under 1 male/km^2 (Silva and Pinto 2006).

As in the case of Spain, the result of this first Portugal-wide census cannot be directly compared with previous estimates due to methodological differences, and thus it was difficult to be confident of the population trend (Silva 2010). However, important land-use changes have recently occurred in the Alentejo (spread of irrigation and olive groves, grassland-use intensification, Faria 2015), and the inference of a decline, at least since the early 2000s, was fully justified. Moreover, a second national census conducted in 2016 provided directly comparable results and confirmed the negative population trend (Silva et al. 2018): the current population estimate is 8900 breeding males (95% CI: 5008–12,836), which corresponds to an overall 49% decline in that country between 2005 and 2016. Interestingly, the percentage decline was greater outside SPAs, but the absolute reduction in male density was higher within those protected areas (Silva et al. 2018). These alarming figures are consistent with and directly comparable to those obtained for Spain. Moreover, since these censuses are simultaneous to that performed in France, it is now possible to estimate the size and global trend of the entire Iberian and Western Europe's population as presented in Table 3.

Little Bustard Trends in France

France is probably the country where the little bustard decline can be most reliably tracked. National censuses were carried out in 1978, 1985, 1995, 2000, 2004, 2008, 2012 and 2016, along with several distribution surveys published in the French Bird Atlas in 1976, 1994 and 2015. Moreover, published records date back to the late nineteenth century (Ternier 1892). According to Yeatman (1976), the species was common, although irregularly distributed, over most of the country's agricultural plains until at least the mid-1930s, with declines as well as population recoveries at least until the 1950s. In other words, the species was not rare in France during the first half of the twentieth century, although it experienced population fluctuations and probably local extinctions and recolonizations, trends that may partly reflect its status as a game species (Yeatman 1976). The little bustard was distributed over all main agricultural plains, including Poitou-Charentes, Beauce and Champagne, respectively, from western to central and northern parts of the country, the pastures of Limagne and les Causses, close to the Massif Central and the lower valleys of Rhône and Durance in the Mediterranean arch that ranges from Narbonne to Camargue (Languedoc and Provence regions, Yeatman 1976; see also map reproduced in Issa and Muller 2015).

A national census of the species was carried out a few years later, in 1978–1979, and provided the first quantitative estimate of the little bustard population in France, yielding 7200 singing males (André 1985). This study confirmed the species' distribution in two main nuclei whose populations differed in their ecology and migration behaviour: a northern population extending from central-western to

Table 3 Summary of national and regional little bustard population estimates across the species' world range in number of breeding males (western sub-range) or total individuals (breeding and wintering estimates in eastern sub-range). Bolds indicate totals for each su-range and the world

Country/ region	Current estimate		Percentage of world total[a]	Trend	References
	males	individuals			
Spain	38,856	57,118	13–39	Negative (48% in 11 years)	García de la Morena et al. (2018)
Portugal	8900	13,083	2–8	Negative (49% in 13 years)	Silva et al. (2018)
France	2455	3609	1.1–1.6	Stable in last 8 years Negative in last 16 years	Gendre et al. (2018)
Italy	352	517	0.2	Stable	Nissardi and Zucca (2011)
Total western Europe	50,563		16–49	–	
North Africa	Extinct?		–	–	Palacín and Alonso (2009)
Western range total	**50,563**		16–49	**Negative**	
Russia		17,000-121,000	5–54	Positive	Collar et al. (2017) and Oparin et al. (2018)
Ukraine		100	0.03–0.05	Unknown	Andryushchenko and Stadnichenko (1999) and BirdLife International (2020)
Kazakhstan and Kyrgyzstan		20,000	6–9	Positive	Collar et al. (2017) and BirdLife International (2020)
China		<1000	0.32–0.50	Unknown	Collar et al. (2017)
Other Asian countries		100	0.03–0.05	Unknown	BirdLife International (2020)
Eastern range total (breeding)		**38,200–142,200**	**12–64**	**Positive**	
Eastern range total[b] (wintering)		**171,700–207,000**	**54–93**	**Positive**	
World total[c]		**222,900–317,000**			

Total number of individuals for the western range has been calculated using the median proportion of 0.47 adult females plus female-like individuals provided by Serrano-Davies et al. (under review). The estimate of world total uses breeding individuals from western range plus wintering individuals for the eastern range, as the most reliable in each case. The corresponding minimum-maximum ranges of this world estimate percentage are also provided, along with population trends

[a]Conservatively calculated to maximize possible range of variation

[b]Based on winter counts in Azerbaijan and Iran (Heiss 2013 and Yousefi et al. 2017, respectively)

[c]Based on the sum of Western range breeding total and Eastern range wintering total

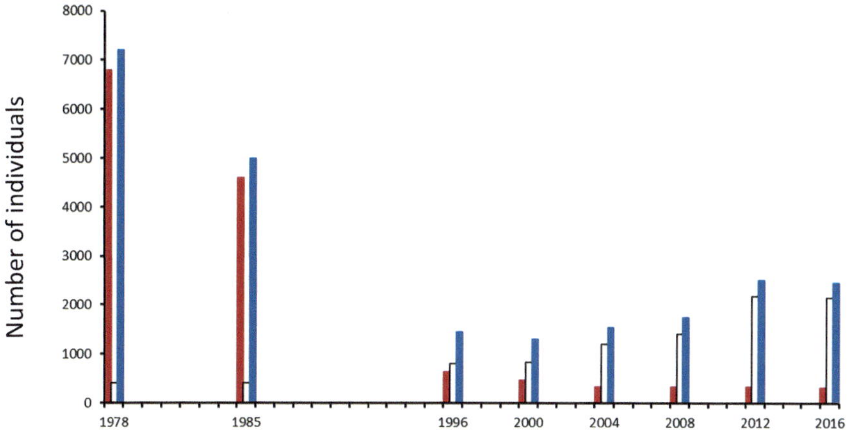

Fig. 4 Trends in number of breeding little bustard males in France from 1978 to 2016. Red bars: number of males in the western population; white bars: number of males in the Mediterranean population (values from 2004 to 2016 correspond to the median of the minimum-maximum interval); blue bars: total number of males in France (values from 2004 to 2016 correspond to the median of the minimum-maximum interval). Based on data from Jolivet and Bretagnolle (2002) and Gendre et al. (2018)

eastern France, consisting of migratory birds inhabiting intensive cereal farmland, which we will simply call 'western population', and a southern sedentary one located in the Languedoc and Provence regions and occupying dry Mediterranean grasslands, which will be referred to as 'Mediterranean population'. According to these census results, the Mediterranean population accounted for 400 breeding males, while the western population comprised 6800 males. A second estimate yielded a total of 5000 breeding males in 1985, 400 of which belonged to the Mediterranean population (Jolivet and Bretagnolle 2002), while a new national census carried out in 1995–1996 allowed a national estimate of 1460 breeding males (Jolivet and Bretagnolle 2002, revised in Gendre et al. 2018), 648 of which belonged to the western and 812 to the Mediterranean population. A third national census performed in 1999–2000 yielded a total of 1300 singing males, 460 of which belonged to the western population and 840 to the Mediterranean one (Jolivet and Bretagnolle 2002). To summarize, the little bustard population of western France had experienced a 94% drop in 22 years, while at the national scale the decrease was 82%, the steepest decline of a bird species ever documented in France (Jolivet and Bretagnolle 2002). The Mediterranean population, however, remained fairly stable over that period and even slightly increased at the end of it (Fig. 4).

After year 2000, and thanks to a combination of conservation measures (from agri-environmental contracts with farmers to population reinforcement) implemented since 1997 in the context of different EU LIFE programmes (see chapter "Little bustard and humans: conservation and management"), the French western population seems to be stabilized at around 340 breeding males, as indicated by the last three national censuses carried out in 2004, 2008 and 2012 (Gendre et al. 2018). However, the Mediterranean population has experienced a steady and marked increase from the 1114–1283 males estimated in 2004 to the 1329–1483 and 2021–2331 males calculated for 2008 and 2012, respectively (Fig. 4). Total minimum-maximum ranges were thus 1453–1626, 1664–1830 and 2360–2674, respectively, for 2004, 2008 and 2012. Such positive growth of the Mediterranean population is probably associated with the abandonment of traditional vineyards and other agricultural crops that occurred in that region over the last decade due to EU agricultural policy, as well as to land expropriation for the building of a high-speed railway communicating Nimes and Montpellier (Devoucoux 2014). The latest national census, carried out in 2016, supports those trends, although a slight decrease in relation to 2012 has occurred in both the Mediterranean and western populations, while the species has practically disappeared from elsewhere in France (Gendre et al. 2018). According to the results of this survey, the total number of breeding little bustard males in France is 2455, 313 of which belong to the western population and 2142 to the Mediterranean population (Gendre et al. 2018, see Fig. 4).

Little Bustard Trends in Italy

Historically, little bustards were locally common in the main flat areas of central and southern continental Italy and the two largest islands of Sardinia and Sicily. In the continent, the species was consistently found in the region of Apulia, which hosted the bulk of the continental population, as well as in Abruzzo and Molise, where the species had locally nested until the end of the 1960s, but it was also irregularly recorded further north, in the eastern Padanian plains (Petretti 2007). The total range occupied by the species in continental Italy was around 50,000 ha in 1950, but decreased to only 1200 ha by the 1980s (Petretti 1985). In Apulia, population was roughly estimated at between 10,000 and 100,000 individuals in the early twentieth century, although it began to decline when extensive and transhumant livestock rearing was replaced by intensive farming during the 1940s, so that by the 1960s the population had dropped to around 1000 individuals (Petretti 2006) distributed over the plateau area known as Tavoliere of Gargano (in Gargano National Park), close to the Adriatic coast of the Manfredonia Gulf. That was already the last population of the species in continental Italy, which continued to decrease to 1000–500 birds during the 1970s, and over the remaining years of the twentieth century, to the 15–20 little bustards reported in 2005 (Petretti 2006). Little bustard densities reported in this population during the 1980s and 1990s ranged from 1.4 to 2.0 males/km^2 (Petretti 1991). Nowadays this relict population, totally isolated from the rest of the species' range, can be considered virtually extinct, since there have been no sightings of birds since 2010 (G. Palumbo pers. comm.).

The little bustard has also vanished from Sicily, where it was considered extinct in the late 1960s (Snow and Perrins 1998; Massa and La Mantia 2007). As a result, Sardinia hosts the only significant little bustard population remaining in the country, which was estimated at 1500–2000 individuals in the mid-1980s (Schenk and Aresu 1985). These numbers are distributed over several localities in central-western Sardinia, all characterized by extensive pastoral and arable landscapes, in both coastal lowlands and inner plateaus (Santangeli and Dolman 2011). The Sardinian population seems to have declined during the last decades to only 352 breeding males (Nissardi and Zucca 2011) due to, among other factors, land-use intensification (Santangeli et al. 2010). In these populations, little bustard density varies from 0.1–0.2 to 2.7–3.4 males/km^2 (95% confidence intervals, Santangeli and Dolman 2011).

Little Bustard Trends in Northern Africa

According to historical reports, the little bustard was a common breeder in cultivated plains of north-western Africa, ranging from Tunisia through northern Algeria to north-western Morocco (Cramp and Simmons 1980; Schulz 1985). The species was also a common winter visitor throughout the same range, at least until the 1970s (Cramp and Simmons 1980; Palacín and Alonso 2009). For example, Thévenot et al. (2003) reported the presence of 1000 wintering birds near Larache (Morocco) in 1964. However, during the last decades of the twentieth century little bustards became extremely scarce in the region, both as breeders and winter visitors, although the exact trend is difficult to track due to the paucity of data. According to Palacín and Alonso (2009), the decline can be attributed to both land-use transformations (agricultural intensification, irrigation, infrastructures) and hunting pressure. These authors extensively reviewed the species' trends in north-west Africa, highlighting the speed and extent of its decline throughout the region. It became extinct both as a breeder and wintering visitor in Algeria during the 1990s (Isenmann and Moali 2000), and nowadays it can also be considered extinct in Tunisia (Isenmann et al. 2005).

As regards Morocco, the little bustard historical range encompassed two disjoint areas, one in the north-west between Tangier in the north and El Jadida in the south, and another in the north-east close to the Algerian border and presumably connecting with historical nuclei in that country (Palacín and Alonso 2009). There are no recent records of the species in the latter area, suggesting it is probably extinct there. However, little bustards were still present until 2005 in five localities in the north-western range, four of them between the Hachef and Loukos rivers, with the fifth some kilometres south of the latter river (Palacín and Alonso 2009). Later records are extremely scarce and no reports have been published since 2009. The numbers provided by Palacín and Alonso (2009) for those five localities in the pre-breeding season (early March–early April) total 47 birds observed in 4 consecutive years (1999–2003). These authors also provided an estimated trend for birds wintering around the locality of Larache for the last four decades of the twentieth century. The steepest fall occurred during the 1960s and 1970s, when the number of observed wintering birds dropped from 1000 to 100. This number was maintained through the 1980s but finally collapsed to extinction in the 2000s. In summary, the little bustard in northern Africa is on the brink of extinction, although such an assertion would require confirmation by further fieldwork.

Little Bustard Trends in the Eastern Range

Here we will refer to those regions within the eastern range of the little bustard for which at least some information is available, including southern Russia and Ukraine, central Asian countries and Turkey. As already mentioned, available information on little bustard numbers and trends in the species oriental range is scarce. Moreover, breeding and wintering estimates largely differ. Winter ones rely on the counting of birds migrating across main flyway passes and staying at winter quarters. These counts yield much larger numbers than breeding estimates.

Historically, the little bustard in Russia distributed across the country's steppe belt, penetrating even in the forest-steppe (Spangenberg 1951). However, over the twentieth century the species disappeared from many regions, so that the current range mainly comprises the Volga–Don interfluve, the Volga–Ural interfluve and eastern Cis-Caucasia (see review in Oparin et al. 2018). For Russia, BirdLife International (2020) provides the figure of 14,000–17,000 individuals only for the Orenburg oblast, in the south-eastern limit of European Russia (Antonchikov 2011). This area, together with the Saratov oblast, sited further west in the Volga Valley and for which Shlyakhtin et al. (2004) report 5900 birds, seems to hold the bulk of the species' population in Russia.

Shlyakhtin et al. (2004) also provide a range of little bustard breeding density values in the Saratov oblast, varying from 0.01 ± 1.1 individuals/km^2 in cereal fields, perennial grass prairies and northern steppe to 0.8 ± 0.3 individuals/km^2 in southern steppes. According to these authors, the population there seems to have been increasing since the early 1980s. Such an increase is consistent with reports from elsewhere in Russia over roughly the same period, involving the re-colonization of large areas by the species (Collar et al. 2017). For example, Fedosov et al. (2017) reported a population of 75,000 individuals in the eastern part of the Orenburg oblast (see reference in Collar et al. 2017) and Korovin (2014) refers to significant increases in breeding male densities from 0 to 5 individuals/km^2 between 1990 and 2010 further east, in the trans-Ural region of Chelyabinsk. Although the reasons for this are unclear, the abandonment of intensive agriculture after the fall of the Soviet Union may have contributed (Kamp et al. 2011). A recent survey conducted in fall (October), i.e. just after breeding and previous to migration, in the Saratov oblast concluded that about 25,000 birds (95% confidence interval 15,015–46,225) were present (Oparin et al. 2018). Since counts were repeated (although irregularly) between 1997 and 2016, maximum numbers at the level of the Saratov oblast peaked at c. 50000 individuals in 2014 (Oparin et al. 2018). Therefore, a significant increase in the little bustard population occurred there in the first decade of the current century. Oparin et al. (2018) attribute this population growth to an increase of 20% in land area of secondary virgin territory suitable for nesting in 1990s–2000s. In addition, there was a significant decrease in the grazing load in the Saratov Trans-Volga region. Overall, if little bustard numbers from the core of the species' range in Russia (i.e. Saratov and Orenburg) are summed, the population size range would be 17,000–121,000 individuals, with a median value of around 69,000 individuals.

The situation in Ukraine is unclear but discouraging. The population estimate provided by Andryushchenko and Stadnichenko (1999) and referenced in BirdLife International (2020) is for 100–110 individuals remaining in the Crimea. However, this estimate is most probably outdated, and no trend figure is currently available for the country. For Kazakhstan and Kyrgyzstan, BirdLife reports a joint figure of 20,000 individuals widespread over those countries' steppe areas (Wassink and Oreel 2007; Wassink 2016), while the breeding numbers in north-west China, Iran and Turkey are residual (5–50 males, BirdLife International 2020). Collar et al. (2017) reviewed recent reports from China and concluded that probably fewer than 1000 birds remained, restricted to Xinjiang province and Ningxia Hui region in the north-west and north of the country, respectively. Iran historically hosted a small breeding population, spread across the lowlands south-east of the Caspian Sea (Evans 1994; Scott 1995), but although the presence of breeding birds has not been confirmed in recent times, careful prospection is needed before their disappearance from the country can be presumed (Sehhatisabet et al. 2012). As regards Turkey, breeding presence has been recorded in Inner Anatolia near Tuz lake (Eken and Magnin 2000; Kirwan et al. 2008). Observations consisted of a few and displaying males and some female, so the species has been considered as breeder in the area (Eken and Magnin). Some birds were observed in post-breeding period in eastern Anatolia, although breeding there was not confirmed, and a few sporadical observations were made in the early 2000s in the Black Sea coast and the Mediterranean district, all out of the breeding season (Kirwan et al. 2008). No population estimates or trends are available.

Nowadays, however, the most reliable overall account of the eastern population probably comes from wintering population estimates. As mentioned above, the main wintering grounds for eastern little bustards are located south of the Caspian Sea in the Transcaucasian region and northern Iran. Azerbaijan hosts the bulk of this wintering population, with 150,000 birds occupying shrub (*Artemisia* and Chenopodiaceae-dominated) and grass steppes, as well as cereal fallows from the Caspian coastal plains to the Greater Caucasus foothills (Gauger 2007). Historically, Azerbaijan seems to have harboured up to 200,000–300,000 wintering individuals before 1930, as well as some breeding nuclei persisting up to the early twentieth century (see revision in Patrikeev 2004). Winter little bustard densities in these areas can be quite high, with an average value of 55 birds/km^2, but rising to 100 in the smaller concentration sites, and flock size may reach several tens of thousands of individuals, although the most frequent size is under 100 birds (Gauger 2007). The area of Besh Barmag Mountain, close to the Azerbaijan Caspian coast, is an important bottleneck for migratory eastern Palearctic birds (Heiss 2013). In the particular case of the little bustard, Heiss (2013) estimated that 105,895 to 123,491 birds crossed this bottleneck in the autumn migration of winter 2011–2012. According to this author, that figure represents 41–48% of the species' world population and 71–82% of the contingent using this flyway (however, the number of little bustards crossing the bottleneck in the spring migrations is only of a few thousands, Heiss 2013; see also chapter "Migration, movements and non-breeding ecology"). The number of little bustards wintering in Azerbaijan seems to have increased during the early 2000s (Gauger 2007), which could reflect the population recovery reported in breeding areas after the abandonment of intensive agriculture following the Soviet Union's breakdown.

Wintering grounds in Iran are located in the north of the country, along the Caspian coast, but also in its north-west end, close to the Azerbaijan border and the north-eastern regions bordering Turkmenistan and Afghanistan (Yousefi et al. 2017), although historical observations from more central parts of the country have also been reported (Sehhatisabet et al. 2012). In spite of this more widely spread winter range, Iran harbours a wintering population preliminarily estimated by Sehhatisabet et al. (2012) as 5000–10,000 individuals. According to these authors, wintering little bustards occupy a variety of habitats including farmland and grazed grasslands, although halophytic shrub steppes associated with wetlands are preferred. The area surveyed in Iran has increased more recently and Yousefi et al. (2017) report 57,000 birds. In addition, habitat changes may have favoured shifts of wintering quarters. Therefore, the magnitude of little bustard winter numbers in Iran may require further assessment. As in Azerbaijan, the Iranian wintering population seems to have recovered in recent times, both in numbers and geographical range, after a period of decrease in the mid- and late twentieth century. Sehhatisabet et al. (2012) attributed this recovery to the hunting ban enforced by the Iranian government from 2006 to 2009 as a measure to prevent an avian Influenza outbreak, although the agrarian changes on the breeding grounds have most probably also contributed.

Summing up the Iranian and Azerbaijan contingents yields 155,000–160,000 little bustards overwintering south of the Caspian Sea and provides a minimum estimated size of the species' eastern population. This figure exceeds the summed estimate for the Iberian Peninsula, so far considered the species' global stronghold, as well as that of the entire western range (Table 3). Furthermore, while most western populations are in decline, eastern little bustard numbers seem to have increased as a result of a de-intensification of agricultural practices which are now, however, being re-intensified.

Conclusions

The different regional trends reviewed in this chapter clearly show that the conservation status of the little bustard in the world has markedly changed in the last two decades. According to the current EU Action Plan for the species (Iñigo and Barov 2010), south-western Europe still was thought to harbour 60% of the world population, mostly concentrated in the Iberian Peninsula. However, the updated information reviewed here indicates that the species is rapidly decreasing throughout its western range, particularly in Iberia, a trend that was already suggested in the Action Plan (Iñigo and Barov 2010). The recent results of the last Spanish and Portuguese national censuses have confirmed that trend, with an alarming reduction of the Iberian population size (Table 3). This review also suggests that the eastern population might be in better health than previously thought (at least in terms of population size), a conclusion based on increasing winter counts in Azerbaijan and Iran. In fact, the global population account presented here strongly suggests, based on winter estimates, that the eastern population might have outnumbered the western one (Table 3), although more precise breeding population estimates based on standard and comparable methods are required for eastern countries (basically Russia and Kazakhstan).

Before the Soviet Union's collapse, ploughing was the major threat to the maintenance of Russian and central Asian steppes. For example, in 1990 almost one million hectares of Russian steppe were converted to arable land (Smelansky and Tishkov 2012). However, the breakdown of socialist economies favoured the abandonment of agricultural activities in natural steppe regions of former Soviet countries (Kamp et al. 2011), particularly in the less productive areas, so that between 1991 and 2007 more than 26 million hectares of cultivated steppe were abandoned in Russia (Smelansky and Tishkov 2012). But since 2007 this tendency has changed and nearly one million hectares have been put back into cultivation in that country, affecting some regions considered key for the little bustard, such as Orenburg province (Smelansky and Tishkov 2012). Therefore, it is critical to assess current land-use changes and agricultural trends in those countries, in order to forecast future availability of suitable habitat for the eastern population of the species, as well as to evaluate their impact on population trends, which could rapidly turn into declines if habitat changes mirror those in the west. In summary, a re-thinking of both the world and national strategies for the species' conservation is required.

Acknowledgements We sincerely thank Nigel Collar and Eladio García de la Morena for reviewing and commenting on this chapter.

References

André R (1985) Some aspects of reproduction in the Little Bustard and a contribution to an estimate of its population in France in 1978–1979. Bustard Stud 2:153–159

Andryushchenko YL, Stadnichenko IS (1999) The current state of bustard, little bustard, and stone curlew in the south of the leftbank Ukraine. Branta 2:135–151

Antonchikov A (2011) Little Bustard: current status and conservation prospects. Steppe Bull 31:33–38. (in Russian)

BirdLife International (2015) European red list of birds. Office for Official Publications of the European Communities, Luxembourg

BirdLife International (2020) Species factsheet: *Tetrax tetrax*. Downloaded from http://www.birdlife.org on 29/01/2020

Bretagnolle V, Denonfoux L, Villers A (2018) Are farming and birds irreconciliable? A 21–year study of bustard nesting ecology in intensive agrosystems. Biol Conserv 228:27–35

Brotons L, Mañosa S, Estrada J (2004) Modelling the effects of irrigation schemes on the distribution of steppe birds in Mediterranean farmland. Biodivers Conserv 13:1039–1058

Casas F, Mougeot F, Arroyo B, Morales MB, Hervás I, García de La Morena EL, Fagan WF, Viñuela J (2019) Opposing population trajectories in two Bustard species: a long-term study in a protected area in Central Spain. Bird Conserv Int 29:308–320. https://doi.org/10.1017/S0959270918000254

Collar NJ (1996) Family Otididae (bustards). In: Del Hoyo J, Elliot A, Sargatal J (eds) Handbook of the birds of the world, Hoatzin to Auks, vol 3. Lynx Edicions, Barcelona, pp 240–273

Collar NJ, Garcia EFJ, De Juana, E (2016) Little bustard (*Tetrax tetrax*). In: del Hoyo J, Elliott A, Sargatal J, Christie DA, De Juana E (eds) Handbook of the birds of the world alive. Lynx Edicions, Barcelona. http://www.hbw.com/node/53736. Accessed 21 July 2016

Collar NJ, Baral HS, Batbayar N, Bhardwaj GS, Brahma N, Burnside RJ, Choudhury AU, Combreau O, Dolman PM, Donald PF, Dutta S, Gadhavi D, Gore K, Goroshko OA, Hong C, Jathar GA, Jha RRS, Jhala YV, Koshkin MA, Lahkar BP, Liu G, Mahood SP, Morales MB, Narwade SS, Natsagdorj T, Nefedov AA, Silva JP, Thakuri JJ, Wang M, Zhang Y, Kessler AE (2017) Averting the extinction of bustards in Asia. Forktail 33:1–26

Cramp S, Simmons KEL (1980) The birds of the western Palearctic, vol 2. Oxford University Press, Oxford

De Juana E (2009) The dramatic decline of the little bustard *Tetrax tetrax* in Extremadura (Spain). Ardeola 56:119–125

De Juana E, Martínez C (1996) Distribution and conservation status of the little bustard *Tetrax tetrax* in the Iberian Peninsula. Ardeola 43:157–167

De Juana E, Martínez C (2001) Action Plan for the conservation of Little Bustard (*Tetrax tetrax*). BirdLife International and the European Commission, Strasbourg

Delgado MP, Traba J, Morales MB (2011) Climate niche constraints in two coexisting steppe birds: the little and the great bustards. Ardeola 58:223–238

Devoucoux P (2014) Conséquences et impacts prévisibles d'une perte d'habitat majeure sur une espèce menacée aux exigences écologiques complexes: effets de la mise en place du contournement ferroviaire à grande vitesse Nîmes-Montpellier sur la dynamique de la population d'outarde canepetière des Costières de Nîmes. Dissertation, University of Poitiers

Eken G, Magnin G (2000) A preliminary biodiversity atlas of the Konya Basin, central Turkey. Dogal Hayati Koruma Dernegi, Istanbul

Estrada A, Delgado MP, Arroyo B, Traba J, Morales MB (2016) Forecasting large-scale habitat suitability of European bustards under climate change: the role of environmental and geographic variables. PLoS One 11(3):e0149810. https://doi.org/10.1371/journal.pone.0149810

Evans M (1994) Important bird areas in the Middle East. BirdLife International, Cambridge

Faria N (2015) Implications of dry grassland management in the ecology and conservation of grassland birds in south Portugal. Dissertation, Autónoma University of Madrid

Faria N, Morales MB, Rabaca JE (2016) Exploring nest destruction and bird mortality in mown Mediterranean dry grasslands: an increasing threat to grassland bird conservation. Eur J Wildl Res 62:663–670

Fedosov VN, Fedosov AV, Antonchikvo A (2017) Surveys of Little Bustard in Orenburg Province and species conservation prospects. Steppe Bull 49:55–58. (in Russian)

García de la Morena E (2015) Ecología y movimientos migratorios del sisón del sisón común (*Tetrax tetrax*) fuera del periodo reproductor. Dissertation, Autónoma University of Madrid

García de La Morena E, De Juana E, Suárez F, Morales MB, Martínez C (2003) Sisón común (*Tetrax tetrax*). In: Martí R, del Moral JC (eds) Atlas de las aves reproductoras de España. SEO/ Birdlife-Ministerio de Medioambiente, Madrid, pp 232–233

García de La Morena E, De Juana E, Suárez F, Morales MB, Martínez C (2004) Sisón común (*Tetrax tetrax*). In: Madroño A, Martí R (eds) Libro Rojo de las Aves de España. SEO/BirdLife-Ministerio de Medioambiente, Madrid, pp 202–207

García de la Morena EL, Bota G, Ponjoan A, Morales MB (2006) El sisón común en España. I censo nacional. SEO/BirdLife, Madrid

García de La Morena EL, Morales MB, Bota G, Silva JP, Ponjoan A, Suárez F, Mañosa S, De Juana E (2015) Migration patterns of Iberian little bustards *Tetrax tetrax*. Ardeola 62:95–112

García de la Morena EL, Bota G, Mañosa S, Morales MB (2018) El Sisón Común en España. II Censo Nacional (2016). SEO-Birdlife, Madrid

Gauger K (2007) Occurrence, ecology and conservation of wintering little bustards *Tetrax tetrax* in Azerbaijan. Arch Nat Landsch 46:5–27

Gendre N, Éraud C, Bretagnolle V, Dalloyau S (2018) L'Outarde canepetière Tetrax tetrax en France en 2012 et 2016: effectifs et répartition. Ornithos 25-5:290–302

Goriup P (1994) Little Bustard *Tetrax tetrax*. In: Tucker GM, Heath MF (eds) Birds in Europe: their conservation status. BirdLife International, Cambridge, pp 236–237

Gudka M, Santos CD, Dolman PM, Abad-Gómez JM, Silva JP (2019) Feeling the heat: elevated temperature affects male display activity of a lekking grassland bird. Plos One. https://doi.org/10.1371/journal.pone.0221999

Hagemeijer WJM, Blair MJ (1997) The EBCC Atlas of European breeding birds: their distribution and abundance. T. and A. D. Poyser and European Bird Census Council, London

Heiss M (2013) The importance of Besh Barmag bottleneck (Azerbaijan) for Eurasian migrant birds. Acta Ornithol 48:151–164

Huntley B, Green RE, Collingham YC, Willis SG (2007) A climatic atlas of European breeding birds. Lynx Edicions, Barcelona

Inchausti P, Bretagnolle V (2005) Predicting short-term extinction risk for the declining Little Bustard (*Tetrax tetrax*) in intensive agricultural habitats. Biol Conserv 122:375–384

Iñigo A, Barov B (2010) Action plan for the little bustard *Tetrax tetrax* in the European Union. SEO/BirfLife and BirdLife International for the European Commission

Isenmann P, Moali A (2000) Birds of Algeria. Société d'Etudes Ornithologiques de France, Paris

Isenmann P, Gaultier T, El Hili A, Azafzaf H, Dlensi H, Smart M (2005) Oiseaux de Tunisie/Birds of Tunisia. SEOF Editions, Paris

Issa N, Muller Y (2015) Atlas des oiseaux de France métropolitaine; nidification et présence hivernale. Delachaux & Niestlé, Paris

Jolivet C, Bretagnolle V (2002) L'outarde canepetière en France: évolution récente des populations, bilan des mesures de sauvegarde et perspectives d'avenir. Alauda 70(2002):93–96

Kamp J, Urazaliev R, Donald PF, Hölzel N (2011) Post-Soviet agricultural change predicts future declines after recent recovery in Eurasian steppe bird populations. Biol Conserv 144:2607–2614

Kirwan GM, Byla K, Castell P, Demirci B, Özen M, Welch H, Marlow T (2008) The birds of Turkey. Helm, London

Korovin VA (2014) Restoration of the little bustard population in the northern Steppe Trans-Urals. Biol Bull 41:856–861. https://doi.org/10.1134/S1062359014100057

López Ávila P, Hidalgo de Trucios S (1998) Revisión del status del sisón: evolución en Extremadura. In: Conservación de la naturaleza y los espacios protegidos de Extremadura. Junta de Extremadura, Mérida, pp 115–121
Mañosa S, Bota G, Estrada J, Cuscó F (2015) Una oportunidad para el sisón en Cataluña. Quercus 356:24–35
Marcelino J, Moreira F, Mañosa S, Cuscó F, Morales MB, García de la Morena EL, Bota G, Palmeirim JM, Silva JP (2018) Long term tracking study of the little bustard in Iberia shows high anthropogenic mortality. Bird Conserv Int 28:509–520
Massa B, La Mantia T (2007) Forestry, pasture, agriculture and fauna correlated to recent changes in Sicily. Foresta 4:418–438
Morales MB, Bretagnolle V, Arroyo B (2005) Viability of the endangered little bustard *Tetrax tetrax* population of western France. Biodivers Conserv 14:3135–3150
Morales MB, García de la Morena EL, Delgado MP, Traba J (2007) Tendencia reciente y viabilidad futura de las poblaciones de sisón común *Tetrax tetrax* en la Comunidad de Madrid. In: De la Puente J, Pérez-Tris J, Juan M, Bermejo A (eds) Anuario Ornitológico de Madrid 2006. SEO-Monticola, Madrid, pp 40–55
Morales MB, Traba J, Delgado MP, García de la Morena EL (2013) The use of fallows by nesting little bustard Tetrax tetrax females: implications for conservation in mosaic cereal farmland. Ardeola 60:85–97
Morales MB, Arroyo B, Traba J (2015) El declive del sisón en el centro de España. Quercus 356: 36–43
Nissardi S, Zucca C (2011) Risultati del primo censimento completo dei maschi territoriali di gallina prataiola tetrax tetrax in Sardegna (2009–2011). Atti del Convegno Italiano di Ornitologia, Cervia
Oparin ML, Kondratenkov IA, Oparina OS, Mamayev AB (2018) Abundance of the Little Bustard *Tetrax tetrax* L. 1758 (Otididae, Aves) in Saratov Oblast. Biol Bull 45(10):1165–1173
Palacín C, Alonso JC (2009) Probable population decline of the Little Bustard *Tetrax tetrax* in north-west Africa. Ostrich 80:165–170
Petretti F (1985) Preliminary data on the status of the Little Bustard in Italy. Bustard Stud 2:165–170
Petretti F (1991) Status of lowland dry grasslands and birds in Italy. In: Goriup PD, Batten L, Norton JA (eds) The conservation of lowland dry grassland birds in Europe, Reading, pp 69–76
Petretti F (2006) Conservation of steppe birdlife in Italy. In: Gafta D, Akeroyd J (eds) Nature conservation. Concepts and practice. Springer, Heidelberg, pp 235–239
Petretti F (2007) The situation of the little bustard in Italy. In: Leitão D, Jolivet C, Rodríguez M, Tavares J (eds) Bustard conservation in Europe in the last 15 years: current trends, best practice and future priorities. RSPB/BirdLife, Bedfordshire, pp 65–69
Patrikeev M (2004) The birds of Azerbaijan. Pensoft, Sofia
Santangeli A, Dolman PM (2011) Density and habitat preferences of male little bustard across contrasting agro-pastoral landscapes in Sardinia (Italy). Eur J Wildl Res 57:805–815
Santangeli A, Aresu M, Cardillo A, Vitale M (2010) Nota sullo status della gallina prataiola Tetrax tetrax nella Zona di Protezione Speciale Pianadi Semestene, Bonorva, Macomer e Bortigali (Sardegna) (2007–2009). Alula 18:142–144
Schenk H, Aresu M (1985) On the distribution, number and conservation of Little Bustard in Sardinia (Italy). Bustard Stud 2:161–164
Schulz H (1985) A review of the world status and breeding distribution of the Little Bustard. Bustard Stud 2:131–152
Scott DA (1995) A directory of wetlands in the Middle East. IUCN and IWRB, Slimbridge
Sehhatisabet ME, Abdi F, Ashoori A, Khaleghizadeh A, Khani A, Rabiei K, Shakiba M (2012) Preliminary assessment of distribution and population size of wintering Little Bustards Tetrax tetrax in Iran. Bird Cons Int 22(279):287
SEO/BirdLife (2018) SACRE. https://www.seo.org/sacre/. Last accessed 31 Oct 2018

Shlyakhtin GV, Tabachishin VG, Khrustov AV, Zav'yalov EV (2004) Ecological segregation of bustards (Otididae) in the north of the lower Volga region: evolutionary and adaptive aspects. Russ J Ecol 35:247–253

Silva JP (2010) Factors affecting the abundance of the little bustard Tetrax tetrax: implications for conservation. Dissertation, University of Lisbon

Silva JP, Pinto P (2006) Final report of Action 2 of the Project LIFE02NAT/P/8476: Conservation of the little bustard in Alentejo. Institute for Nature Conservation. Available via http://portal.icnb.pt/NR/rdonlyres/783CA43B-7C0E-4FA4-8286-F4EB120C4A49/0/Life_SISAO.pdf. Accessed 17 Jan 2011

Silva JP, Santos M, Queirós L, Leitão D, Moreira F, Pinto M, Leqoc M, Cabral JA (2010) Estimating the influence of overhead transmission power lines and landscape context on the density of little bustard *Tetrax tetrax* breeding populations. Ecol Model 221:1954–1963

Silva JP, Catry I, Palmeirim JM, Moreira F (2015) Freezing heat: thermally imposed constraints on the daily activity patterns of a free-ranging grassland bird. Ecosphere 6:article number 119. https://doi.org/10.1890/ES14-00454.1

Silva JP, Correia R, Alonso H, Martins RC, D'Amico M, Delgado A, Sampaio H, Godinho C, Moreira F (2018) EU protected area network did not prevent a country wide population decline in a threatened grassland bird. PeerJ 6:e4284. https://doi.org/10.7717/peerj.4284

Smelansky IE, Tishkov AA (2012) The steppe biome in Russia: ecosystem services, conservation Status, and actual challenges. In: Werger MJA, van Staalduinen MA (eds) Eurasian steppes. Ecological problems and livelihoods in a changing world. Springer, Heidelberg, pp 45–101

Smith AC (1868) A sketch of the birds of Portugal. Ibis 10:428–460

Snow DW, Perrins CM (1998) The Birds of the Western Palearctic (Concise Edition) vol 1. Non-Passerines. Oxford University Press, Oxford

Spangenberg EP (1951) Order Otidiformes. In: Ptitsy Sovetskogo Soyuza (Birds of the Soviet Union), vol 2. Sovetskaya Nauka, Moscow, pp 139–168

Suárez-Seoane S, Osborne PE, Alonso JC (2002) Large-scale habitat selection by agricultural steppe birds in Spain: identifying species–habitat responses using generalized additive models. J Appl Ecol 39:755–771

Suárez-Seoane S, Garcia de la Morena EL, Morales MB, Osborne PE, De Juana E (2008) Maximum entropy niche-based modelling of seasonal changes in little bustard (Tetrax tetrax) distribution. Ecol Model 219:17–29

Tarjuelo R, Barja I, Morales MB, Traba J, Benítez-López A, Casas F, Arroyo B, Delgado MP, Mougeot F (2015) Effects of human activity on physiological and behavioral responses of an endangered steppe bird. Behav Ecol 26:828–838

Ternier L (1892) La sauvagine en France. In: Oiseaux de mer, de rivières et de marais. Firmin-Didot, Paris

Thévenot M, Vernon R, Bergier P (2003) The birds of Morocco. BOU checklist No. 20. British Ornithologists' Union and British Ornithologists' Club, London

Villers A, Millon A, Jiguet F, Lett JM, Attié C, Morales MB, Bretagnolle V (2010) Tracking of wild and captive-bred Little Bustards Tetrax tetrax from western France, and implications for reinforcement programmes. The Ibis 152:254–261

Wassink A (2016) The new birds of Kazakhstan. Arend Wassink, privately published, Texel

Wassink A, Oreel GJ (2007) The birds of Kazakhstan. De Cocksdorp, Texel

Yeatman L (1976) Atlas des oiseaux nicheurs de France. Société Ornithologique de France/ Ministère de la Qualité de la Vie Environnement, Paris

Yousefi M, Kafash A, Malakoutikhah S, Ashoori A, Khani A, Mehdizade Y, Ataei F, Ilanloo SS, Rezaie HR, Silva JP (2017) Distance to international border shapes the distribution pattern of the growing Little Bustard Tetrax tetrax winter population in northern Iran. Bird Conserv Int 28: 499–508

Breeding Biology and Demographic Traits: Population Parameters, Reproduction and Survival

S. Mañosa, G. Bota, A. Villers, Vincent Bretagnolle, and Manuel B. Morales

Introduction

The fate of any living population is ultimately determined by the balance between reproduction, emigration, immigration and death. Understanding these processes, in both quantitative and qualitative terms, is essential for the appropriate management and conservation of animal populations. It is however surprising how little quantitative information exists about the vital parameters of the little bustard, despite being such an unusual bird in terms of its mating system, and how much evolutionary and conservation biologists can learn from it. While the details of the complex breeding

S. Mañosa (✉)
Departament de Biologia Evolutiva, Ecologia i Ciències Ambientals, Institut de Recerca de la Biodiversitat (IRBio), Universitat de Barcelona, Facultat de Biologia, Barcelona, Catalonia, Spain
e-mail: smanosa@ub.edu

G. Bota
Centre de Ciència i Tecnologia Forestal de Catalunya (CTFC), Solsona, Catalonia, Spain
e-mail: gerard.bota@ctfc.es

A. Villers
Office Français de la Biodiversité, Direction de la Recherche et de l'Appui Scientifique, Unité Avifaune Migratrice, Villiers-en-Bois, France
e-mail: alexandre.villers@ofb.gouv.fr

V. Bretagnolle
Centre d'Etudes Biologiques de Chizé, CNRS & La Rochelle Université, Beauvoir sur Niort, France
e-mail: vincent.bretagnolle@cebc.cnrs.fr

M. B. Morales
Terrestrial Ecology Group, Departamento de Ecología & Centro de Investigación en Biodiversidad y Cambio Global (CIBC), Universidad Autónoma de Madrid, Madrid, Spain
e-mail: manuel.morales@uam.es

V. Bretagnolle et al. (eds.), *Little Bustard: Ecology and Conservation*, Wildlife Research Monographs 5, https://doi.org/10.1007/978-3-030-84902-3_5

behaviour and mating system of the little bustard are dealt with in other chapters of this book, here we summarize what is known about some aspects of the breeding biology and vital rates of the species. Although early studies on this topic considered eastern populations in Asia, most recent quantitative information on breeding and demographic parameters comes from western populations in France and the Iberian Peninsula. Our aim is to provide an overview of the variation of the vital parameters of the species across its distribution range, an essential basis for understanding the varying fate and life history strategies of extant little bustard populations, as well as for identifying gaps in our knowledge that future research can plug.

Breeding Densities

Most information about little bustard breeding densities involves densities of displaying males, which may not necessarily reflect the real status of the populations, since male and female densities can be very different in some populations (see below). However, owing to their secretive behaviour the actual density of females, which is a key (if not the most important) parameter in terms of demography, is unknown in most cases (but see Devoucoux et al. 2019 for an exception). It is in any case important to use the concept of density with caution in this species, as it is known to exhibit a clearly aggregated distribution pattern, resulting from underlying habitat and social factors (see chapters "Habitat selection and space use" and "Behavioural ecology of the little bustard: sexual selection and mating systems", respectively). This means that density estimates are very sensitive to the spatial scale of the study and the statistical tools used to compute it, which are therefore crucial to consider when comparing different sites. Density estimates tend to increase exponentially as the sampled area in the studies becomes smaller (Table 1 and Fig. 1), so that comparing male densities between study sites is not always straightforward. This is probably because studies based on small sampled areas reflect a focus on the most favourable habitats for the species and also because studies based on progressively larger areas may include areas of unfavourable habitats. The effect is accentuated by the lekking behaviour of little bustard males, which causes them to congregate. For the same reason, extrapolating breeding density estimates obtained at a local scale to a larger geographical area will result, in most cases, in an unrealistic evaluation of population size, particularly when extrapolations are done regardless of habitat suitability or within well-defined distribution ranges.

Data on male densities have been estimated in several western European populations, most of them in farmland areas. The lowest absolute male densities (not corrected for the size of the sampled area), fluctuating between 0.13 and 1.5 males/km^2 have been recorded in farmland areas of western-central France (Boutin and Metais 1995; André 1985; Jiguet et al. 2000). The highest are those reported in farmland areas in the Nîmes region in south-eastern France (Devoucoux et al. 2019), where up to 23.4 males/km^2 have been observed, compared to those of 5–14 males/km^2 in traditional pastures and 'montados' (open green oak forest) of the Alentejo

Table 1 Male densities in several regions of the little bustard's western range, in relation to the size of the area sampled in each study site

Site	Area (ha)	Males/ km^2	References
1. Sardinia	250	2.4	Petretti (1993)
2. Nîmes (SE France)	409	23.44	Devoucoux et al. (2019)
3. Apulia (SE Italy)	520	1.5	Petretti (1993)
4. Castro Verde (S Portugal)	900	7.50	Silva et al. (2014)
5. Valdetorres (C Spain)	1500	1.93	Morales et al. (2008a)
6. Campo Real (C Spain)	1600	2.40	Morales et al. (2008a)
7. Bellmunt (NE Spain)	1794	4.24	Tarjuelo et al. (2013)
8. Belianes (NE Spain)	1839	4.02	Tarjuelo et al. (2013)
9. Abbasanta (Sardinia)	4314	3.02	Santangeli and Dolman (2011)
10. Cabrela (S Portugal)	4653	0.50	Faria et al. (2012a)
11. Crau (SE France)	6080	2.70	Wolff et al. (2001)
12. Campeda (Sardinia)	8712	0.13	Santangeli and Dolman (2011)
13. Ciudad Real (C Spain)	10,698	2.34	Tarjuelo et al. (2013)
14. Campidano (Sardinia)	11,281	0.31	Santangeli and Dolman (2011)
15. Crau (SE France)	16,050	2.00	Wolff et al. (2001)
16. LTSER Zone Atelier Plaine & Val de Sèvre (CW France; 1996)	34,000	0.15	Jiguet (2001)
17. LTSER Zone Atelier Plaine & Val de Sèvre (CW France; 2018)	43,000	0.065	Bretagnolle (unpubl.)

In the site LTSER Zone Atelier Plaine & Val de Sèvre, two values are reported (one in 1996, one in 2018), since this site has seen strong variation in population size (see Bretagnolle et al. 2018)

(Portugal) (Schulz 1985; Silva et al. 2014). Most Iberian traditional cereal farmlands hold intermediate male densities ranging from 1.9 to 4.0 males/km^2 (Morales et al. 2014). In Crau, SE France, densities range from 1.0 to 8.0 males/km^2 (Cheylan 1985; André 1985; Wolff et al. 2001), while in Sardinia 0.8–3.1 males/km^2 have been reported (Schenk and Aresu 1985; Petretti 1993). In the virtually extinct population in Apulia (southern Italy), Petretti (1993) reported a local density of 1.5 males/km^2 in *Asphodelus* pastures. The residuals of the regression points in Fig. 1 (how far each point is from the regression line in the vertical axis and whether it is above (higher density than predicted) or below (lower density than predicted)) can be used as a comparable relative density estimate between the points in the graph. Although somewhat simplistic, this approach may tentatively indicate that the populations above the line—some areas in Ciudad Real and Catalonia (in Spain), some in Sardinia (in Italy), Castro Verde (in Portugal) and the French Mediterranean area—actually hold relatively dense populations in comparison to the populations below the line—Apulia, Sardinia (in Italy), Cabrela (in Portugal) and Deux-Sèvres (in France).

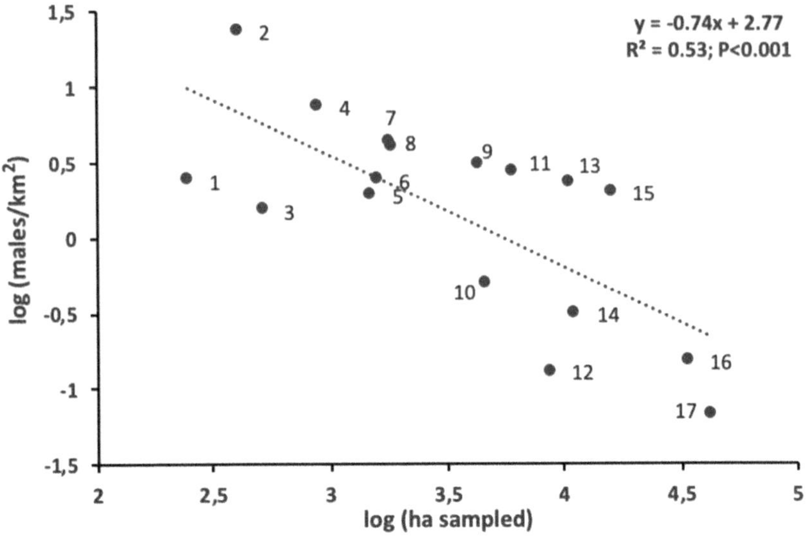

Fig. 1 Relationship between recorded male densities and the size of the area sampled in each study site. Point numbers refer to region, as in Table 1

Although eastern European and Palearctic populations represent a huge part of the world distribution range of the species, much less information exists about male densities there. In this region, the little bustard breeds mainly on natural or pastural steppes and in arable (mainly fallow) land. In Russia, according to the surveys conducted in 1975–1980 by Belik (1992) in the Rostov Oblast (south-eastern Russia), densities could vary from 0.1 male/km^2 in the steppes between the Sal and Don River to 6–10 males/km^2 on sandy steppes, or 6–20 males/km^2 in alfalfa fields. Amazingly high densities of 60 males/km^2 were reported in 1985 in the Saratov region (Moseykin 1992), but this probably reflects a very local estimate.

In Kazakhstan, mean breeding density in Karaganda and Kostanay was estimated at 0.30 birds/km^2 (81% males), with big differences depending on habitat. Thus, 1.35 birds/km^2 were estimated in old and recently mown perennial grass fields, 0.53 birds/km^2 in abandoned arable fields, 0.26 birds/km^2 in pristine steppe and only 0.07 birds/km^2 in active arable fields (Koshkin 2011). A more recent survey conducted in the Aktobe region in 2017 (S. Mañosa, F. Cuscó, G. Bota and R. Urazaliyev, unpublished data) by means of distance sampling revealed an average density of males (±SE) of 0.138 ± 0.45 males/km^2.

Because of the difficulty of counting females in the field, very little information exists on female densities. Devoucoux et al. (2019) reported local female densities of 24.9 females/km^2 but, at larger spatial scales, a usual female density estimate range of between 0.25 and 6 females/km^2 or even lower (Table 2). At plot scale the density of females and nest has been found to be associated with the density of displaying males (Wolff et al. 2002; Tarjuelo et al. 2013; Devoucoux et al. 2019). This

Table 2 Female little bustard densities reported in several western European study sites (in central-western France site, minimum-maximum density values presented, depending on year)

Site	Area (ha)	Females/ km^2	References
Cabrela (S Portugal)	1590	0.25	Faria et al. (2012a, b)
Crau (SE France)	≈ 1800	≈ 0.5–6.0	Wolff et al. (2002)
Campo Real (C Spain)	275	1.4	Morales et al. (2008b)
Valdetorres (C Spain)	300	0.54	Morales et al. (2008b)
Alentejo (S Portugal)	12	7.3	Silva et al. (2014)
Nîmes (SE France)	409.51	24.9	Devoucoux et al. (2019)
Lleida Plain (NE Spain)		0.05–2.70	Bota et al. (2015)
LTSER Zone Atelier Plaine & Val de Sèvre (CW France)	43,000	0.028–0.056	Bretagnolle unpubl.

relationship is globally true, as well, at regional scale. In the Lleida Plain, in Catalonia, little bustard counts conducted in seven different Special Protection Areas in 2002, 2009 and 2014 revealed that spring female densities also tend to be higher at sites where male densities are high (Fig. 2), although in some cases low male density areas may hold relatively high female densities.

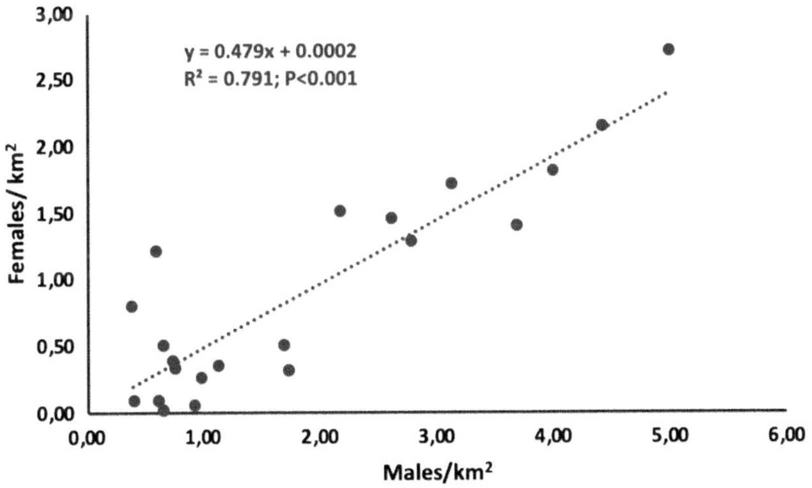

Fig. 2 Relationship between recorded male and female densities in seven sites in the Lleida Plain (Catalonia, NE Spain) in 2002, 2009 and 2014. Each point represents a site and year. Drawn from data taken from Bota et al. (2015)

Breeding Phenology, Laying Dates and Re-nesting

Breeding phenology may differ between populations, depending on latitude and climate, as well as on the specific migratory behaviour in each one. In the migratory population of western-central France, males arrive first, between March and April, with some individuals displaying directly upon arrival, while females arrive from April onwards (Bretagnolle and Inchausti 2005). Then males remain in the study area and display until early July, although there are seasonal movements by some of the males, presumably tracking sites with females. Display rates by males are almost stable between early May and late June (Jiguet and Bretagnolle 2001). In the migratory Atlantic population of Galicia (north-west Spain), the earliest sightings take place in early May (Rivas et al. 2005). In the mainly sedentary population of Catalonia (north-east Iberia), wintering groups break up progressively from late January to early April, and males settle on their breeding territories and start displaying in early April, with maximum displaying activity reported in late April and early May. Depending on their age and/or social status (Ponjoan et al. 2012), males remain on their lek all through the breeding season, and leave for the post-reproductive grounds in late May or early June. In this area, it is not rare to observe some males still displaying in late June or early July, chasing females with chicks. In central and western Spain, display periods have been reported during April and May (Martínez 1994) with maximum intensity in May (Delgado et al. 2010), similar to what is described in southern Portugal (Faria et al. 2012a; Moreira et al. 2012). In Sardinia, bustard displaying activity is concentrated between April and June (Petretti 1993). However, inter-annual phenological displacements may occur within sites, as at Valdetorres (central Spain, Delgado et al. 2010), which probably reflect inter-annual differences in weather conditions (Fig. 3). Indeed, it has recently been shown that little bustard activity level is strongly determined by temperature (Silva et al. 2015).

Egg-laying starts in late April (Lett et al. 2000; Moseykin 1992; Bretagnolle et al. 2018; Lapiedra et al. 2011; Cuscó et al. 2021), but may continue up to late July (Bretagnolle et al. 2018). The earliest start of laying has been reported in the Lleida cereal plains, in Catalonia, where radio-tracking and GPS tagging have revealed that laying takes place between 17 April and 13 July (first and replacement clutches included) (Lapiedra et al. 2011; Cuscó et al. 2021), but concentrates in May. In the migratory populations of western-central France, Bretagnolle et al. (2018) recorded laying dates for 105 nests between 1997 and 2009, ranging from 2 May to 22 July, but with a median on 11 June, which is consistent with previous records for three radio-tracked females (20 May–25 June; Lett et al. 2000). However, they found that egg-laying showed a very clear bimodal pattern, with 42% of clutches being laid from 1 May to 15 June, and 58% from 15 June to 22 July. The second mode is likely constituted by replacement clutches. Using a 5-year sliding window they further showed that in the early years the second mode was higher, while more recently the first mode was higher, suggesting a trend in laying to occur earlier every year (Bretagnolle et al. 2018). The most retarded reported dates for the start of laying

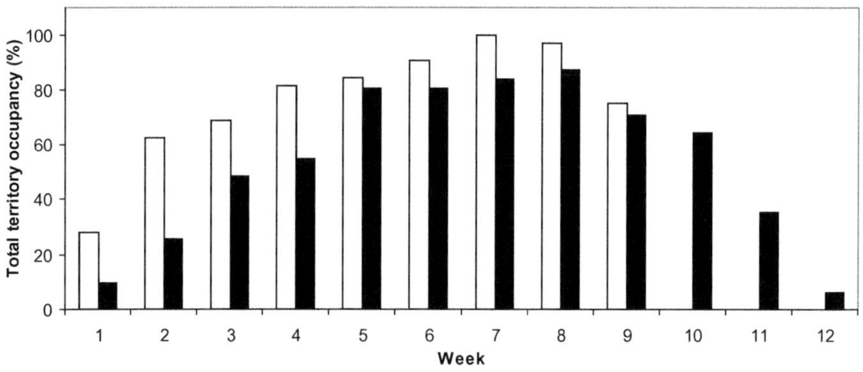

Fig. 3 Territory occupancy of male little bustards in Valdetorres (Central Spain) in 2001 and 2002 (white and black bars, respectively). Week 1 corresponds to the last week of March [Based on Delgado et al. (2010)]

have been found in the Saratov region, in south-western Russia (Moseykin 1992), where the egg-laying period covered 17 May–19 June, which suggests that the nesting period can be considerably concentrated in some regions or latitudes.

Re-nesting is common after clutch loss (but probably not after chick loss) and late clutches (June–July) reported in many areas are probably in most cases replacement clutches, although limited nest-site availability in farmlands as the season progresses may reduce the possibility of re-nesting. Lett et al. (2000) reported a radio-tracked female laying a first clutch on 1 June and a replacement clutch on 24–25 June. According to Boutin and Metais (1995), replacement clutches can be found as late as early August in France. In Lleida Plains, as many as three successive clutches can be laid by a single female (Cuscó et al. 2021). In this area, a radio-tracking study by Lapiedra et al. (2011) found that 2 of 9 little bustard females (22%) that lost their first clutch produced a second clutch. Subsequent research based on GPS tagging in the same area found that among 16 failed first and second clutches, 10 were replaced (63%), while failed third clutches were never replaced (Cuscó et al. 2021).

Taken globally, all this information suggests earlier laying in the southern and sedentary populations as compared to northern and migratory ones. On the other hand, the longer laying period reported in western-central France may reflect the prolonged suitable conditions provided by Atlantic weather systems compared to those under Mediterranean or continental climates, where harsh winters and/or extreme summer temperatures may limit the onset and/or the end of the laying season.

Nests

Eggs are laid in concealing vegetation on the ground, on a shallow, 4–9 cm deep round depression, 18–21 cm in diameter, usually lined with a few sticks and fragments from the surrounding vegetation (Plate 1) and frequently also some female feathers (Dementiev and Gladkov 1951; Cramp and Simmons 1980). The nest cup is usually built in flat ground, and females take care not to remove vegetation screening the nest. In most cases, nests are concealed from above by herbaceous cover and shadow (Plate 1a), and are usually just deep enough for the back of the incubating female to be confounded with the surrounding ground. Vegetation cover around the nest has to be tall enough to conceal the nest and protect the eggs and the female from the sun, the rain and aerial predators, but the females seem to avoid too dense swards, which impair walking to and from the nest, which is always accessed by foot (Bretagnolle et al. 2018).

In Deux-Sèvres, 53.8% of the nests were found in alfalfa fields, 20.1% in set-aside fields and 23.4% in other meadows including 12.6% in artificial meadows (such as ray grass) and 10.1% in permanent meadows ($n = 119$ nests in 1997–2009); 70.5% of nests were in either pure or mixed leguminous fields, and only 27.7% were in pure grass meadows (mostly set-asides). The habitat used for breeding represented

a c

b d

Plate 1 Examples of little bustard nests from central Spain; (**a**) well camouflaged nest with four eggs in an old fallow field (Photo by M.B. Morales); (**b**) nest with two hatched eggs in old fallow field (Photo by M.B. Morales); (**c**) nest with two eggs in natural shrub-steppe vegetation (Photo by Jesús Herranz); (**d**) empty nest whose simple structure can be appreciated (a 1 euro coin was placed to provide an idea of dimensions; Photo by M.B. Morales)

a mere 14% of that available (Bretagnolle et al. 2018). Vegetation height at laying was 38.7 ± 19.6 cm (n = 62, range 5–105.8 cm). In the dry extensive pseudo-steppes of the Lleida plains (Catalonia), females positively select fallow land to nest, although most nests are set in dry cereal fields, which is the dominant land cover and are in general less than 40 cm tall and of relatively low stem density (Lapiedra et al. 2011; Cuscó et al. 2021). In central Spain, Morales et al. (2013) found that long-term fallows are the main agricultural habitat type used and positively selected by breeding little bustard females for nesting, followed in importance by young fallows. Cereal fields are also used, but clearly less than expected based on its availability. In more intensive farmland areas in western France, however, most nests are found in alfalfa, set-aside and fallow land (Jolivet 1994; Bretagnolle et al. 2018), probably because cereal crops are taller and thicker than in Mediterranean areas. In Iberian 'páramos' (semi-natural cold steppes in the Iberian highlands), little bustard nests have also been found in natural shrub-steppe vegetation (Plate 1c) (J. Herranz pers. obs.).

In grassland fields in the Alentejo (Portugal), the probability of occurrence of potential nesting females was higher in the largest fields, as well as in grassland 20 cm tall as compared to 40–60 cm tall (Silva et al. 2014), but selection for larger fields has not been detected in other areas. Indeed, size did not have any effect on nesting field selection in Deux-Sèvres (average size of fields with nests was 3.56 ha, range 0.47–26.6 ha; Bretagnolle et al. 2018). In two separate studies conducted in western France (Bretagnolle et al. 2018) and the Lleida Plain (Cuscó et al. 2021) no trend was found to place the nest close to field margins than expected by random. For example, in western France, the average distance from the nest to the nearest margin was 26.1 m, range 0–100 m, which did not differ from random.

Male territories contain resources that may be used by the females (Traba et al. 2007), although this does not seem to increase male attractiveness (Jiguet et al. 2002). Although male and female little bustards have been described as selecting different microhabitats during the breeding season (Morales et al. 2008a; Silva et al. 2014), this seems to be more the result of their different roles in reproduction, rather than of different feeding or physiological requirements. This difference is probably not sufficient to generate a spatial segregation at meso-scale, so that nests and families are more frequently found inside rather than outside male territories (Morales et al. 2013) and nests were found closer to displaying males than random in Portuguese grasslands (Faria et al. 2012a; Silva et al. 2014). Lett et al. (2000) also reported that, after mating, females nest in general 100–1000 m from the display sites, but Bretagnolle et al. (2018) provided several cases in which females laid at 5–10 km from males. Indeed, GPS tracking in Lleida Plain (unpublished data) indicates that some females travel more than a hundred kilometres in the days prior to laying, suggesting the possibility that laying may occur far from mating sites on some occasions.

Eggs and Incubation

Eggs are unspotted variable olive-brown due to biliverdin deposition (Plate 1) and have a variable, but always very light, brownish patterning. They are 47–58.2 mm long, 35–41 cm wide and weigh around 41 g (Dementiev and Gladkov 1951; Cramp and Simmons 1980; see also chapter "Captive breeding, handling and care, and the impact of releases on wild populations"). Clutch size is 1–6 eggs, usually 3–4 (Cramp and Simmons 1980). An average clutch size of 3.09 eggs (range 2–5) has been reported in western France (Bretagnolle et al. 2018). Boutin and Metais (1995) reported an average clutch size of 3.7 eggs in France. A one-egg clutch was found also in Deux-Sèvres (France), but it was probably a replacement, and was finally abandoned. Extreme clutches of 9 and 11 eggs, supposedly laid by a single female, have also been reported (Dementiev and Gladkov 1951), but they are probably the result of intraspecific brood parasitism. A captive hen laid 11 eggs consecutively (Von Frish 1976), but recent captive breeding has shown some females can lay up to 20 (chapter "Captive breeding, handling and care, and the impact of releases on wild populations"). Clutches of 5 eggs were reported as the commonest in Portugal (Labitte 1955; Ferguson-Lees 1967 in Cramp and Simmons 1980). Some field evidence suggests that 1-year-old females may lay smaller clutches (one egg less on average) than older ones (Bretagnolle and Inchausti 2005). Labitte (1955) reported replacement clutches of 2, which suggest smaller size than first clutches, but in Catalonia replacement clutches of 3, the usual size of first clutches in the area, have been recorded (S. Mañosa and F. Cuscó unpublished data).

Eggs are probably laid at 2-day intervals (Von Frish 1976); this is supported by observations on captive females (chapter "Captive breeding, handling and care, and the impact of releases on wild populations"). Incubation lasts for 20–22 days and, as no pair bond exists, is conducted exclusively by the female. Although it has been suggested that incubation may begin before the last egg is laid (Cramp and Simmons 1980), a study conducted in Catalonia, where 28 clutches were monitored by means of GPS tagging (Cuscó et al. 2021) showed that the period of full nest attention lasts for about 21–22 days, suggesting that incubation begins with the last egg. According to this study the females spend on average 87% of the daytime on the nest. During the day, breaks from incubation were more frequent in the morning and early evening, and nest attendance was maximum over midday and during the late evening. A similar pattern was also observed in western France as indicated by temperature recorders placed in dummy eggs (Villers and Bretagnolle unpublished data). Temperature constraints and predation risk are probably the factors that shape this behavioural pattern. During incubation, females do not go far from the nest. In central France, Lett et al. (2000) reported Minimum Convex Polygon ranges of 11–12 ha during the laying and incubating phase, but most activity (90–95%) was concentrated on 2–3 ha around the nest, a value similar to the average (±SD) of 1.4 ± 2.0 ha ($n = 27$) obtained in the Lleida Plains (Catalonia) by GPS tagging (Cuscó et al. 2021).

Laying Rates, Nest Survival and Hatching

In a given breeding season not all females necessarily lay eggs. In a radio-tracking study conducted in Lleida Plains (Catalonia) in 2006–2008, Lapiedra et al. (2011) found that as many as 5 of 22 (22.7%) hens monitored by VHF transmitters alive at the start of the breeding season (1 April for convention) did not lay (one because it died and the other four for no apparent reason). A similar 16% value was obtained for a different set of 19 females monitored by GPS in the same area in 2010–2016 (Cuscó et al. 2021), indicating a relative consistency of this parameter. Females that do not lay may be younger and less experienced, older or simply in poorer body condition.

In the stable population of La Crau, southern France, where females nest in fallow and steppe, Wolff (2001) estimated that nest survival from laying to hatching was about 65%. In the smaller populations of the agricultural region of Poitou-Charentes, in western-central France, the estimated nest survival was 58.4–64% (Bretagnolle et al. 2011, 2018). Of the total nest losses, 93% were due directly or indirectly to farming practices (i.e. harvest), causing the destruction of the nest or desertion by the incubating female, and 7.5% to predation (Bretagnolle et al. 2011, 2018). This means that, in this area, 33.5% of all nests are lost to harvest operations. This proportion was even higher (50%) between 1997 and 2004, before the establishment of an agri-environmental scheme (AES) in the area. In Portugal, where little bustards nest mainly in natural or semi-natural pasture, Boutin and Metais (1995) report that of 25 nests only 11 (44%) hatched. In this case, the main cause of nest losses was nest trampling by livestock (50%), followed by predation (36%) and farming practices (14%). These authors reported that nest destruction was commoner in pasture than in cereal crops. In the cereal pseudo-steppes of Lleida, in Catalonia, nest survival is also relatively low, at 32.1–47.1% (Lapiedra et al. 2011; Cuscó et al. 2021). In this area, however, farming operations account for a lesser proportion of nest losses, only 33–44% of all of them, the remainder being attributed to predation or other causes not related to human intervention (Cuscó et al. 2021).

Chick Growth and Care

Eggs hatch synchronously, all in the same day, allowing the chick and the hens to leave the nest. Chicks are precocial and nidifugous, and are brooded and cared for exclusively by the hen. Before 5 days old, they are fed bill-to-bill by the female. Chick food consists exclusively of arthropods (mainly coleopterans and orthopterans), at least until they are a few weeks old (Jiguet 2002). According to measurements obtained in captivity (Bretagnolle et al. 2018; Ivanova 1992; Alex Llopis, unpublished data), chicks grow slowly until 5 days old, but their daily weight gain increases thereafter to an average rate of 13.2 g/day (range 2.1–26.2 g/day for two Lleida chicks) (Fig. 4). Although full size is not attained until 50–55 days (Fig. 5),

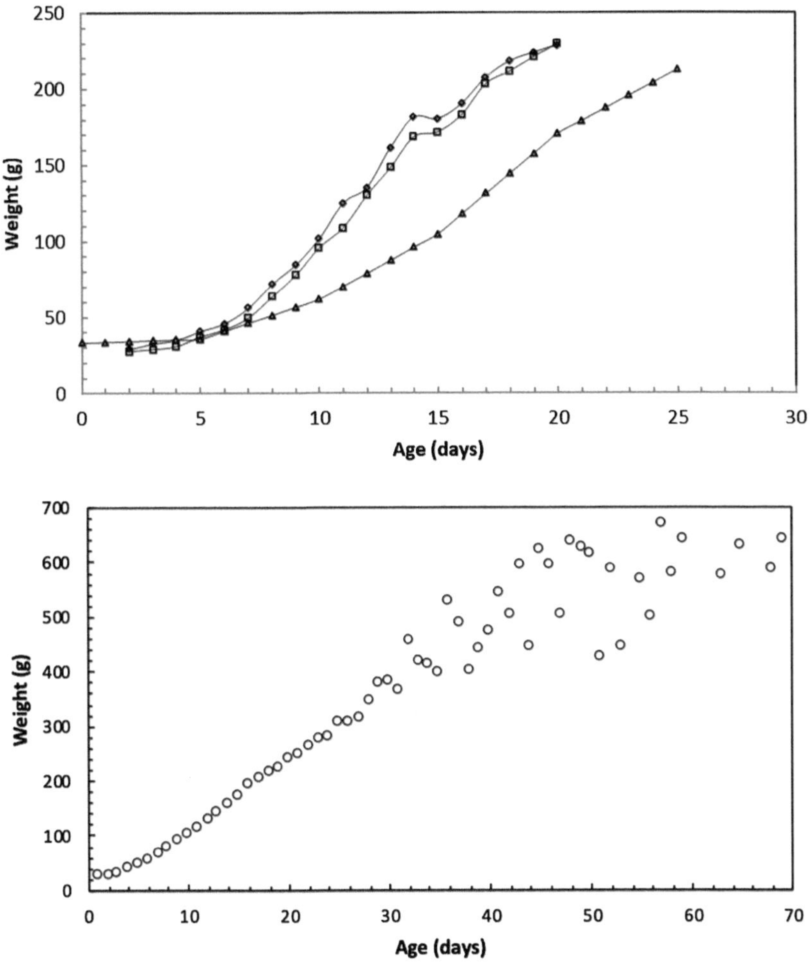

Fig. 4 Weight gain of little bustard chicks. (Top) Drawn from data in Ivanova (1992), assuming an average egg weight of 41 g (triangles), and on the measurements of 2 little bustard chicks hatched in captivity in the Vallcalent Wildlife Recovery Center (Lleida, Catalonia) in 2006, raised and measured by Alex Llopis (diamonds and squares). (Bottom) Based on measurements of 41 little bustard chicks measured in captivity in 1997–2004 [Based on Bretagnolle et al. (2018)]

they can fly as early as 25–30 days old because tarsus length and wing length are attained much quicker (Fig. 5).

During the rearing period, if feeding conditions are adequate for both the chicks and the hen, little bustard families can remain in a very small area of only a few hectares. However, if they are disturbed (e.g. by farming operations) or the availability of food is low, the families are obliged to move from patch to patch, covering an overall much larger range. Chicks can walk relatively large distances from a very early age, although this may have a survival cost (Bretagnolle et al. 2011; Lapiedra et al. 2011). In central France, Lett et al. (2000) reported Minimum Convex Polygon

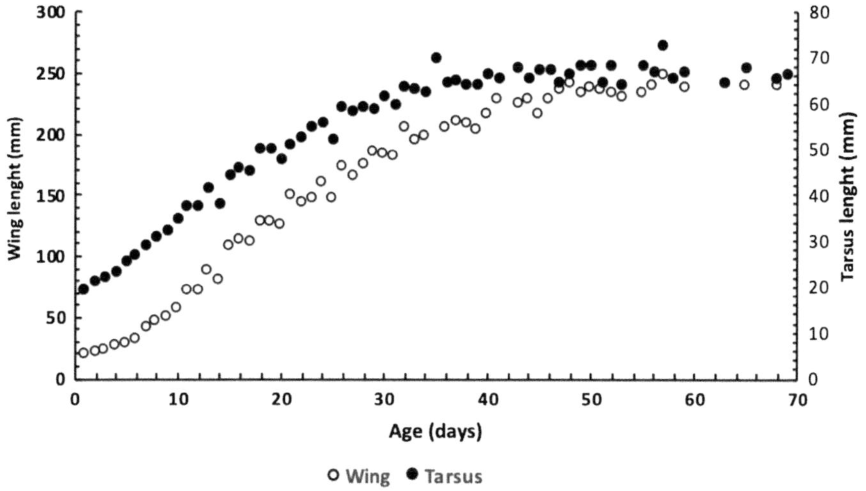

Fig. 5 Tarsus and wing length growth (mm) of 41 little bustard chicks measured in captivity in 1997–2004 [Based on Bretagnolle et al. (2018)]

(MPC) ranges of 20–25 ha during the brooding phase, but most activity (95%) was concentrated within a 10–15 ha range ($n = 3$). In the Lleida cereal plains (Catalonia), families had much larger average home ranges (MPC) of 136.9 ha (range 10–432 ha, $n = 8$; Lapiedra et al. 2011). These ranges included 1–3 activity areas ranging from 2.7 to 106 ha each, separated 290–676 m from each other.

Brood and Chick Survival

Chick mortality is probably high during the first days of life. Although the average clutch size is between 3 and 4 eggs, it is usual to see families of only one or two fully grown chicks. In southern Portugal, brood size reduced from 2.3 chicks in 2-week-old broods to 2.1 chicks in 4-week-old broods (Boutin and Metais 1995). In the Poitou-Charentes region of western-central France in 1997–2001, average family sizes also showed strong evidence of brood reduction (Bretagnolle et al. 2011): the average (±SD) number of chicks per family (including total brood failure) averaged 1.32 ± 1.35 for families 7 days old, 0.77 ± 1.20 at under 15 days of age, 0.51 ± 1.00 at 15–30 days, 0.49 ± 1.00 at 30–45 days, and 0.36 ± 0.97 at more than 45 days (see also Bretagnolle et al. 2018 for updated figures). In this area, no evidence of chick predation was found, and all the chicks that were recovered dead showed signs of starvation. In spite of this, in about 60% of broods at least one chick survived from hatching to fledging in this area (Berthet et al. 2012). Anecdotal observations have shown that chicks can be predated by raptors such as marsh harrier *Circus aeruginosus* (S. Mañosa and F. Cuscó, pers. obs.), Montagu's harrier *C. pygargus* and hen harrier *C. cyaneus* (B. Arroyo, pers. com.).

In the cereal pseudo-steppes of Lleida (Catalonia), 33–62.5% of broods survived to fledge some chicks (Lapiedra et al. 2011; Cuscó et al. 2021). In this area, earlier broods had higher survival probabilities than later ones, which seems to be related to the decline of the availability of cover as the season progresses and cereal harvesting advances, leaving the families with less protection and food and forcing them to perform long displacements between the few small remaining patches of suitable vegetation (Lapiedra et al. 2011). Families keeping within a smaller area during the whole rearing period successfully fledged twice as many young as families forced to wander, which made faster and longer movements that cause a progressive increase in the area used through the rearing period (Lapiedra et al. 2011). In fact, chicks are able to travel long distances early after hatching, as has been found in western France by Bretagnolle et al. (2011). When grasshopper abundance was low, radio-tracked females with chicks younger than 10 days old could travel 1–2 km a day in search of food. This resulted in losing one chick every 2 days and sometimes in the loss of the whole brood.

Final Breeding Success and Productivity

Breeding success can be defined as the percentage of females present in a given area at the start of the breeding season that manage to fledge at least one young. Reported figures for this parameter oscillated between 7 and 61% on seven sites monitored in 1997–2000 in western France (Morales et al. 2005), where the inter-population median average was 33.2%. In the Lleida Plains (Catalonia) breeding success among 42 breeding females tracked in 2006–2016 was even lower, at only 15–23% (Lapiedra et al. 2011; Cuscó et al. 2021).

Very few data exist on the productivity (average number of chicks fledged per breeding female) of the little bustard. This is because it is very difficult to estimate breeding success in such a shy, precocial species. Estimates based on family counts have been used in some studies, but the values obtained are only an indication of the real production of young per successful female. In this way, in western-central France, between 1991 and 1994, the average size at fledging of 30 families (thus excluding complete failure) was 2.2 fledglings per family, similar to results found in Portugal, where values of 2.1 fledglings per family were reported (Boutin and Metais 1995). Post-breeding counts conducted in 2008 in Iberian cereal pseudo-steppes gave much lower average values, of 1.11 chicks/family in Ciudad Real, and 1.30–1.83 in Lleida Plains (Tarjuelo et al. 2013).

On eight study sites monitored in western France between 1997 and 2002, Morales et al. (2005) and Bretagnolle and Inchausti (2005) reported average productivity values ranging from 0.108 to 1.028 fledglings/female (thus including complete failure), with a median of 0.501 and 0.384, respectively. These figures were obtained by dividing the number of fledglings by the number of females recorded in each study site. In the agricultural region of Poitou-Charentes, in western France, the estimated productivity based on a sample of 87 families recorded between 1997 and 2008 was 0.1 chicks/female (Bretagnolle et al. 2011). Using similar calculations, Morales et al. (2008b) obtained equivalent values of 0.3 and 0.4 between 2005 and 2006 in Campo Real and Valdetorres (Madrid, central Spain), respectively. By following the fate of nests discovered before hatching in one of these areas between 1997 and 2009, Bretagnolle et al. (2018) determined productivity values of 0.62 fledging per female and year, but this included active clutch protection through a breeding centre (see chapter "Captive breeding, handling and care, and the impact of releases on wild populations"). Using an unbiased methodology based on radio-tracked or GPS-tagged females conducted in the Lleida population in Catalonia, the average number of young raised per female was 0.27 ($n = 22$) and 0.25 ($n = 20$) in two separate studies conducted in 2006–2008 and 2010–2016 (Lapiedra et al. 2011; Cuscó et al. 2021).

It has been suggested that, on an individual basis, the number of chicks produced may be partially related to age, younger birds producing on average 15% fewer fledglings than older ones (Bretagnolle and Inchausti 2005). Average (±SD) fecundity values for younger females was assumed to be 0.28 ± 0.18 fledglings per female for 1-year-old females and 0.44 ± 0.26 for older females (Inchausti and Bretagnolle 2005). On a population basis, the number of fledglings raised in a given year seems to be related to grasshopper abundance in the western-central French population (Bretagnolle et al. 2011).

Sex Ratio

Considering the polygynous mating system of the little bustard, the proportions of females to males in the population is a parameter of outstanding demographic relevance (Morales et al. 2005). However, this ratio is difficult to establish, because females are difficult to observe and count. Boutin and Metais (1995), by counting the number of males that followed overflying females, reported a ratio of 209 males for 66 females, that is 0.32 females per male in Deux-Sèvres in the 1990s (western France). Although this method may bias estimates in favour of males, Cheylan (1985) obtained a similar 0.33 females per male value by counting pre-nuptial groups in La Crau, south-eastern France. Both are, indeed, very low female sex-ratios if, according to the modelling approach of Morales et al. (2005), the mid-term viability of little bustard populations declines sharply when the proportion of females for every male in the population is less than 1.0. A less unbalanced sex-ratio of 0.68 ± 0.08 females for every male was estimated in the Poitou-Charentes region, in western-central France, in 1997–2003, consistently similar across years and sites (Inchausti and Bretagnolle 2005; Jiguet and Bretagnolle 2014). By systematically searching 0.25 ha plots in spring in two study areas in Madrid province (central Spain), Morales et al. (2008b) obtained sex-ratio values of 0.72 and 0.54 females per male, respectively. In the Saratov region, based on data collected in 1985, Moseykin (1992) reported a sex-ratio of 0.4 female per male. However, in the Alentejo (Portugal), where some of the highest local male and female breeding densities of little bustard have been recorded in the world, 1.3 females per male ratio was reported (Silva et al. 2014). Spring counts over 400 1 ha quadrats in the Nîmes area (southern France) gave equal densities of male and females, suggesting a more or less balanced sex-ratio (Devoucoux et al. 2019). In a recent compilation of sex-ratio data across several western European populations in France, Spain and Portugal, Serrano-Davies et al. (submitted) found that most populations had an unbalanced sex-ratio in favour of males, and that this low proportion of females positively correlates to low young ratios in the populations.

It thus seems that demographically healthy populations (i.e. Alentejo, Mediterranean France) exhibit more or less balanced sex-ratios or even female biased ratios, while the typical pattern in declining populations is a more or less pronounced lack of females (Serrano-Davies et al. submitted). However, the reasons for this male-biased sex-ratio in some populations are unknown. The secondary sex-ratio, that is, the proportion of sexes at hatching, is usually assumed to be 1:1 (Inchausti and Bretagnolle 2005). However, this might not be the case, as Von Frish (1976), in La Crau, reported only three females among 17 chicks artificially raised up to maturity. If we assume 1:1 sex-ratio at hatching, then biased tertiary sex-ratios observed in some populations should be related to differential survival of the sexes (see below). Alternatively, this sexual bias may be due to an uneven spatial distribution of the sexes, deriving from the different ecological requirements of males and females, or to the peculiar social aggregation patterns and mating system of the species.

Maturity, Survival and Senescence

Females are sexually mature and can breed in their first spring, whereas males are not mature usually before their second year, when they acquire their breeding plumage for the first time (Schulz 1985; Bretagnolle and Inchausti 2005). However, immature males have been observed trying to copulate with decoy females during trapping sessions (Eraud and Villers, pers. obs.) or displaying (G. Bota, personal observation). In a necropsy of a first-year male with juvenile plumage the gonads were found to be active (G. Bota, personal observation).

Very little published information exists on annual survival rates for the little bustard. The oldest recorded birds in the wild are two GPS-tagged males, one in Deux Sèvres (France), which was at least 18 years old in 2019 since it was captured already as adult, and a 14-year-old male observed in Lleida (Mañosa et al. 2018) which was seen again in spring 2019 (Roger Guillem, personal communication) as a 15-year-old bird. However, lifespan may usually go no further than 10 years. Indeed, Inchausti and Bretagnolle (2005) calculated that no more that 5% of the little bustards in a stable population in western-central France would be older than 9 years. Although some authors have assumed that birds younger than 1 year may have a very high mortality rate, of up to 75% (Morales et al. 2005), the few data available on this aspect in western-central France, based on a sample of 35 ringed and radio-tagged adult birds between 1998 and 2002 (Inchausti and Bretagnolle 2005), indicated that the best fit to the survival data was obtained by assuming constant survival over time. However, in many bird species immature birds survive less well than adult birds. Annual survival rates of released fledglings in western France (2002–2013) was c.45%, which gives a survival probability of c.55% once taking into account recapture probability (calculations made by R. Fay using V. Bretagnolle's unpubl. data). Therefore 1-year-old birds have slightly lower survival rates than adults. Inchausti and Bretagnolle (2005) assumed that annual survival rates for birds older than 8 years was 60%, slightly lower than that the average 70% obtained for younger age classes. Data collected in the same study area (Inchausti and Bretagnolle 2005) indicated that average (±SD) annual survival rate estimates were lower for females (68% ± 11%) than for males (72% ± 11%), although this difference was not significant. These data are the only available to suggest that female survival is lower than male survival in this species. Reduced female survival could be associated with their higher exposure to predation or mowing during incubation or to carry-over effects of reproductive costs. This was confirmed by more recent estimates based on a larger sample of birds (2002–2018), with adult survival rates slightly higher, 80% for adult males versus 72% for adult females (calculations made by R. Fay using V. Bretagnolle, unpubl. data).

In Iberia, an adult survival rate estimate based on a larger sample size of 151 radio-tagged adult birds captured in Spain and Portugal between 2001 and 2013 (Marcelino et al. 2017) produced a value of 67%, only slightly lower than the estimate produced in France. In any case, both values are below the minimum of ca. 0.80 threshold in annual adult survival necessary to ensure population mid-term survival (Morales et al. 2005), especially for females, and well below the estimates obtained in stable or increasing populations of other bustard species, such as the great bustard *Otis tarda* (90%, Alonso et al. 2004) or the African houbara bustard *Chlamydotis undulata* (75–86%, Hardouin et al. 2015).

Therefore, in spite of the very different conditions in the breeding grounds of the Iberian and central-western France populations, both seem to have similar adult survival rates. It is worth noting that bustards from the western-central France population overwinter in Iberia (Villers et al. 2010), where they might be subject to similar causes of mortality, and that costs associated with migration might be limited for these individuals, since they usually reach their wintering grounds in a couple of night flights of ~300–400 km per night (Villers et al. 2010). As estimated by Marcelino et al. (2017), half of this mortality might be of anthropogenic origin, mainly poaching and powerline collision casualties. The elimination of this mortality would probably raise survival to acceptable values, guaranteeing population viability.

Concluding Remarks

Although the amount of available quantitative information on the demographic parameters of the little bustard has increased in recent years, it comes from relatively few areas and very little is known about how these parameters change in response to environmental variables. Information is, however, far scarcer in the eastern populations of Russia and Kazakhstan. Current evidence indicates that most studied populations in western Europe exhibit low breeding success and male-biased sex-ratios. They also present survival rates lower than expected from potential longevity, which is consistent with their current declines. Further studies are needed to increase our knowledge on the breeding biology and demographic traits of the species and to better understand how these parameters vary within and between populations, as a prior step to understanding how we can ensure the viability of the little bustard populations across its large distribution range. In particular, unbiased productivity estimates in different areas and over many years are needed to better understand which factors determine the temporal and spatial variation in breeding success and productivity values. Studies of survival rates are also needed to provide further insight into mortality and causes of death, and particularly to identify, and eventually understand, the origin of sex biases in survival rates.

Acknowledgements Thanks to Alex Llopis for keeping, measuring and providing the data on little bustard chick weight increase in Vallcalent (Lleida). Thanks are also due to Tomás Santos for his useful review of the manuscript. Nigel J. Collar kindly reviewed the English and further commented on the manuscript.

References

Alonso JC, Martin CA, Alonso JA, Palacin C, Magaña M, Lane SJ (2004) Distribution dynamics of a great bustard metapopulation throughout a decade: influence of conspecific attraction and recruitment. Biodivers Conserv 13:1659–1674

André R (1985) Some aspects of reproduction in the little bustard and a contribution to an estimate of its population in France in 1978/1979. Bustard Stud 2:153–160

Belik VP (1992) Distribution, numbers and some features of the ecology of the Little Bustard in the southeast of the European part of the USSR. Bustard Stud 5:73–77

Berthet ETA, Bretagnolle V, Segrestin B (2012) Analyzing the design process of farming practices ensuring little bustard conservation: lessons for collective landscape management. J Sustain Agric 36:319–336

Bota G, Sardà F, Cardador L (2015) Cens poblacional de sisó (*Tetrax tetrax*) en les zones de secà de laPlana de Lleida: seguiment quinquenal de la població de sisó CTFC-Infraestructures.cat. Solsona

Boutin JM, Metais M (1995) L'Outarde Canepetière. Eveil Editeur, Saint Yrieix

Bretagnolle V, Inchausti P (2005) Modelling population reinforcement at a large spatial scale as a conservation strategy for the declining little bustard (*Tetrax tetrax*) in agricultural habitats. Anim Conserv 8:59–68

Bretagnolle V, Villers A, Denonfoux L, Cornulier T, Inchausti P, Badenhausser I (2011) Rapid recovery of a depleted population of Little Bustards *Tetrax tetrax* following provision of alfalfa through an agri-environment scheme. Ibis 153:4–13

Bretagnolle V, Denonfoux L, Villers A (2018) Are farming and birds irreconciliable? A 21–year study of bustard nesting ecology in intensive agrosystems. Biol Conserv 228:27–35

Cheylan G (1985) Le Statut de la canepetière *Tetrax tetrax* en Provence. Alauda 53:90–99

Cramp S, Simmons KEL (eds) (1980) The birds of the Western Paleartic, vol 2. Oxford University Press, London

Cuscó F, Bota G, Llovet A, Mañosa S (2021) Nesting and incubation behaviour of the little bustard Tetrax tetrax and its relation to hatching success. Ardeola 68:95–122

Delgado MP, Traba J, García de la Morena EL, Morales MB (2010) Habitat selection and density-dependent relationships in spatial occupancy by male Little Bustards *Tetrax tetrax*. Ardea 98: 185–194

Dementiev GP, Gladkov NA (eds) (1951) Birds of the Soviet Union, vol 2. Israel Program for Scientific Translations, Jerusalem

Devoucoux P, Besnard A, Bretagnolle V (2019) Sex-dependent habitat selection in a high-density Little Bustard *Tetrax tetrax* population in southern France, and the implications for conservation. Ibis 161:310–324

Faria N, Rabaça JE, Morales MB (2012a) The importance of grazing regime in the provision of breeding habitat for grassland birds: the case of the endangered little bustard (*Tetrax tetrax*). J Nat Conserv 20:211–218

Faria N, Rabaça JE, Morales MB (2012b) Linking plant composition and arthropod abundance to establish little bustard breeding requirements in pastureland dominated landscapes. Biodivers Conserv 21:2109–2125

Hardouin LA, Hingrat Y, Nevoux M, Lacroix F, Robert A (2015) Survival and movement of translocated houbara bustards in a mixed conservation area. Anim Conserv 18:461–470

Inchausti P, Bretagnolle V (2005) Predicting short-term extinction risk for the declining Little Bustard (*Tetrax tetrax*) in intensive agricultural habitats. Biol Conserv 122:375–384

Ivanova VS (1992) Early growth of the little bustard when reared in captivity. Bustard Studies 5: 180–185

Jiguet F (2001) Défense des ressources, choix du partenaire et mécanismes de formation des leks chez l'Outarde canepetière (Tetrax tetrax), une espèce menacée des plaines céréalières. Unpublished PhD. Université Pierre et Marie Curie, Paris

Jiguet F (2002) Arthropods in diet of Little Bustards *Tetrax tetrax* during the breeding season in western France. Bird Study 49:105–109

Jiguet F, Bretagnolle V (2001) Courtship behaviour in a lekking species: individual variations and settlement tactics in male little bustard. Behav Process 55:107–118

Jiguet F, Bretagnolle V (2014) Sexy males and choosy females on exploded leks: correlates of male attractiveness in the little bustard. Behav Process 103:246–255

Jiguet F, Arroyo B, Bretagnolle V (2000) Lek mating systems: a case study in the Little Bustard *Tetrax tetrax*. Behav Process 51:63–82

Jiguet F, Jaulin S, Arroyo B (2002) Resource defence on exploded leks: do male little bustards, *Tetrax tetrax*, control resources for females? Anim Behav 63:899–905

Jolivet C (1994) Recherches sur la biologie de l'Outerde canepetière *Tetrax tetrax* dans le canton de Montreuil-Bellay (Maine-etLoire) en 1993-1994. Bull Gr Angevin Ét Orn LPO Anjou 22:45–53

Koshkin MA (2011) Habitat preferences of steppe breeding birds in Central Kazakhstan, in relation to different forms of land use. MSc dissertation, University of East Anglia

Labitte A (1955) La reproduction d'*Otis tetrax* dans la partie nord du départément d'Eure-et-Loir. ORFO 25:144–147

Lapiedra O, Ponjoan A, Gamero A, Bota G, Mañosa S (2011) Brood ranging behaviour and breeding success of the threatened little bustard in an intensified cereal farmland area. Biol Conserv 144:2882–2890

Lett JM, Guyomarch JC, Boutin JM (2000) Occupation de l'espace par les femelles reproductrices chez l'outarde canepetière (*Tetrax tetrax*). Chantiers d'Éthologie 20:1–20

Mañosa S, Guillem R, Bota G (2018) Rècord de longevitat del sisó *Tetrax tetrax*. L'Abellerol 56:23

Marcelino J, Moreira F, Mañosa S, Cuscó F, Morales MB, García de la Morena E, Bota G, Palmeirim J, Silva JP (2017) Tracking data of the Little Bustard *Tetrax tetrax* in Iberia shows high anthropogenic mortality. Bird Conservation International 28:509–520

Martínez C (1994) Habitat selection by the little bustard *Tetrax tetrax* in cultivated areas of Central Spain. Biol Conserv 67:125–128

Morales M, Bretagnolle V, Arroyo B (2005) Viability of the endangered little bustard *Tetrax tetrax* population of western France. Biodivers Conserv 14:3135–3150

Morales MB, Traba J, Carriles E, Delgado MP, García de la Morena E (2008a) Sexual differences in microhabitat selection of breeding little bustards *Tetrax tetrax*: ecological segregation based on vegetation structure. Acta Oecol 34:345–353

Morales MB, García de la Morena E, Delgado MP, Traba J (2008b) Tendencia reciente y viabilidad futura de las poblaciones de sisón común (*Tetrax tetrax*) en la Comunidad de Madrid. Anuario Ornitológico de Madrid 11:40–55

Morales MB, Traba J, Delgado MP, García de la Morena EL (2013) The use of fallows by nesting Little Bustard *Tetrax tetrax* females: implications for conservation in mosaic cereal farmland. Ardeola 60:85–97

Morales MB, Casas F, García de la Morena E, Ponjoan A, Calabuig G, Martínez-Padilla J, Garcia JT, Mañosa S, Viñuela J, Bota G (2014) Density dependence and habitat quality modulate the intensity of display territory defence in an exploded lekking species. Behav Ecol Sociobiol 68: 1493–1504

Moreira F, Silva JP, Estanque B, Palmeirim JM, Lecoq M, Pinto M, Leitao D, Alonso I, Pedroso R, Santos E (2012) Mosaic-level inference of the impact of land cover changes in agricultural landscapes on biodiversity: a case-study with a threatened grassland bird. PLoS One 7:e38876

Moseykin VN (1992) Ecology and protection of the little bustard in the Saratov region. Bustard Stud 5:78–91

Petretti F (1993) Notes on the lek behaviour of the little bustard in Italy. Avocetta 17:19–22

Ponjoan A, Bota G, Mañosa S (2012) Ranging behaviour of little bustard males, *Tetrax tetrax*, in the lekking grounds. Behav Process 91:35–40

Rivas O, de Castro A, Llaneza L, Otero J (2005) Population size, phenology and habitat preferences of little bustard *Tetrax tetrax* in an isolated area of the northwestern Iberian península. Ardeola 52:147–149

Santangeli A, Dolman PM (2011) Density and habitat preferences of male little bustard across contrasting agro-pastoral landscapes in Sardinia (Italy). Eur J Wildlife Res. https://doi.org/10.1007/s10344-011-0492-3

Schenk H, Aresu M (1985) On the distribution, number and conservation of the Little Bustard in Sardinia (Italy), 1971-1982. Bustard Stud 2:161–164

Schulz H (1985) A review of the world status and breeding distribution of the Little Bustard. Bustard Stud 2:131–151

Silva JP, Estanque B, Moreira F, Palmeirim JM (2014) Population density and use of grasslands by female little bustards during lek attendance, nesting and brood-rearing. J Ornithol 155:53–63

Silva JP, Catry I, Palmeirim JM, Moreira F (2015) Freezing heat: thermally imposed constraints on the daily activity patterns of a free-ranging grassland bird. Ecosphere 6:119

Tarjuelo R, Delgado MP, Bota G, Morales MB, Traba J, Ponjoan A, Hervas I, Mañosa S (2013) Not only habitat but also sex: factors affecting spatial distribution of Little Bustard *Tetrax tetrax* families. Acta Ornithologica 48:119–128

Traba J, Morales MB, García de la Morena EL, Delgado MP, Krištín A (2007) Selection of breeding territory by little bustard (*Tetrax tetrax*) males in Central Spain: the role of arthropod availability. Ecol Res 23:615–622

Villers A, Millon A, Jiguet F, Lett JM, Attie C, Morales MB, Bretagnolle V (2010) Migration of wild and captive-bred Little Bustards *Tetrax tetrax*: releasing birds from Spain threatens attempts to conserve declining French populations. Ibis 152:254–261

Von Frish O (1976) Zur Biologie der Zwergtrappe *Tetrax tetrax*. Bonn Zool Beitr 27:21–38

Wolff A (2001) Changements agricoles et conservation de la grande avifaune de plaine: étude des relations espèce-habitats à différents échelles chez l'outarde. PhD Thesis. Montpellier. France

Wolff A, Paul JP, Martin JL, Bretagnolle V (2001) The benefits of extensive agriculture to birds: the case of the little bustard. J Appl Ecol 38:963–975

Wolff A, Dieuleveut T, Martin JL, Bretagnolle V (2002) Landscape context and little bustard abundance in a fragmented steppe: implications for reserve management in mosaic landscapes. Biol Conserv 107:211–220

Habitat Selection and Space Use

Juan Traba, Manuel B. Morales, João Paulo Silva, Vincent Bretagnolle, and Pierrick Devoucoux

Introduction

The distribution of a species is the result of evolutionary, ecological and/or anthropogenic processes that operate at different spatial and temporal scales. The way a species uses and selects a habitat influences its survival and reproductive performance and, therefore, its spatial persistence. This field has been largely studied in the little bustard. Thus, we start this chapter by establishing definitions of terms used here. Habitat refers to a distinctive set of physical environmental factors that a species uses for its survival and reproduction (Block and Brennan 1993). Habitat selection refers to the hierarchical process of behavioural responses by which a species chooses between distinct available resources (Johnson 1980), and which result in the disproportionate use of particular habitats, thus influencing survival and fitness of individuals (Block and Brennan 1993).

J. Traba (✉) · M. B. Morales
Terrestrial Ecology Group, Departamento de Ecología, Universidad Autónoma de Madrid (TEG-UAM), Madrid, Spain

Centro de Investigación en Biodiversidad y Cambio Global (CIBC-UAM), Universidad Autónoma de Madrid, Madrid, Spain
e-mail: juan.traba@uam.es; manuel.morales@uam.es

J. P. Silva
Centre for Applied Ecology "Prof. Baeta Neves", ISA, Centre for Environmental Biology, FCUL Faculty of Sciences, University of Lisbon, Lisbon, Portugal

CIBIO-InBIO, Universidade do Porto, Campus de Vairão, Vairão, Portugal
e-mail: jpsilva@cibio.up.pt

V. Bretagnolle · P. Devoucoux
Centre d'Etudes Biologiques de Chizé, CNRS & La Rochelle Université, Beauvoir sur Niort, France
e-mail: vincent.bretagnolle@cebc.cnrs.fr

V. Bretagnolle et al. (eds.), *Little Bustard: Ecology and Conservation*, Wildlife Research Monographs 5, https://doi.org/10.1007/978-3-030-84902-3_6

Two other aspects related to the habitat selection process (Jones 2001) should be considered when reading this chapter in the framework of habitat selection theory: (1) it is a decision-making process, and individuals may face other possibilities that differ in structure and/or quality; (2) it is an adaptive process, and in general we assume that individuals' choices are related to increased fitness in certain habitats. Unfortunately, researchers often lack the information to evaluate habitat selection through measures or indicators of increased fitness. Besides, this process may not be independent of many other non-habitat factors which can influence habitat selection patterns: predation, intra- and interspecific competition, conspecific attraction, site fidelity, food availability, personality traits. To integrate all previously mentioned aspects, in this chapter we consider habitat selection as a complex process including all the responses of an individual when using a particular habitat.

Habitat selection is typically a process involving decisions at several spatial scales (Morris 1987). This multi-scalar process ranges from the macroscale to macro- and microhabitat. Macroscale habitat refers to continental and regional factors that determine, to a large extent, species distribution. Macrohabitat is taken here as the area in which organisms carry out their biological functions (Morris 1987; see also the section Macrohabitat below, and all the references therein). Macrohabitat is therefore related to landscape configuration and composition. On the other hand, we refer here to microhabitat as the horizontal and vertical structural characteristics, and food availability of those locations actually perceived and finally chosen by an organism (Morales et al. 2008). Finally, we review variation in habitat selection according to sex, as well to seasonal and regional differences.

Factors Influencing Little Bustard Macroscale Distribution

When analysing a species' distribution pattern, both ecological and evolutionary factors have to be considered. These two types of drivers of species-specific space occupancy reflect the influence of local, mainly biotic, fine-grained factors, which modulate the effect of those others acting on a much greater scale (Ricklefs 1987). Species distribution at a macroscale (continental, for example) seems to be influenced by abiotic, mainly climatic, factors (Thuiller et al. 2005).

Delgado et al. (2011) carried out a complete analysis of the climate niche of little (and great) bustard at a Western Palearctic scale (see chapter "The little bustard around the world: distribution, global conservation status, threats and population trends" for distribution range). These authors revealed a relatively constrained climate niche for the species, due to its association with both temperature and precipitation-related variables (see also Estrada et al. 2016). The little bustard seems to prefer somewhat oceanic climates, with lower temperature range and milder winters than those preferred by the great bustard. The strong negative association between little bustard presence and temperature range suggests the avoidance of marked continental climate, at least for the Iberian population (Delgado et al. 2011). Mean temperature during the breeding season has a non-linear but predominantly positive effect on the distribution pattern of the species, suggesting a preference for warm springs, although this variable had relatively lesser importance. In addition to pure thermic constraints, little bustard presence seems to be associated with soil moisture (measured by the SMI index which describes the amount of soil water available), which is related to primary production (Delgado et al. 2011). The relationship between little bustard presence and soil moisture showed a typical peaking pattern, with an apparent optimum at mean values of this variable, which suggests the avoidance of both extremely dry and wet soils (see also De Juana et al. 1988). This result, however, contrasts with that for annual rainfall, which showed a U-shaped response, the slope being negative for the lower and positive for the higher extreme values of the variable. These results could be indicating the combined effect of both temperature and precipitation in shaping little bustard distribution, although some methodological biases related to the greater number of presence cells in Iberia should be considered. However, the effect of climate on habitat configuration could be behind these results, as the optimal habitat for the species, namely treeless, open herbaceous vegetation, can appear in both high precipitation areas, like southern Russia, or drier ones, such as the Mediterranean.

Climate modulates habitat, and consequently, plant growth and invertebrate food availability (Noy-Meir 1973). Therefore, species' distribution constraints are modulated by large-scale climate factors. Large-scale predictive models built for continental Spain have shown the effect of habitat type and topography on the presence of little bustard during the breeding season (Suárez-Seoane et al. 2002). In general, the species' probability of occurrence decreases as overall greenness increases, suggesting some kind of preference for arid areas, at least in the Iberian context. A complementary explanation suggests a preference for areas where the habitat is suitable throughout the whole breeding season, such conditions being coincident with areas where agriculture remains extensive (i.e. avoiding intensification). In summary, little bustard selects areas dominated by dry crops and pastures with low values of greenness (see also Suárez-Seoane et al. 2008). The effect of climate has been also found at smaller spatial scale, as precipitation during the breeding season and the previous months seems to have positive effects on population size, probably due to its effect on primary productivity and food availability (Delgado et al. 2009).

The species' probability of occurrence seems to show a typical optimum peak in the 300–700 m range, with a long tail through higher altitudes (Suárez-Seoane et al. 2002, 2008), although this could be context-dependent. For instance, French little bustard populations are essentially found near sea level in southern France (i.e., 99% under 100 m in Costière Nîmoise and in La Crau; pers. obs.). However, this last effect could presumably reflect a proxy variable for climate, which ultimately affects plant productivity and invertebrate availability for the birds. Little bustards prefer gently undulating terrain (Suárez-Seoane et al. 2002; Osborne and Suárez-Seoane 2007), although this can be also a habitat-related feature, as grass-steppes and cultivated areas are usually located on flat or near to flat areas.

Suárez-Seoane et al. (2002) also found some effect of anthropogenic disturbance on little bustard distribution, with birds preferring areas far from roads, although the importance of this in determining little bustard occurrence was much lower than that of the above-mentioned habitat-related variables (see however Silva et al. 2004; Arcos and Salvadores 2005; Osborne and Suárez-Seoane 2007). Again, this seems to be a context-dependent effect, as some populations seem to be highly impacted by human infrastructures. This is the case of Costière Nîmoise (southern France), an area with high densities of little bustards where preliminary studies reveal that the effects of railways, roads and paths are even higher than those of habitat composition (Devoucoux 2014).

Other factors influence the species' distribution at a large scale. Historical restrictions, behavioural constraints such as conspecific attraction, and local processes such as the availability of optimal habitat, and local extinction and colonization, limit the space occupied by a species (Estrada et al. 2016). Indeed, when other factors are included in the analyses, climate continues to be the main driver of the present distribution of the little bustard, but other non-climatic factors also contribute, namely a preference for areas dominated by dry crops, the avoidance of high human population density (Estrada et al. 2016; see also Suárez-Seoane et al. 2002, 2008, for a lower-scale analysis) and intra- and intersexual relationships (Devoucoux et al. 2018; see below). However, the importance of geographic predictors, i.e. the proximity between occupied locations, was, by far, the most relevant factor in explaining the current distribution of little bustard at a European scale (Estrada et al. 2016). Specifically, geographic factors could reflect its absence from some regions, perhaps due to direct human pressure or habitat loss over the past 2000 years. Besides, the species has a tendency to aggregate in reproductive nuclei (see chapter "Behavioural ecology of the little bustard: sexual selection and mating systems"), showing strong philopatry. This suggests that the current distribution of little bustard at this scale is spatially constrained beyond the effect of environmental variables. As a consequence, dispersal constraints and site fidelity may restrict the future distribution as climate change pushes the species' optimal climate conditions northwards (Estrada et al. 2016). Finally, a simpler distribution pattern has been found for winter, as only habitat and climate variables explain the species' distribution at a regional scale (Suárez-Seoane et al. 2008).

Macrohabitat Selection

Although the original habitat of the little bustard is natural herbaceous steppes (Cramp and Simmons 1980), in Western Europe the species preferentially uses both extensive croplands and grasslands (García de la Morena et al. 2006). It occupies many habitats resembling its original steppe-like habitat, such as the extensive sheep and cattle pastures of Sardinia, southern France, southwest Iberia and Central Europe, extensive cultivated areas in Spain, but also more intensive croplands in France (Salamolard and Moreau 1999; Bretagnolle et al. 2018).

The statement that little bustards prefer areas dominated by dry crops may sound rather general. Fortunately, this is a well-investigated pattern, as habitat preference studies have usually been centred on the type of land cover or crop used by the species. More specifically, little bustards usually avoid cultivated cereals but select legume crops, specifically alfalfa, and more or less long-term abandoned fields and fallows in Spain and Portugal, set-aside or natural vegetation in France (Martínez 1994; Salamolard and Moreau 1999; Wolff et al. 2001; Morales et al. 2005, 2006; Moreira et al. 2012; see however Devoucoux et al. 2018), and mainly flat landscapes (Tarjuelo et al. 2014). There is an overall tendency for little bustards to select extensive agricultural farmland, with higher male densities in such habitats (Wolff et al. 2001), though local male density may not be always the best proxy for optimal habitat (see, e.g., García et al. 2007). Devoucoux et al. (2018) in southern France mention probably the highest little bustard density in Western Europe (mean male density: 23.44/100 ha; mean female density: 24.91/100 ha; totalling up to 50 birds per 100 ha) linked to small and heterogeneous fields and land abandonment.

In general terms, the abundance of little bustard males seems to be positively correlated with heterogeneous landscapes with small fields and high habitat diversity or some degree of mosaic structure (Martínez 1994; Wolff et al. 2001; Morales et al. 2006; Delgado et al. 2010; Santangeli and Dolman 2011; Devoucoux et al. 2018), which may reflect habitat complementarity or habitat supplementation. In the Iberian range, the preference for a typical mosaic of habitats created by extensive cereal farming, sometimes called pseudo-steppes (Suárez et al. 1997), could be simply a selection of optimal habitats, but restricted to the few, small scattered fields remaining in the landscape. Wolff et al. (2001) suggest three processes, not mutually exclusive, behind habitat complementation: (1) natural habitats may be scarce, or do not provide sufficient resources during the breeding season, so males need to extract food from adjacent agricultural habitats; (2) for the same reason, natural habitats may be a poor nesting habitat for females, prompting males to display in the vicinity of better nesting habitats where they are likely to attract more females; (3) other agricultural land uses may be selected outside the breeding season. Despite this consideration, bird density may also reach very high values in Western Europe within landscapes dominated by large fields (Moreira et al. 2012). Silva et al. (2010) indeed detected a preference for large homogeneous grassland fields in southern Portugal, perhaps because where little optimal habitat remains, little bustards may include other, suboptimal habitats in their territories. Alternatively, large, apparently homogeneous grassland fields may be considerably heterogeneous in terms of vegetation structure, owing to extensive grazing, thus also providing microhabitat complementary resources for breeding birds (Silva et al. 2010), more similar to the original true steppes. In these last cases, little bustards use stable habitats (old or young fallows), which they will keep using over the years, leading to a typical territorial system that, in turn, will attract satellite males (Silva et al. 2007), consequently leading to a high-density system with overlapping ranges.

There is some constancy in the pattern of habitat selection between years, which suggests a strong habitat fidelity to permanent and semi-permanent habitats (Delgado et al. 2010). Both the polygynous mating system and the marked philopatry of the little bustard could be behind this inter-annual repeatability of observations in similar locations, even though land cover may greatly change between years (Delgado et al. 2010). Indeed, if the habitat is inter-annually stable, they may show habitat and site fidelity at a very local scale, using the same locations as in previous years (Silva et al. 2007). When the habitat changes between years, males slightly modify locations from year to year, in order to find the optimal habitats that meet their requirements (Morales et al. 2005; Delgado et al. 2010). At certain spatial scales (below 1500 ha), the space use of male little bustards could be driven largely by habitat preferences rather than by factors associated with its typical mating system, such as presence of other males. This pattern may be associated with low density of males, when the number of birds is not large enough for all the optimum areas to be occupied, and the area is not saturated (Delgado et al. 2010). In contrast, changes in male displaying locations between consecutive years after site destruction by a High Speed Railway were less than 500 m, thus leading to an intensification of intrasexual competition due to high male density (Devoucoux et al. unpubl). In this case, when males reach high density and optimal habitat is saturated, they are forced to use suboptimal habitats, and spatial distribution seems to be driven by female presence and other males' density (Devoucoux et al. 2018).

Most habitat selection studies have considered the focal location of the bird to establish habitat preferences (see for instance Martínez 1994; Salamolard and Moreau 1999; Wolff et al. 2001; Santangeli and Dolman 2011; Delgado et al. 2010). However, little bustard territories are greater than the precise location of the display ground (Jiguet et al. 2000), and therefore may include a mosaic of crops and field types. For this reason, findings on habitat selection of little bustard males in highly cultivated landscapes, when just the precise location of the male is considered, have to be considered with caution.

Although fallows and old fallows show great inter-annual stability in plant structure, which explains why they are preferentially used by little bustards, they also depend on grazing or cutting to prevent plant succession and excessive vegetation height (see below, Microhabitat selection). Finally, in relation to human disturbance, whereas some studies reveal no particular response of little bustards to roads, villages or inhabited houses (Martínez 1994; Faria and Rabaça 2004), other studies suggest the contrary (Suárez-Seoane et al. 2002; Silva et al. 2004, 2010; Arcos and Salvadores 2005; Osborne and Suárez-Seoane 2007; García et al. 2007; Santos et al. 2016). For instance, comparing locations of displaying males with random points without males, Arcos and Salvadores (2005; NW Spain) found significant differences in distance to roads (576.93 $\pm$ 68.85 m with males, 287.54 $\pm$ 77.86 m without males) and to paths (141.4 $\pm$ 14.4 m with males, 65.62 $\pm$ 12.1 m without males). Moreover, Silva et al. (2004; South Portugal) found that little bustards are sensitive to inhabited houses and road proximity.

Regional Differences in Habitat Selection

Little bustards occupy extensive (non-intensive) grasslands or cereal farmland, with varying degrees of land-use mosaics, depending on the prevailing landscape type

(Morales et al. 2005, 2006; Silva et al. 2010). In this sense, habitat selection is modified by regional differences in habitat availability. Unfortunately, most information comes from studies in Western Europe, while information from farther East is very scarce. A brief review is presented here.

In France, displaying males prefer extensive agricultural habitats (Wolff et al. 2001; Devoucoux et al. 2018); in intensive farmland habitats, they need alfalfa and set-aside, though they also use spring crops (flax, sunflower) or even wheat very early in the spring season (Salamolard and Moreau 1999; Bretagnolle et al. 2018), although these latter crops rank low on the list of preferred habitats for Spanish birds (Delgado et al. 2010). In summer, more intensive crops, such as cereals or oilseed rape, are always avoided (Salamolard and Moreau 1999; Wolff et al. 2001). Sunflower and maize are also avoided when tall, although they might be used and actually preferred transiently in their early stages of development, when lower than 30 cm (Rivas et al. 2005). Little bustards also seem to avoid hedges and woody areas in some parts of France where hedges against winds effects on crops (3–5 m in height, mainly cypress) are numerous (Devoucoux 2014).

In the northwest Iberia population, whose cereal farmland presents some similarity with that of western France, the species selects semi-permanent crops such as raygrass and alfalfa as well as pastures (Arcos and Salvadores 2005; Rivas et al. 2005), but also uses more intensive fields, such as maize crops (Rivas et al. 2005), although retaining a preference for short vegetation. In the central part of Iberia the species uses the mosaic landscape associated with extensive cereal farming, selecting mainly fallows and extensive pastures (Martínez 1994; Suárez et al. 1997; Morales et al. 2006; Silva et al. 2014; Tarjuelo et al. 2013, 2014), although some more intensive cereal-dominated farmlands also hold important populations, as in Lleida plains (Lapiedra et al. 2011). The species can tolerate around 20% of dry woody cultures and orchards (Tarjuelo et al. 2014).

In Sardinia, where the only Italian population remains (see chapter "The little bustard around the world: distribution, global conservation status, threats and population trends"), the species selects fallows and extensive pasturelands during the breeding season, and avoids either arable, hay meadow or others land uses (Santangeli and Dolman 2011), although it may also avoid fallows in the summer (Santangeli and Cardillo 2012). This discrepancy has been explained by the differential effect of livestock grazing through the year: in spring, grazing shapes habitat structure for displaying males, while in summer it depletes the already limited food resources (i.e. green plants) available (Santangeli and Cardillo 2012). However, this is a site-specific process, depending on the traditional grazing management carried out in each region (Silva et al. 2007).

In Portugal, males select large homogeneous fields in contrast to the mosaic landscape that can constitute male territories elsewhere (Silva et al. 2010; Moreira et al. 2012). Female preference for Fabaceae-dominated sites has been recorded (Faria et al. 2012), which seem to be related to the high nutrient content of these wild plants, largely consumed by little bustards (Bravo et al. 2017).

Finally, in southern Russia, where some extension of natural steppe remains, little bustard populations occupy natural steppes, but also occupy farmland areas where they prefer fields sown with winter crops or stubble fields (Shlyakhtin et al. 2004). A similar habitat use pattern has been described for little bustard populations in Kazakhstan (Kamp et al. 2011). In Iran, a high concentration of wintering little bustards seems to be related to a combination of available winter crops close to non-hunting areas (Yousefi et al. 2017).

Seasonal Differences in Macrohabitat Selection

Some changes in habitat selection pattern may occur over the year. During summer, there are no clear overall patterns regarding little bustard macrohabitat selection, although some preference for melon fields has been observed (Silva et al. 2007). In central Spain, little bustard females and families seem to prefer semi-permanent vegetation and stubbles while ploughed lands are avoided (Tarjuelo et al. 2013). During Mediterranean summers, little bustards (as other birds) experience food shortages, which may explain why they perform post-reproductive movements towards areas with greater food availability (detailed in chapter "Migration, movements and non-breeding ecology"). However, they may be still forced to occupy suboptimal habitats, resulting in daily movements between areas with high cover of green plants and dry fallow lands. In this sense summer habitat selection may be quite distinct from the other seasons.

Autumn habitat selection has been studied by Faria and Silva (2010) in southern Portugal. During this season, males prefer cereal stubbles and fallows and, in general, more productive fields. Cereal stubbles, with little or no herbicide treatment, generally show a fast growth of weeds after the first autumn rains. Fallows also show a higher availability of green plants, suggesting the importance of feeding requirements above others once the breeding period has ended.

Microhabitat Selection

At a fine scale, vegetation height, as well as vegetation structure and floristic composition within preferred grassland fields (mainly pastures and fallows of varying ages) are the main traits that affect little bustard occurrence (Morales et al. 2008; Faria et al. 2012; Silva et al. 2014; Traba et al. 2015). Vegetation height selection patterns have been studied in detail in Devoucoux et al. (2018), who found that vegetation height affected both males and females probability of presence, with a significant negative correlation between vegetation height and presence. However, the effect of vegetation height differed between sexes: whereas females showed a quadratic response to vegetation with an optimal height estimated at around 30 cm, males showed a strict decrease in presence probability with vegetation height.

The preference of little bustard for pastures, legume crops and fallows has been related to the greatest availability of trophic resources, especially cultivated legumes and wild herbs, and invertebrates, in these fields (Martínez 1994; Wolff et al. 2001; Santangeli and Dolman 2011), but also to both vertical and horizontal vegetation structure (Morales et al. 2008; Morales and Traba 2009; Traba et al. 2015). In addition, a few studies have focused on plant composition in the habitat selection by little bustard (Faria et al. 2012). Despite the importance of food in any habitat selection process, only a few studies have estimated differences in food availability between land uses and/or between little bustard territories and control locations (Martínez 1998; Salamolard and Moreau 1999; Traba et al. 2008; Faria et al. 2012; Devoucoux et al. 2018), and only two have directly evaluated little bustard diet through stomach and faeces analyses (Jiguet 2002; Bravo et al. 2017). Several microhabitat studies have highlighted the importance of arthropod availability (Martínez 1998; Jiguet et al. 2002; Traba et al. 2008; Faria et al. 2012; see however Devoucoux et al. 2018). For instance, alfalfa fields and natural grasslands have been considered to harbour the highest number and diversity of invertebrates (Salamolard and Moreau 1999).

Tarjuelo et al. (2019) studied the relationships between horizontal and vertical plant structure and food availability (weed and arthropod biomass). To do so, in spring 2013, they sampled 30 random points in one typical extensive farming area of central Spain and recorded variables about horizontal and vertical structure and complexity of the vegetation, and vegetal and arthropod biomass. They found that (1) bare ground cover is significantly and negatively correlated with all the items of food availability; (2) as weed and litter cover increases in any particular location, weed biomass (food source for adult little bustards) increases as well. In contrast, as vertical complexity of vegetation is greater, only a weak relationship with weed biomass exists; (3) however, arthropod biomass is not related to any of the plant structure variables, neither positively nor negatively, which seems to suggest that arthropod availability is modulated by field management (herbicide and pesticide treatment, ploughing, etc.) more than by vegetation structure (Table 1 and Fig. 1).

Several studies have analysed differences between intensive and extensive agricultural landscapes (see, for instance, Guerrero et al. 2011). Extensive, traditional agricultural habitats hold higher plant diversity and arthropod biomass and diversity, providing food for both adults and chicks (Emmerson et al. 2016). In addition, extensive agriculture has lower, less uniform and/or less dense vegetation, which might better match the trade-off little bustards need between refuge and visibility (Morales and Traba 2009). In summary, vegetation structure seems to be more important than the type of vegetation (plant composition), or even the type of agrarian use (i.e. fallow land or pasture) (Morales et al. 2008; Silva et al. 2010), although its role is always modulated by the intensity of field management.

Finally, a few microhabitat studies have focused on the importance of local plant composition during the breeding season (Martínez 1998; Faria et al. 2012), but they consistently suggest that little bustards select sites with high floristic richness (Martínez 1998) and an abundance of Fabaceae species, although with some differences between the sexes (see below).

Table 1 Spearman correlations (and *p*-values in parenthesis) between vertical and horizontal plant structure and food availability variables, in 30 microhabitat locations in Central Spain

	Fresh plant biomass	Dry plant biomass	Fresh weed biomass	Arthropod biomass
Bare ground cover	**−0.459 (<0.05)**	**−0.449 (<0.05)**	**−0.789 (<0.001)**	−0.12 (0.552)
Green vegetation cover	**0.687 (<0.001)**	**0.679 (<0.001)**	**0.596 (<0.001)**	0.072 0.721
Litter cover	**0.457 (<0.05)**	**0.450 (<0.05)**	**0.681 (<0.001)**	0.16 0.426
Weed cover	0.241 (0.200)	0.210 (0.266)	**0.849 (<0.001)**	0.127 (0.528)
Number of contacts <5 cm	0.355 (0.054)	0.328 (0.077)	**0.738 (<0.001)**	−0.038 (0.851)
Number of contacts 5-10 cm	**0.589 (<0.01)**	**0.557 (<0.01)**	**0.600 (<0.01)**	0.100 (0.621)
Number of contacts 10-30 cm	**0.651 (<0.001)**	**0.645 (<0.001)**	**0.654 (<0.001)**	0.165 (0.412)
Number of contacts >30 cm	**0.924 (<0.001)**	**0.923 (<0.001)**	0.062 (0.749)	0.369 (0.058)
Maximum vegetation height	**0.706 (<0.001)**	**0.713 (<0.001)**	**0.463 (<0.05)**	0.300 (0.129)

Based on unpublished data from J. Traba and M.B. Morales, and Tarjuelo et al. (2019)
Bold indicates significant relationships

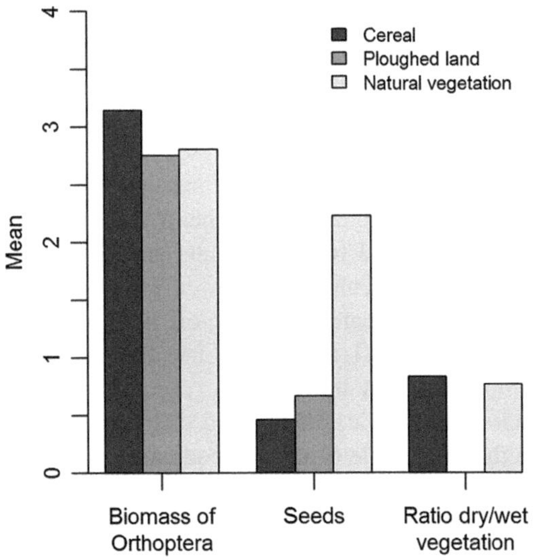

Fig. 1 Mean value of three potential food resources for the little bustard—Biomass of orthoptera (mg), seed volume (mm^3) and ratio of dry/wet vegetation—measured in different agricultural habitats (cereal, ploughed land and natural vegetation formed by fallows of different ages and short shrubs). Based on unpublished data from R. Tarjuelo, J. Traba, L. Arribas and M. B. Morales, and Tarjuelo et al. (2019)

Sexual Differences in Microhabitat Selection

Breeding Season

Whereas the habitat preferences of breeding male little bustards have been relatively well studied, thanks to their conspicuousness, habitat selection by females has been addressed in only a few studies. These all confirm the different choices of the sexes, with males seeking habitats with high detectability and females looking for secluded areas to place their nests, feed their chicks and avoid both predators and sexual harassment (Morales et al. 2008, 2013; Tarjuelo et al. 2013; Traba et al. 2015). Even so, Devoucoux et al. (2018) found only vegetation height differs between the sexes at a microhabitat scale, although these results may be influenced by the extremely high density of birds, perhaps the highest ever recorded for this species at around 50 individuals per 100 ha of suitable habitat.

During the breeding season, males court females and defend their territories from other males, both activities requiring habitats with good visibility, defendability from other males and rapid detection of females (Salamolard and Moreau 1999; Morales and Traba 2009; Traba et al. 2015), plus abundant food (Wolff et al. 2001; Traba et al. 2008). Both sexes seem to select similar feeding microhabitat features related to food availability (Morales and Traba 2009; Traba et al. 2015).

The driver of habitat selection by displaying males is the physical architecture of the location site, specifically the height of vegetation with crop height being more important than crop type (Martínez 1998; Moreira 1999; Salamolard and Moreau 1999; Morales et al. 2008; Devoucoux et al. 2018), as calling males usually avoid cereal crops since plant height is already too tall during the breeding season, reducing visibility. On the other hand, ploughed fields are also poorly represented in display sites since, although they offer high visibility, they provide neither sufficient protection nor food resources (Morales et al. 2008; Morales and Traba 2009; Traba et al. 2015). Calling males use different kinds of crops in the early stages of growth, and avoid them when the vegetation height exceeds 30–40 cm (Salamolard and Moreau 1999), preferring heights around 15–20 cm (Martínez 1994; Moreira 1999).

Male territories include a high diversity of crops, some of them scarce at the landscape scale, as alfalfa crops or long-term fallows (Wolff et al. 2001; Delgado et al. 2010). These kinds of fields offer a high content in plants and invertebrates, especially nutritive for both adults and fledglings. As already said, invertebrate abundance in alfalfa fields is higher than in other cultivated crops such as cereals or sunflower (Salamolard and Moreau 1999). Short- and long-term fallows offer high wild invertebrate diversity (Traba et al. 2008), and they are frequently incorporated into the core areas of displaying males. During breeding, little bustard males adopt a loose aggregation system (exploded leks, see chapter "Behavioural ecology of the little bustard: sexual selection and mating systems"). This arrangement could allow males to hold large territories, so they can include not only the preferred displaying habitat but also optimal conditions for female foraging and nesting (Jiguet et al. 2000; Jiguet 2002; Morales et al. 2008, 2013; Morales and Traba 2009). Although the home ranges of adult males during the breeding season may be as large as 40 ha (Ponjoan et al. 2012; 27–83 ha, in Schulz 1985; 19 ha, Jiguet et al. 2000), core areas, which can be considered as exclusive male territories, are highly variable and site-specific differences have been found, from 1.5 ha (Delgado et al. 2010) to 9 ha (Ponjoan et al. 2012). Finally, during the breeding season, males usually prefer to occupy open areas on low rises, which provide greater visibility for the reproductive display (Silva et al. 2004).

For their part, females are cryptic and secretive, and thus their habitat selection has been little studied (but see Salamolard and Moreau 1999; Morales et al. 2008, 2013; Lapiedra et al. 2011; Tarjuelo et al. 2013; Devoucoux et al. 2018; Bretagnolle et al. 2018). In contrast to males, females need shelter, but with sufficient visibility to allow surveillance (vegetation height around 30 cm, Devoucoux et al. 2018; over 30 cm in Morales et al. 2008; under 42 cm in Silva et al. 2014), and food for brood rearing, thus preferring habitats with higher vegetation cover and greater amounts of litter and green weeds (Morales et al. 2008, 2013; Lapiedra et al. 2011; Traba et al. 2015). In the vineyard plains of southern France, females tend to select vegetation height around 30 cm (Devoucoux et al. 2018).

During the pre-nesting, nesting and post-nesting stages, little bustard females select sites of male concentration (Wolff et al. 2002; Silva et al. 2014; Tarjuelo et al. 2013). That is, females visit places located in areas that increase the probability of an encounter between sexes (Jiguet et al. 2002). Besides, male territories provide more food (large beetles) than other (with no male) areas (Martínez 1998; Traba et al. 2008; Faria et al. 2012), which suggests that, although males do not provide parental care, females locate their nests in sites with a potentially higher fledging survival. Recently, local habitat selection by both sexes has been shown to be strongly positively correlated with the abundance of the other sex (Devoucoux et al. 2018). In this sense, the exploded lek mating system attributed to the little bustard, where males provide only their genes to the females, could, under some circumstances (e.g., low density), be a resource-defence polygyny system (Delgado et al. 2010; Morales et al. 2013), with males offering and actively defending good nesting places.

In summary, females prefer set-aside and long-term fallows more than males do (Salamolard and Moreau 1999; Morales et al. 2013; Tarjuelo et al. 2013). In any case, females are more mobile than males during the breeding season, and they can nest several kilometres outside the courtship area.

Nesting Microhabitat Preferences

Either near or far from male display territories, nests are located in habitats with low risk of predation and high food availability for the precocial downy young (Lapiedra et al. 2011; Bretagnolle et al. 2018), characterized by a mixed herb–grass vegetation (Shlyakhtin et al. 2004). Vegetation height is however highly variable between nests, from less than 20 cm at laying up to about 1 m (Bretagnolle et al. 2018), suggesting weak selection for this habitat parameter. Two post-hatching space-use patterns have been described in little bustard families: sedentary, that is centred around the nesting place, or wandering, which seems to depend on the quality of nesting habitats and their food availability (Lapiedra et al. 2011). In Portugal, nesting and brood-rearing females prefer large grassland fields, with vegetation height around 40 cm and plant cover around 75% (Silva et al. 2014), avoiding bare ground (Moreira 1999). This is not necessarily the case in intensive farming habitat in western France, where families are often seen in stubble (Bretagnolle et al. 2018) or alfalfa at any vegetation height (Bretagnolle et al. 2011). Under certain circumstances, such as very extensive traditional cereal farming on unproductive soils, or conversely in highly intensive farming, females could nest in cereal fields (Delgado and Moreira 2000; Bretagnolle et al. 2018). In central Spain, somewhat shorter plant height is preferred by females (up to 30 cm), avoiding dense vegetation above this height (Morales et al. 2008). These vegetation features are found in both short- and long-term fallows, which provide high floristic and invertebrate diversity and protection due to the vertical heterogeneity of the vegetation (Morales et al. 2008, 2013; Tarjuelo et al. 2013). Since the choice of nesting habitat also depends on landscape configuration, in areas of Iberia where agricultural intensification has been greater than elsewhere, females nest more frequently in cereal fields, as reported for Catalonia (Lapiedra et al. 2011). In the intensively cultivated cereal plains of central France, females nest mostly in multi-annual crops such as alfalfa fields as well as grasslands and set-asides (Salamolard and Moreau 1999; Berthet et al. 2012; Bretagnolle et al. 2018).

The ecological conditions required for nesting are different from those needed to rear chicks (Lapiedra et al. 2011), which adds more evidence to the need of mosaic-complex landscapes or vegetation heterogeneity for the conservation of this (and other) farmland species. During the brood-rearing period, families avoid habitats subjected to higher levels of disturbance, such as sheep-grazed pastures, fruit-tree orchards (Tarjuelo et al. 2013) or harvested cereal fields, where visibility is higher (Lapiedra et al. 2011), although the last seems to be site-specific, as females with families select stubbles during summer in Central Spain (Tarjuelo et al. 2013) and in southern (P. Devoucoux, pers. obs.) and western France (V. Bretagnolle, pers. obs). In this period, families (females and chicks) are also found near the previous male concentration sites (Tarjuelo et al. 2013), which suggests a preferential utilization of these areas, in accordance with their higher food availability (Traba et al. 2008). In cereal-dominated sites, as in the Catalonian Lleida plains, broods preferentially use cereals before harvest. After harvest, broods increase the use of field margins and fallows (Lapiedra et al. 2011) and stubbles (Tarjuelo et al. 2013), where the arthropod availability is higher than in cereal crops (Traba et al. 2008), although this result depends on the management intensity in each field. Certainly, the need to include both high-quality feeding and nesting habitats must be crucial in the final configuration of brood ranges, which can vary between 3.1 ha ($n = 1$, Schulz 1985) to 10–430 ha (mean = 137 ha; $n = 8$, Lapiedra et al. 2011), although these huge differences are partially due to different home range estimation methodologies.

Sexual Differences in Plant Structure and Composition and Between Seasons

In relation to plant composition, males prefer high-diversity sites, especially of Asteraceae (Martínez 1998; Faria et al. 2012), but when these are scarce they use sites dominated by *Ornithopus compressus* and some clover (*Trifolium* spp.) species (Fabaceae) (Faria et al. 2012). Females select Fabaceae-dominated places (Faria et al. 2012). These intersexual differences in plant composition reflect the differential selection related to the structural characteristics of the vegetation (Morales et al. 2008) and are probably due to contrasting sexual behaviours and breeding requirements (Morales and Traba 2009). When such natural vegetation is absent, little bustards may alternatively use alfalfa fields, which resemble the architecture of natural vegetation (Salamolard and Moreau 1999; Jiguet et al. 2002).

In relation to plant structure, the occurrence of males increases with shorter vegetation and greater legume and green herb cover (Moreira 1999; Santangeli and Dolman 2011; Faria et al. 2012). This preference is not confined to the reproductive period. During summer, males are associated with areas with short vegetation, below 30 cm tall, but with high cover of green plants, which possibly means food resources for the little bustard (Rivas et al. 2005; Silva et al. 2007). During autumn, males still prefer short vegetation around or below 10 cm (Faria and Silva 2010), which is even shorter than that described for the winter and breeding seasons (10–30 cm; Silva et al. 2004; below 40 cm; Silva et al. 2010; Morales et al. 2008), although this could be an effect of the range of plant height available during this season. More details on habitat selection out of the breeding season can be found in chapter "Migration, movements and non-breeding ecology".

Sexual Differences in Arthropod Biomass in Occurrence Areas

As we have mentioned above, fields selected by little bustards seem to be associated with high biomass of arthropods, although with some differences between the sexes (Faria et al. 2012). Both sexes use sites with high abundance of Acrididae and Formicidae (Faria et al. 2012), although little bustards seem to reject ants from their diet (Jiguet 2002). Males select territories which include large beetles such as carabids (Traba et al. 2008), or large orthopterans such as bush crickets (Tettigoniidae, Faria et al. 2012) and grasshoppers (Acrididae, Martínez 1998). Females prefer locations with high abundances of Coleoptera (Faria et al. 2012) or grasshoppers (Badenhausser and Bretagnolle 2005; Badenhausser et al. 2009).

In summary, breeding males select territories with enough food to cope with energy-consuming courtship activity, but also for use by potentially nesting females, thus contributing to increased mating success (Morales et al. 2008). This kind of microhabitat selection pattern remains constant both over the entire breeding season and among study sites, suggesting that little bustard preferences in plant structure are independent of habitat composition at the landscape level (Morales et al. 2008).

Once breeding has ended, plant food requirements increase in importance and males occupy the bottom of hills, which seems to be related to the availability of green plants as food (Silva et al. 2007; more details in chapter "Migration, movements and non-breeding ecology").

Less is known about the preferences of females. As mentioned above, during the breeding season they prefer sheltered sites, i.e. locations with vegetation cover and spatial density at low to medium heights (up to 30 cm above ground), and high litter and green weed cover, thus suggesting that the females microhabitat selection is determined by two main constraints: the need for shelter and anti-predator surveillance, and the need for food for them and their young (Morales et al. 2008).

Concluding Remarks

Some general trends on macro- and microhabitat selection emerge from this review (Fig. 2). First, the strong relationship of little bustard distribution with temperature-related variables suggests the sensitivity of the species to temperature changes (Delgado et al. 2011). Besides, its dependence on both conspecific attraction and

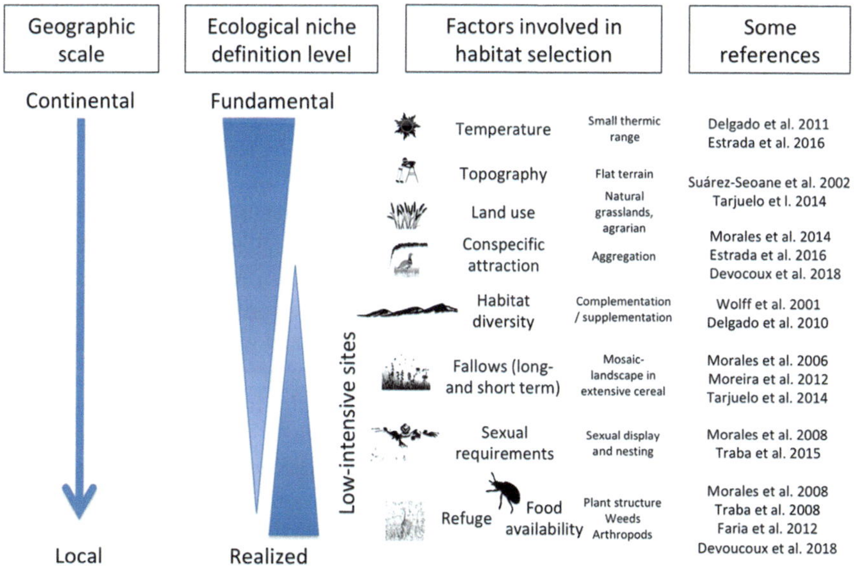

Fig. 2 Schematic representation of hierarchical multiscale process of little bustard habitat selection. Scales from geographic to local are represented, which resembles the transition from fundamental to realized niche. Drivers of species-specific space occupancy are also shown, reflecting the influence of local biotic, fine-grained factors, which modulate the effect of others, mainly abiotic, acting on a much greater scale. Some references supporting each of the items are also provided

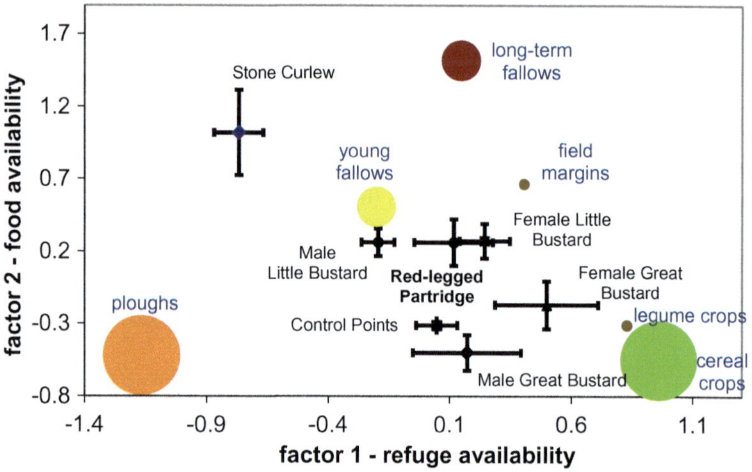

Fig. 3 Steppe bird species located by their preference in the space defined by the two main microhabitat (ecological) gradients in extensive agrosystems of central Spain. Control points (indicating availability) are also shown. See Morales and Traba (2009) and Traba et al. (2015) for more details on statistics. Error bars indicate ± SE around PCA mean values

land use makes it especially vulnerable to climate change (Estrada et al. 2016), due to its strong philopatric tendency.

Second, this chapter provides support for the role of long-term fallows as a critical habitat for little bustards, and reinforces the role of young fallows as an alternative habitat, especially when they offer inter-annual stability. In a summarizing work, Traba et al. (2015) examined microhabitat preferences of little bustard and other coexisting medium- to large-sized steppe birds in typically extensive agrarian landscape mosaics in central Spain. They found two main ecological gradients, one related to variation in cover and the other related to food availability. The position in such habitat space of both the agrarian habitats and the species is shown, providing a link from micro- to macrohabitat in relation to each species' preferences (Fig. 3). Both little bustard male and female preferences are close to the centroid of a triangle formed by the three main field types in these extensive agrosystems: cereal crops, ploughed fields and long-term fallows. Such results highlight the importance of mosaic dry cereal farmland for this species, at least in microhabitat terms.

The strong relationship between little bustard and other steppe birds and extensive farmland, and their negative response to agricultural intensification, were established when the transformation of fallows and low-yield cereal crops into intensive crops dramatically reduced the population of little bustard. This intensification produced a rapid change, as fallows and old fallows vanished from the agricultural landscape, being replaced by more productive land uses. As a consequence, since the 1970s and 1980s, habitat loss and degradation has been the most important factor affecting little bustard abundance in Europe, especially in France (Schulz 1985; Goriup and Batten 1990; Jolivet 1997: Traba and Morales 2019). The management of fallows is a critical point for optimum habitat conservation, as both over- and undergrazed fallows and pastures can become unsuitable as breeding habitat for little bustard females (Faria et al. 2012). In central Spain, sheep grazing has practically disappeared from many areas, leading to vegetation succession in old fallows and thus decreasing their suitability for the species. In this situation, young fallows remain the only (even if suboptimal) habitat available for nesting and brood-rearing (Morales et al. 2013).

However, in recent years a new, rather steep population decline has been observed in Spain and Portugal (see chapter "The little bustard around the world: distribution, global conservation status, threats and population trends"), attributed at least partially to a second phase of the intensification process (Morales et al. 2015; Traba and Morales 2019). This includes increased use of herbicides and other agricultural pesticides (Martínez and Tapia 2002; Morales et al. 2015), affecting not only crops but also stubbles and young fallows. These effects on the wild flora and invertebrate fauna typical of farmlands are not sufficiently known, but seem to affect landscape for long periods of time (Geiger et al. 2010). Despite the scarcity of species-specific studies about the effect of intensification on little bustard density (but see Wolff et al. 2001; García et al. 2007; Bretagnolle et al. 2018), the general conclusion is that little bustard density is higher in low-intensive sites than in intensive ones (Wolff et al. 2001; Martínez and Tapia 2002; Moreira et al. 2012).

Third, this review has also shown that little bustards select a rather specific range of vegetation structure as a result of several trade-offs between anti-predator vigilance, food requirements and sexual constraints. This selection pattern goes beyond landscape composition, and relies on microhabitat characteristics, independent of particular land use. However, with increasing farming intensification, vegetation structure becomes more uniform across fields, and thus the chance to find optimum microhabitats decreases. In this sense, the loss of optimal habitat diversity in intensive areas could facilitate the predation of chicks, as the spatial heterogeneity associated with fallows and pastures seems to be important in reducing nest predation (Martínez and Tapia 2002). The breeding success of little bustard, and consequently the total productivity of the species, could be decreasing in parallel with the decrease of long-term fallow lands, the preferred nesting place for the species. Thus, the conservation of the little bustard is inevitably linked to the conservation of its habitats. Managing young fallows extensively could still meet this challenge, by providing an adequate plant structure for nesting and feeding, and minimizing agrochemical inputs to avoid undesirable trophic effects on weeds and insects (Moreno et al. 2010; Traba and Morales 2019).

Acknowledgements We are thankful to Francisco Moreira for his useful comments on an earlier version of this chapter, and to Nigel Collar for his thorough English review.

References

Arcos F, Salvadores R (2005) Habitat selection by territorial males of Little Bustard *Tetrax tetrax* in a NW Iberian population. Ardeola 52:151–157

Badenhausser I, Bretagnolle V (2005) Grasshopper abundance in grassland habitats in Western France. In: Lillak R, Viiralt R, Linke A, Geherman V (eds) Integrating efficient grassland farming and biodiversity. Grassland Science in Europe, Tartu, pp 445–448

Badenhausser I, Amouroux P, Lerin J, Bretagnolle V (2009) Acridid (Orthoptera: Acrididae) abundance in Western European grasslands: sampling methodology and temporal fluctuations. J Appl Entomol 133:720–732

Berthet ETA, Bregtanole V, Segrestin B (2012) Analyzing the design process of farming practices ensuring little bustard conservation: lessons for collective landscape management. J Sustain Agric 36:319–336

Block WM, Brennan LA (1993) The habitat concept in ornithology: theory and applications. Curr Ornithol 11:35–91

Bravo C, Cuscó F, Morales MB, Mañosa S (2017) Diet composition of a declining steppe bird the Little Bustard (*Tetrax tetrax*) in relation to farming practices. Avian Conserv Ecol 12(1):3

Bretagnolle V, Villers A, Denonfoux L, Cornulier T, Inchausti P, Badenhausser I (2011) Rapid recovery of a depleted population of Little Bustards *Tetrax tetrax* following provision of alfalfa through an agri-environment scheme. Ibis 153:4–13

Bretagnolle V, Denonfoux L, Villers A (2018) Are farming and birds irreconcilable? A 21-year study of bustard nesting ecology in intensive agroecosystems. Biol Conserv 228:27–35

Cramp S, Simmons KEL (1980) Handbook of the birds of Europe, the Middle East and North Africa. Vol. 2. Hawks to Bustards. RSPB, Oxford University Press, Oxford

De Juana E, Santos T, Suárez F, Tellería JL (1988) Status and conservation of steppe birds and their habitat in Spain. In: Goriup PD (ed) Ecology and conservation of grassland birds. ICPB, Cambridge, pp 113–123

Delgado A, Moreira F (2000) Between-year variations in Little Bustard *Tetrax tetrax* population densities are influenced by agricultural intensification and rainfall. Ibis 152:633–642

Delgado MP, Morales MB, Traba J, García de la Morena EL (2009) Determining the effects of habitat management and climate on the population trends of a declining steppe bird. Ibis 151: 440–451

Delgado MP, Traba J, García de la Morena EL, Morales MB (2010) Habitat selection and density-dependent relationships in spatial occupancy by male Little Bustards *Tetrax tetrax*. Ardea 98: 185–194

Delgado MP, Traba J, Morales MB (2011) Climate niche constraints in two coexisting steppe birds: the little and the great bustard. Ardeola 58:223–238

Devoucoux P (2014) Conséquences et impacts prévisibles d'une perte d'habitat majeure sur une espèce menacée aux exigences écologiques complexes: Effets de la mise en place du Contournement ferroviaire á grande vitesse Nîmes-Montpellier sur la dynamique de la population d'Outarde canepetière des Costiéres de Nîmes. Life Sciences. Université de Poitiers. PhD Dissertation

Devoucoux P, Besnard A, Bretagnolle V (2018) Sex-dependent habitat selection in a high-density Little Bustard *Tetrax tetrax* population in southern France, and the implications for conservation. Ibis. https://doi.org/10.1111/ibi.12606

Emmerson M, Morales MB, Oñate JJ, Batáry P, Berendse F, Liira J, Aavik T, Guerrero I, Bommarco R, Eggers S, Pärt T, Tscharntke T, Weisser W, Clement L, Bengtsson J (2016) How agricultural intensification affects biodiversity and ecosystem services. Adv Ecol Res 55: 43–97

Estrada A, Delgado MP, Arroyo B, Traba J, Morales MB (2016) Forecasting large-scale habitat suitability of European bustards under climate change: the role of environmental and geographic variables PLOS One 9/57

Faria N, Rabaça JE (2004) Breeding habitat modelling of the Little Bustard *Tetrax tetrax* in the site of community importance of Cabrela (Portugal). Ardeola 51:331–343
Faria N, Silva JP (2010) Habitat selection of the Little Bustard during the beginning of an agricultural year. Ardeola 57:363–373
Faria N, Rabaça JE, Morales MB (2012) Linking plant composition and arthropod abundance to establish Little Bustard breeding requirements in pastureland dominated landscapes. Biodivers Conserv 21:2109–2125
García J, Suárez-Seoane S, Miguélez D, Osborne PE, Zumalacárregui C (2007) Spatial analysis of habitat quality in a fragmented population of little bustard (Tetrax tetrax): implications for conservation. Biol Conserv 137:45–56
García de la Morena EL, Bota G, Ponjoan A, Morales MB (2006) El sisón en España. I Censo Nacional (2005). SEO/BirdLife, Madrid
Geiger F, Bengtsson J, Berendse F, Weisser WW, Emmerson M, Morales MB, Ceryngier P, Liira J, Tscharntke T, Winqvist C, Eggers S, Bommarco R, Pärt T, Bretagnolle V, Plantegenest M, Clement LW, Dennis C, Palmer C, Oñate JJ, Guerrero I, Hawro V, Aavik T, Thies C, Flohre A, Hänke S, Fischer C, Goedhart PW, Inchausti P (2010) Persistent negative effects of pesticides on biodiversity and biological control potential on European farmland. Basic Appl Ecol 11:97–105
Goriup PD, Batten L (1990) The conservation of steppic birds: a European perspective. Oryx 24: 215–223
Guerrero, I., Morales, M. B., Oñate, J. J., Aavik, T., Bengtsson, J., Berendse, F., Clement, L.W., Dennis, C., Eggers, S., Emmerson, M., Fischer, C., Flohre, A., Geiger, F., Hawro, V., Inchausti, P., Kalamees, A., Kinks, R., Liira, J., Meléndez, L., Pärt, T., Thies, C., Tscharntke, T., Olszewski, A. And Weisser, W. W. 2011. Taxonomic and functional diversity of farmland bird communities across Europe: effects of biogeography and agricultural intensification. Biodivers Conserv, 20: 3663–3681
Jiguet F (2002) Arthropods in diet of Little Bustards *Tetrax tetrax* during the breeding season in western France. Bird Study 49:105–109
Jiguet F, Arroyo B, Bretagnolle V (2000) Lek mating systems: a case study in the Little Bustard *Tetrax tetrax*. Behav Process 51:63–82
Jiguet F, Jaulin S, Arroyo B (2002) Resource defence on exploded leks: do male little bustards, *Tetrax tetrax*, control resources for females? Anim Behav 63:899–905
Johnson DH (1980) The comparison of usage and availability measurements for evaluating resource preference. Ecology 61:65–71
Jolivet C (1997) L'Outarde canepetière, *Tetrax tetrax*, en France: le déclin s'accentue. Ornithos 4: 73–77
Jones J (2001) Habitat selection studies in avian ecology: a critical review. Auk 118(2):557–562
Kamp J, Urazaliev R, Donald PF, Hölzel N (2011) Post-soviet agricultural change predicts future declines after recent recovery in Eurasian steppe bird populations. Biol Conserv 144:2607–2614
Lapiedra O, Ponjoan A, Gamero A, Bota G, Mañosa S (2011) Consequences of agricultural intensification on the ranging behavior and breeding success of threatened steppe-land birds: the case of little bustard. Biol Conserv 144:2882–2890
Martínez C (1994) Habitat selection by the little bustard *Tetrax tetrax* in cultivated areas of Central Spain. Biol Conserv 67:125–128
Martínez C (1998) Microhabitat selection of Little Bustard Tetrax tetrax during breeding season. Ardeola 45:73–76
Martínez C, Tapia GG (2002) Density of the Little Bustard *Tetrax tetrax* in relation to agricultural intensification in Central Spain. Ardeola 49:301–304
Morales MB, Traba J (2009) Compromisos adaptativos en la selección de hábitat de aves esteparias. In: Dopazo H, Navarro A (eds) Adaptación y Evolución: 150 años después del Origen de las Especies. SESBE, pp 304–313

Morales MB, García JT, Arroyo B (2005) Can landscape composition changes predict spatial and annual variation of little bustard male abundance? Anim Conserv 8:167–174
Morales MB, Suárez F, García de La Morena E (2006) Réponses des oiseaux de steppe aux différents niveaux de mise en culture et d'intensification du paysage agricole: une analyse comparative de leurs effets sur la densité de population et la selection de l'habitat chez l'outarde canepetiere *Tetrax tetrax* et l'outarde barbue *Otis tarda*. Rev Ecol (Terre et Vie) 61:261–270
Morales MB, Traba J, Carriles E, Delgado MP, García De La Morena EL (2008) Sexual differences in microhabitat selection of breeding little bustards *Tetrax tetrax*: ecological segregation based on vegetation structure. Acta Oecol 34:345–353
Morales MB, Traba J, Delgado MP, García de la Morena EL (2013) The use of fallows by nesting little bustard *Tetrax tetrax* females: implications for conservation in mosaic cereal farmland. Ardeola 60:85–97
Morales MB, Traba J, Arroyo B (2015) El declive del sisón en el centro de España. Quercus 356: 36–43
Moreira F (1999) Relationships between vegetation structure and breeding bird densities in fallow cereal steppes in Castro Verde, Portugal. Bird Study 46:309–318
Moreira F, Silva JP, Estanque B, Palmeirim JM, Lecoq M, Pinto M, Leitao D, Alonso I, Pedroso R, Santos E (2012) Mosaic-level inference of the impact of land cover changes in agricultural landscapes on biodiversity: a case-study with a threatened grassland bird. PLoS One 7(6): e38876
Moreno V, Morales MB, Traba J (2010) Avoiding over-implementation of agri-environmental schemes for steppe bird conservation: a species-focused proposal based on specialist criteria. J Environ Manag 91:1802–1809
Morris DW (1987) Ecological scale and habitat use. Ecology 68:362–369
Noy-Meir I (1973) Desert ecosystems: environments and producers. Annu Rev Ecol Syst 4:25–51
Osborne P, Suárez-Seoane S (2007) Identifying core areas in a species' range using temporal suitability analysis: an example using little bustards *Tetrax tetrax* L. in Spain. Biodivers Conserv 16:3505–3518
Ponjoan A, Bota G, Mañosa S (2012) Ranging behaviour of little bustard males, *Tetrax tetrax*, in the lekking grounds. Behav Process 91:35–40
Ricklefs RE (1987) Community diversity: relative roles of local and regional processes. Science 235:167–171
Rivas O, De Castro A, Llaneza L, Otero J (2005) Population size, phenology and habitat preferences of little bustard *Tetrax tetrax* in an isolated area of the Northwestern Iberian Peninsula. Ardeola 52:147–149
Salamolard M, Moreau C (1999) Habitat selection by Little Bustard *Tetrax tetrax* in a cultivated area of France. Bird Study 46:25–33
Santangeli A, Cardillo A (2012) Spring and summer habitat preferences of little bustard in an agro-pastoral area in Sardinia (Italy). Ital J Zool 79:329–336
Santangeli A, Dolman PM (2011) Density and habitat preferences of male little bustard across contrasting agro-pastoral landscapes in Sardinia (Italy). Eur J Wildl Res 57:805–815
Santos M, Bessa R, Cabral JA, Pacheco FAL, Leitão D, Moreira F, Pinto M, Lecoq M, Silva JP (2016) Impacts of land use and infrastructural changes on threatened Little Bustard *Tetrax tetrax* breeding populations: quantitative assessments using a recently developed spatially explicit dynamic modelling framework. Bird Conserv Int 26:418–435
Schulz H (1985) A review of the world status and breeding distribution of the Little Bustard. Bustard Stud 2:131–151
Shlyakhtin GV, Tabachishin VG, Khrustov AV, Zavyalov EV (2004) Ecological segregation of Bustards (Otitidae) in the north of the Lower Volga Region: evolutionary and adaptive aspects. Russ J Ecol 35:247–253
Silva JP, Pinto M, Palmeirim JM (2004) Managing landscapes for the little bustard *Tetrax tetrax*: lessons from the study of winter habitat selection. Biol Conserv 117:521–528

Silva JP, Faria N, Catry T (2007) Summer habitat selection and abundance of the threatened little bustard in Iberian agricultural landscapes. Biol Conserv 139:186–194

Silva JP, Palmeirim JM, Moreira F (2010) Higher breeding densities of the threatened little bustard *Tetrax tetrax* occur in larger grassland fields: implications for conservation. Biol Conserv 143: 2553–2558

Silva JP, Estanque B, Moreira F, Palmeirim JM (2014) Population density and use of grasslands by female Little Bustards during lek attendance, nesting and brood-rearing. J Ornithol 155:53–63

Suárez F, Naveso MA, De Juana E (1997) Farming in the drylands of Spain: birds of the pseudosteppes. In: Pain DJ, Pienkowski MW (eds) Farming and birds in Europe. The Common Agricultural Policy and its implication for bird conservation. Academic Press, London, pp 297–330

Suárez-Seoane S, Osborne PE, Alonso JC (2002) Large-scale habitat selection by agricultural steppe birds in Spain: identifying species-habitat responses using Generalised Additive Models. J Appl Ecol 39:755–777

Suárez-Seoane S, García de la Morena EL, Morales MB, Osborne PE, De Juana E (2008) Maximum entropy niche-based modelling of seasonal changes in little bustard (*Tetrax tetrax*) distribution. Ecol Model 219:17–29

Tarjuelo R, Delgado MP, Bota G, Morales MB, Traba J, Ponjoan A, Hervás I, Mañosa S (2013) Not only habitat but also sex. Factors conditioning spatial distribution of breeding Little Bustard *Tetrax tetrax* families. Acta Ornithologica 48:119–128

Tarjuelo R, Morales MB, Traba J, Delgado MP (2014) Are species coexistence areas a good option for conservation management? Applications from fine scale modeling in two steppe birds. PLOS One 9:e87847. 9/57

Tarjuelo R, Morales MB, Arribas L, Traba J (2019) Abundance of weeds and seeds but not of arthropods differs between arable habitats in an extensive Mediterranean farming system. Ecol Res 34:624–636

Thuiller W, Richardson DM, Pysek P, Midgley GF, Hughes GO, Rouget M (2005) Niche-based modelling as a tool for predicting the risk of alien plant invasions at a global scale. Glob Chang Biol 11:2234–2250

Traba J, Morales MB (2019) The decline of farmland birds in Spain is strongly associated to the loss of fallow-land. Sci Rep 9:9473

Traba J, Morales MB, García de la Morena EL, Delgado MP, Krištín A (2008) Selection of breeding territory by little bustard (*Tetrax tetrax*) males in Central Spain: the role of arthropod availability. Ecol Res 23:615–622

Traba J, Morales MB, Carmona CP, Delgado MP (2015) Resource partition and niche overlap in a steppe bird assemblage. Community Ecol 16:178–188

Wolff A, Paul JP, Martin JL, Bretagnolle V (2001) The benefits of extensive agriculture to birds: the case of the little bustard. J Appl Ecol 38:963–975

Wolff A, Dieulevent T, Martin JL, Bretagnolle V (2002) Landscape context and little bustard abundance in a fragmented steppe: implications for reserve management in mosaic landscapes. Biol Conserv 107:211–220

Yousefi M, Kafash A, Malakoutikhah S, Ashoori A, Khani A, Mehdizade Y, Ataei F, Ilanloo SS, Rezaei HR, Silva JP (2017) Distance to international border shapes the distribution pattern of the growing Little Bustard *Tetrax tetrax* winter population in Northern Iran. Bird Conserv Int 6:1–10

Migration, Movements, and Non-breeding Ecology

Manuel B. Morales, S. Mañosa, A. Villers, E. L. García de la Morena, and Vincent Bretagnolle

Introduction

The study of species' post-breeding ecology has become a rapidly growing field in avian biology as researchers have realized its relevance, not only in completing our picture of species' annual biological cycles, but also in our understanding of their breeding biology and population dynamics, and thereby for their conservation and management (Sherry and Holmes 1996; Newton 1998). Therefore, the identification of species' requirements and threats outside of the reproductive season is crucial to

M. B. Morales (✉)
Terrestrial Ecology Group, Departamento de Ecología, Universidad Autónoma de Madrid, Madrid, Spain

Centro de Investigación en Biodiversidad y Cambio Global (CIBC-UAM), Universidad Autónoma de Madrid, Madrid, Spain
e-mail: manuel.morales@uam.es

S. Mañosa
Facultat de Biologia, Departament de Biologia Evolutiva, Ecologia i Ciències Ambientals, Institut de Recerca de la Biodiversitat (IRBio), Universitat de Barcelona, Barcelona, Catalonia, Spain
e-mail: smanosa@ub.edu

A. Villers
CEBC, UMR7372, CNRS-Université de la Rochelle, Beauvoir sur Niort, France
e-mail: Alexandre.Villers@ofb.gouv.fr

E. L. García de la Morena
Biodiversity Node, Spain, Spain
e-mail: eladio.garcia@secim.es

V. Bretagnolle
Centre d'Etudes Biologiques de Chizé, CNRS & La Rochelle Université, Beauvoir sur Niort, France
e-mail: vincent.bretagnolle@cebc.cnrs.fr

V. Bretagnolle et al. (eds.), *Little Bustard: Ecology and Conservation*, Wildlife Research Monographs 5, https://doi.org/10.1007/978-3-030-84902-3_7

the development of effective conservation strategies that take those requirements into account (Vickery et al. 2014). Birds inhabiting dynamic and seasonally changing landscapes such as cereal farmland in temperate and Mediterranean climates are faced with rapid oscillation in the availability of limiting resources, and knowing how they cope with these natural and human-induced changes beyond the breeding season is essential if their currently declining trends are to be reversed (Suárez 2004).

Birds and other animals usually respond to environmental changes through their propensity to move to a different area, offering individuals better survival or breeding opportunities (Newton 2010). This tendency to migrate is usually adaptive and thus genetically fixed in most populations. Consequently, migratory behaviour has evolved in many species whenever adequate selection pressures are present (Berthold 1993; Newton 2010). Moreover, experimental evidence has proved that migratory behaviour can evolve in a sedentary population or be lost in a migratory one in a relatively short time period (even decades in short-lived species) if selective advantages leading to enhanced breeding performance and/or survival are strong (Berthold et al. 1992; Pulido 2007; Pulido and Berthold 2010).

In the present chapter, we focus on two main and intrinsically linked aspects of little bustard post-breeding biology. First, we describe the little bustard seasonal movements and migration, highlighting the diversity of movement patterns found across the species' distribution range. Then, we synthesize post-breeding habitat and resource use, which includes habitat selection and diet composition, as well as flocking behaviour and home range features. After reviewing little bustard post-breeding biology, we finish by discussing some of the main threats faced by the species, specifically outside the breeding season, as well as the conservation measures that can be applied to minimize them.

Little Bustard Seasonal Movements

Overview

Animals undertaking a regular annual return movement from reproductive to post-breeding areas are usually regarded as migratory (Bernis 1966a; Sinclair 1983; Terrill and Able 1988). In the case of birds, when this movement occurs after the breeding season in a restricted range of directions (usually north to south and vice versa in the higher latitudes of the Northern Hemisphere), associated with seasonal changes in the conditions for survival and involving variable distances (from tens to thousands of kilometres), it is known as typical or directional migration, and it usually produces massive displacements of birds and a corresponding geographical shift of populations (Newton 2010). However, bird species show a huge range of movement patterns (Berthold 1993; Newton 2010): at the opposite end of that range, populations whose individuals perform much shorter displacements, so that their distributions remain unchanged within and across years, are considered resident or sedentary (Newton 2010). Bird migratory behaviour is largely determined by genetic adaptation to local environmental conditions, which generates an enormous

Table 1 Comparison of the main winter ecological and conservation aspects revised in this chapter between the two little bustard distribution sub-ranges

	Western range	Eastern range
Migration status	Resident, partially and fully migratory	Fully migratory north of Black and Caspian Seas; Resident in the Middle East and Iran?
Migration patterns	Autumn migration, summer migration, nomadic movements, sedentary	Autumn migration, Summer migration? Nomadic?
Main wintering habitat	Extensive cereal farmland Intensive irrigated (alfalfa) farmland	Grazed grass and shrub-steppe, littoral shrub-steppe
Other ecological features	Frequent association with pin-tailed sandgrouse (commensalism)	Unknown
Habitat-related threats	Land-use changes, Intensive irrigation methods	Overgrazing, desertification, others not evaluated
Human-induced disturbances	Legal hunting activity, other leisure activities, poaching, agricultural works	Hunting, poaching
Anthropogenic mortality sources	Shooting, wire collision, wind turbine collision?	Shooting, falconry, others not evaluated

intraspecific variation, particularly in species with large distributions (Alerstam et al. 2003; Pulido 2007). Differences in migration behaviour may be found even within populations so that those where all individuals leave breeding localities are known as annual or fully migratory populations, while those where only some individuals migrate and others do not are called partially migratory (Chapman et al. 2011).

As introduced in chapter "The little bustard around the world: distribution, global conservation status, threats and population trends", the little bustard's world range, comprises both migratory and sedentary populations. In general terms, most populations currently representing the western sub-range are considered sedentary (Snow and Perrins 1998, see Table 1). Little bustard populations of central-western France are the exception since they are mainly fully migratory, using areas in central and southern Iberia as wintering quarters (Villers et al. 2010). Different authors have speculated about the possible wintering of European little bustards in North Africa in historical times, although no evidence based on radio/satellite tracking, ring recovery or direct observation of birds crossing the Mediterranean has ever been provided (see review in Palacín and Alonso 2009).

Remaining little bustard populations from the eastern sub-range, conversely, are considered mainly migratory (Snow and Perrins 1998). Birds from southern Russia and Central Asia have long been known to leave their breeding grounds during the harsh winters typical of those regions (e.g. Shlyakhtin et al. 2004), and large winter concentrations have been observed south of the Black and Caspian Seas (Gauger 2007; Sehhatisabet et al. 2012; Yousefi et al. 2017). In addition, there are historical records of massive autumn migration at bottlenecks along the main eastern flyways (Phillips-Wolley 1881; Heiss 2013). Nothing is known about the seasonal movements of the marginal populations that might still breed in Iran and other parts of the Middle East (see chapter "The little bustard around the world: distribution, global conservation status, threats and population trends"), although similar patterns to those found in western Mediterranean populations might be expected (Table 1).

Nevertheless, recent studies based on radio and satellite tracking have shown that little bustard seasonal movements can be quite diverse and that populations once considered sedentary in fact exhibit a variety of movement patterns associated with different local or regional-scale factors. Some of these populations are far from sedentary (García de la Morena et al. 2015). Such intraspecific diversity of movement has also been described in other bustards and steppe birds (e.g. Morales et al. 2000; Alonso et al. 2001; Palacín 2007; Limiñana et al. 2008; Combreau et al. 2011) and, as mentioned, is generally found in species whose extensive distributions encounter a large variety of seasonal environments (Newton 2010). On the other hand, in spite of the spatial segregation and different behaviour of the sexes during the breeding period (see chapter "Habitat selection and space use"), male and female little bustards form mixed post-breeding flocks and, although much of what we know from their movements are based on tagged males, there is no evidence of differential migration of the sexes, as described for other bustard species (e.g. Morales et al. 2000; Alonso et al. 2000). In this section, we review and discuss, in the framework of bird migration theory, the whole variety of seasonal movement patterns, from typical directional migration to strictly sedentary behaviour, shown by the little bustard across its entire range.

Long-Distance Directional Migration: Eastern Range

As stated above and reviewed in chapter "The little bustard around the world: distribution, global conservation status, threats and population trends", basically all little bustard breeding populations found north of the Black and Caspian Seas, in the species' eastern sub-range, are considered fully migratory (e.g. Martin et al. 2018). This conclusion is based on the following observations: (1) disappearance of birds from breeding grounds during winter to avoid snow cover or heavy frost, (2) the concentration of very large numbers of birds south of those seas, mainly in eastern Azerbaijan (150–200,000 individuals, Gauger 2007), and northern Iran (up to ca. 57,000 in recent years, Yousefi et al. 2017), and (3) the passage of very large numbers of migrating little bustards flying through the Besh Barmag bottleneck of the Caucasus migration flyway (Heiss 2013). Therefore, this seems to be a typical north–south autumn migration between breeding and wintering areas, likely comprising a few thousand kilometres.

In spite of these reports, no attempt to radio or satellite-track little bustards during their migration movements in Central Asia has yet been made. Therefore, their precise migration routes, stopovers, and timing, as well as their degree of breeding and wintering site fidelity and possible pre-migratory movements, are unknown. These birds seem to stay on the Iranian wintering grounds from November to February

(Sehhatisabet et al. 2012), and apparently through March at the Azerbaijani ones (Heiss 2013), although supplementation in this latter area by birds returning from further south or south-east (Iran) during the spring migration cannot be discarded. Information on dates of departure from, and arrival in, breeding areas is scarce. According to Shlyakhtin et al. (2004), little bustards breeding in the Transvolga region of southern Russia undertake their autumn migration from early October to early November, flying south to the northern Caspian region and then south-west to Ciscaucasia (northern Caucasus). These birds would likely join the contingents passing through the Besh Barmag bottleneck (Heiss 2013). Shlyakhtin et al. (2004) also report the arrival of little bustards on their nesting grounds in the second half of April, although this might occur as late as early May in higher latitudes such as northern Kazakhstan (Snow and Perrins 1998). Since these dates may imply some level of timing mismatch between departure from winter quarters and arrival at breeding sites, and the numbers passing through the bottleneck are far fewer in spring, different and less direct spring flyways (e.g. associated with prospecting trips) could exist and be revealed with the use of adequate tracking technology, as described in other Central Asian migratory birds (Terraube et al. 2012; Kessler et al. 2013).

Although the numbers mentioned above clearly suggest that Azerbaijan and Iran hold the bulk of the eastern winter population of the species, additional records indicate that other regions in Central Asia receive smaller but still significant numbers of wintering little bustards. In fact such information has existed since the early and mid-twentieth century (e.g. Zarudny 1915, Ivanov 1940, Maslov 1947), although its limited availability to non-Russian speakers and the real population changes that have occurred since then have not allowed greater clarity on the evolution of wintering numbers in the region. E.A. Kreutzberg (pers. comm.) reports a flock of 1500 little bustards near Termez, Uzbekistan, at the border with Afghanistan in February 2001. This observer reports smaller numbers (120–140 birds) at the same site the following winter and has compiled observations of small flocks (from a few individuals to a few tens) of migrating and wintering little bustards in Uzbekistan during the early twenty-first century. Observations of migrating birds mainly involve central Uzbekistan (Kyzylkum desert region), and wintering flocks tend to be located in the southern Surkhandarya province, close to the Afghan border. In this latter region, more recent observations of large flocks have been reported by A. Ten and V. Soldatov (pers. comm.), who found a winter roost with 1400 little bustards near the Amu Darya river and a flock of 1980 birds north-west of that area in January 2018. Gavrilov and Gavrilov (2005) report on the regular concentration of several hundreds of birds from the second half of August to early May in southern Kazakhstan, close to the Kirgiz border, although there are no more recent records from this area (E.A. Kreuzberg, pers. comm.). Finally, in south-western Tajikistan, the little bustard seemed to winter in relatively large numbers in the Javan valley where 1500–4000 individuals were regularly observed between 2004 and 2010 according to R.S. Muratov (pers. comm.), who reports however that the amount of birds wintering there decreased to 300–500 in later years. This observer also reports migration and irregular wintering in the Sir Darya valley in northern Tajikistan. Overall, these observations indicate the existence of several minor but still significant little bustard wintering areas in Central Asia, as well as migration flyways different from the main one that runs west of the Caspian Sea and through the Caucasus to Azerbaijan and northern Iran, which should be confirmed in the future through the use of satellite telemetry. Nevertheless, it is worth noting that the total latitudinal range of migration movements of little bustards in the eastern range is quite similar to that observed in the west (see below), ranging from around 50°N (e.g. northern Kazakhstan or former breeding areas in north-western Europe) to around 37°N (e.g. wintering grounds in Central Asia or southern Iberia).

Long-Distance Directional Migration: Western Range

In the western sub-range, little bustard populations of central-western France have always been known to desert their breeding areas in autumn (see, for example Snow and Perrins 1998), although some individuals may overwinter close to them (Villers et al. 2010). It is thus reasonable to assume that all populations historically breeding in central-western, northern, and eastern France performed similar movements. In the early twentieth century, little bustards also bred in south-western France, where they are now extinct; their winter movements were unknown. Although French populations were suspected of overwintering in Iberia (e.g. Bernis 1966b), their precise winter quarters were not revealed until radio, and satellite tracking was applied. Using data provided by 32 radio- and satellite-tracked birds tagged between 1997 and 2007, Villers et al. (2010) showed that the great majority of little bustards breeding in central-western France (fitted in Poitou-Charentes and Centre regions) perform a typical directional migration to overwinter in central and south-western areas of the Iberian Peninsula. Additional data from birds tracked after 2010 have confirmed these patterns.

According to Villers et al. (2010) and other cited reports, five main wintering areas of French little bustards in Iberia can be distinguished (Fig. 1). One is the river Duero basin in the Spanish Northern Plateau. This area was also used by birds as stopovers on their further route south (Delgado et al. 2010; Villers et al. 2010). The second is the Spanish Southern Plateau, mainly the valleys of the rivers Tagus and Jarama, but also farther south in La Mancha region (Morales et al. 2002). The third is Extremadura, also in Spain, a large region with different wintering sites, located mainly in the Tagus basin, but also in its south-eastern confines, in La Serena (Delgado et al. 2010). A fourth wintering area is located around the Tagus estuary in Portugal. Finally, a fifth site was found in the Lleida plains in Catalonia (north-eastern Spain). Stopover sites used during both the autumn and spring migrations were also identified (Fig. 1). Stopovers were located mainly in northern Spain (both during autumn and spring migration, Villers et al. 2010), mostly in the Northern Plateau, but also in the Ebro valley (Morales et al. 2002; Delgado et al. 2010; Villers et al. 2010). However, some stopover sites were also identified in northern Extremadura during both autumn and spring migrations and the Southern Plateau during the spring migration (Delgado et al. 2010). A few birds also regularly stop at French

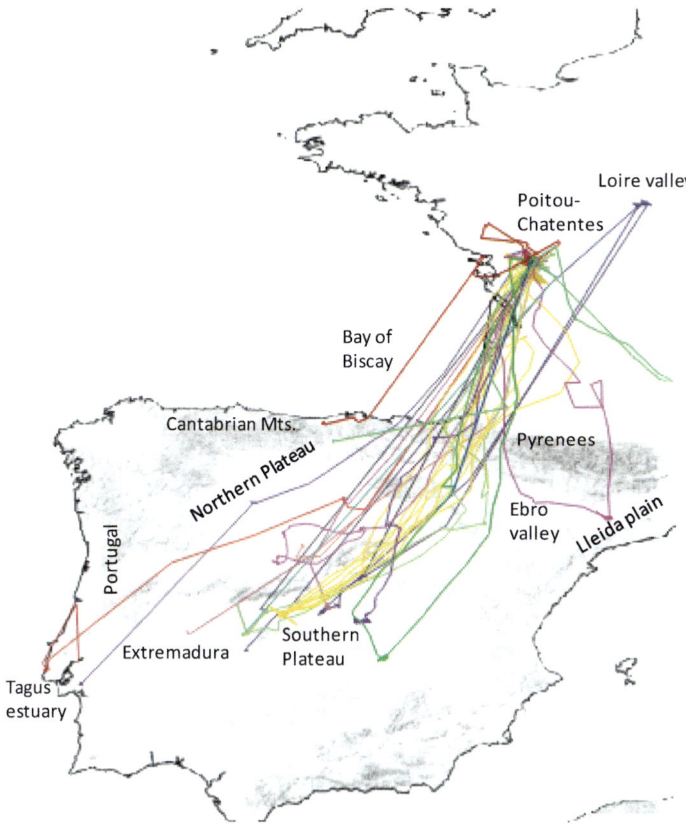

Fig. 1 Precise migration routes, stopovers, and wintering sites of French little bustards wintering in the Iberian Peninsula provided by GPS telemetry. Significant regions and geographical features mentioned in the text are shown. Based on Villers et al. (2010) and updated with unpublished data from V. Bretagnolle and A. Villers

sites before crossing the Pyrenees. During their stay in wintering areas, French little bustards join Iberian birds in wintering flocks (Delgado et al. 2010). At least some French birds perform smaller-scale movements, changing overwintering sites (Morales et al. 2002) in a rather nomadic way (sensu Newton 2010), as also often observed in Iberian birds.

Little bustards from central-western France leave their breeding grounds during October and early November. These departures involve most birds from a given area, which indicates that French little bustards migrate in flocks (Villers et al. 2010) as observed in the eastern populations. Using data from GPS-satellite transmitters, Villers et al. (2010) showed that migration to Iberian wintering quarters is generally accomplished in a single or two consecutive night flights (Villers et al. 2010). The flights were ca. 450 km long if they were direct and 330–380 km if consecutive, with total distances ranging from 600 to 1178 km (Villers et al. 2010). Two main flyways were identified and confirmed by more recent GPS-tagged birds (Fig. 1). One included birds flying over land and using passes in the western Pyrenees. The other included bustards that entered the Peninsula through the eastern Cantabric coast, crossing the Bay of Biscay in about 4 h. One particular individual flew at 64 km/h and at a maximum distance from the coast of 66 km (Villers et al. 2010). The existence of one or more sea flyways is consistent with the relatively high number of historical little bustard observations along the Spanish Cantabric coast compiled by García de la Morena (2015). Although, to date, only four migratory French little bustards have been tracked in at least 2 consecutive years (Villers et al. 2010, V. Bretagnolle and A. Villers unpublished data), data suggest high variability in migratory behaviour both between and within individuals, with some birds using different flyways in consecutive years, while others have crossed the Pyrenees via the exact same pass for 2 consecutive years, only to switch to a different route for another two spring migrations.

Directional migration, however, is not exclusive to French little bustards within the species' western range. Using radio, satellite and GPS-satellite tracking, García de la Morena et al. (2015) have identified different populations in Iberia where little bustards perform a typical north–south autumn migration from their breeding or post-breeding sites to winter quarters several hundred kilometres distant. These populations breed in areas of northern and central Spain where spring and summer productivity is high, but winters are cold. Such localities spread over a large region ranging from Galicia, in north-western Spain, to the Ebro valley in the north-east, and comprise the Spanish Northern Plateau and the high-altitude plains (*páramos*) of the Central and Iberian ranges (Fig. 2). Birds tagged in these populations leave breeding or post-breeding summer grounds during late September, October, and November to spend the winter in thermo and meso-Mediterranean localities in the southern half of Iberia, from Alentejo (southern Portugal) to Extremadura and the Spanish Southern Plateau (Fig. 2a), where they join local populations in relatively large flocks (from tens to several hundreds, historically even a few thousands of individuals, Otero 1985; García de la Morena 2015). Birds return to their breeding sites during March and April (Fig. 3a), although some may reach lekking areas as late as June in Galicia. In this respect, it is worth noting that two adult males captured as breeders in Galicia were located as late as 15 May in Alentejo and Extremadura at lekking sites where local males were calling to attract mates. Their movements in this period, as revealed by GPS locations, suggested that they might be attempting breeding in those localities (own unpubl. data). These two birds later flew north to their capture sites, where other males were at their lekking territories throughout June. This movement behaviour resembles that of common quails (*Coturnix coturnix*), which in western Europe are known to track favourable reproductive environmental conditions (ecosystem productivity) from wintering quarters in northern Africa up to France and through Iberia, breeding successively along the way as those conditions are met (Sardà-Palomera et al. 2012). Nevertheless, these males may have still been too young to be definitely established as territorial breeders.

Little bustards completely or almost completely leave their breeding or post-breeding sites in the northern half of Iberia, where they are basically spring and

a

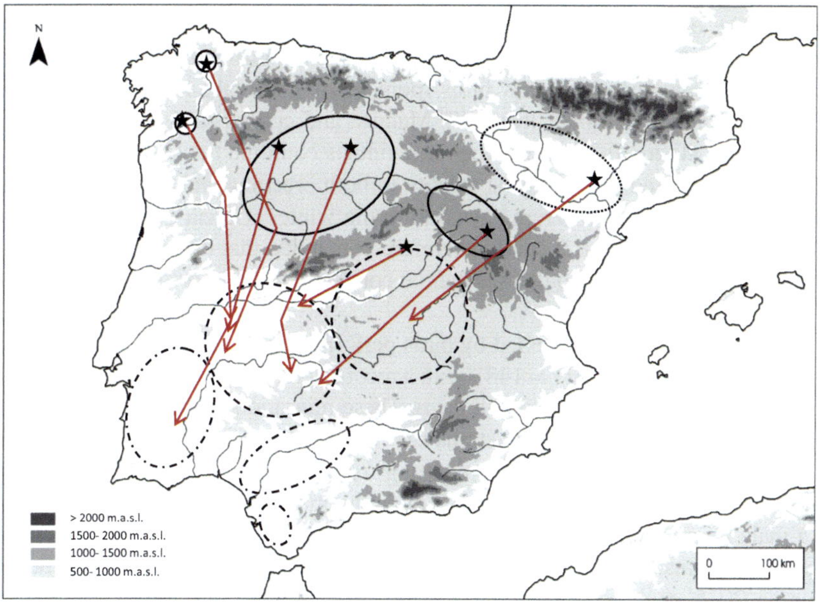

b

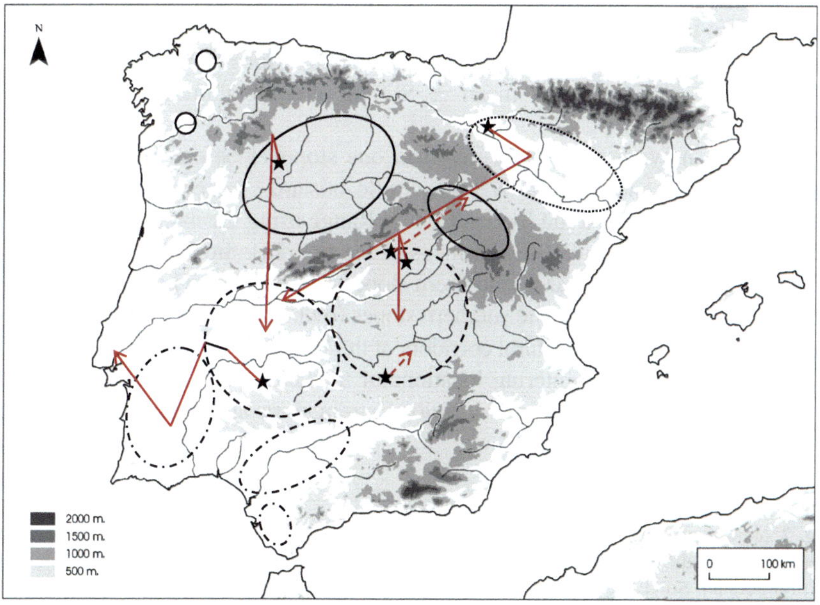

Fig. 2 Simplified migration movements performed by some little bustards radio- and satellite-tracked in Iberia, and representing the two migration types described in this chapter. Stars show locations where breeding birds were captured, and red arrows indicate wintering or summering movements. (**a**) typical autumn migration; (**b**) summer migration by pure summer (dashed arrows)

summer visitors, although, in some localities of the Spanish Northern Plateau, some wintering or stopover visitors may be observed (García de la Morena 2015 see also above). The only reported exceptions are populations in Catalonia, in the eastern Ebro valley, where breeding little bustards overwinter in irrigated alfalfa fields a few kilometres away, and only a few birds sporadically undertake long-distance migration to central and southern Iberia (S. Mañosa and co-workers' unpublished data, Sampietro et al. 2013). In cases when individuals could be tracked for more than a year, they displayed between-year fidelity to both breeding and wintering grounds, although some visited different sites during the winter (García de la Morena et al. 2015, own unpubl. data). It can be therefore concluded that these northern or high-altitude central Iberian populations can be considered totally or partially migratory (Fig. 2a).

The average distance travelled by birds performing long-distance autumn migration in Iberia is 319 km (SD = 148), with a recorded maximum of 565 km (García de la Morena et al. 2015, own unpubl. data). These migration flights occur basically at night. As flapping flyers, little bustards can reach relatively high instant speeds. For example the mean speed of birds migrating from northern Spain to their wintering grounds in south-western Iberia was 85 km/h, but maximum speeds reached 98 km/h in a male flying with a tailwind during spring migration from his wintering site in western Andalusia to his breeding territory on the Iberian high plateau (García de la Morena 2015, authors' own unpubl. data). French little bustards were regularly recorded flying over 90 km/h during migration, and one individual travelled as fast as 137 km/h during spring migration, probably thanks to a strong tailwind. Birds flew at a mean altitude (relative to ground level) of 326 ± 273 m, and flying altitude did not seem to be affected by relief, although variance in relative altitude tended to be lower when birds crossed higher areas (mean ± SD for ground altitudes <500 m. a.s.l. = 326 ± 320 m and mean ± SD for ground altitudes >1000 m.a.s.l. 319 ± 180 m).

Migration movements usually involved one stopover before crossing the main mountain ranges encountered by birds. Little bustards tagged in Galicia and the Northern Plateau crossed the Iberian Central Range using the natural passages of its western sector (altitudes 910–1275 m.a.s.l., see Fig. 2a). Stopovers on the autumn migration were located near the northern slopes of the range, while in spring, birds usually used stopovers near the southern slopes (own unpubl. data). Birds migrating from the Ebro valley used natural corridors of the Iberian System at ca. 1218 m.a.s.l., Fig. 2a). Although the wintering locations of tracked little bustards were concentrated in south-western Iberia (Alentejo, Extremadura and adjacent areas of the

---

Fig. 2 (continued) and winter-summer migrants (solid arrows). Ellipsoids roughly represent the areas where the different migration patterns described for Iberian little bustards are dominant (see text for details). Solid: the dominance of full autumn migration; pointed: the dominance of partial autumn migration; dashed: coexistence of partial autumn and summer migration, receiving winter migrants; dashed and pointed: the dominance of partial summer migration, receiving winter migrants. Based on García de la Morena et al. (2015) and García de la Morena (2015)

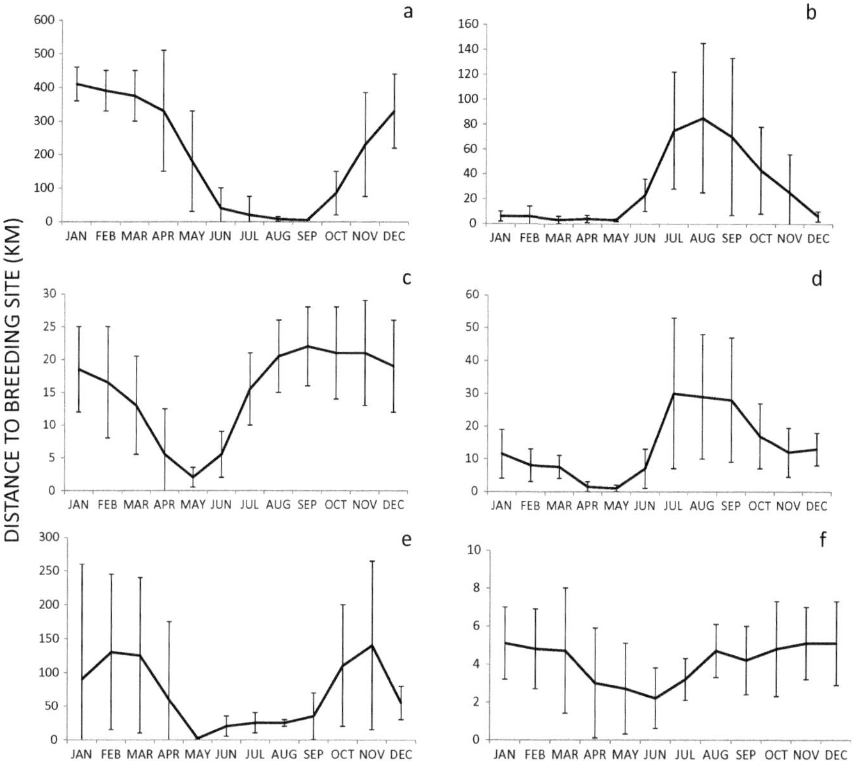

Fig. 3 Migration patterns described for the little bustard in the Iberian Peninsula according to mean distance to breeding sites of 47 radio- and satellite-tracked birds (bars indicate 95% confidence intervals). (**a**) typical autumn migration ($N = 6$); (**b**) pure summer migration ($N = 5$); (**c**, **d**, **e**) summer-autumn migration (see text for differences, $N = 15, 6, 7$, respectively); (**f**) sedentary ($N = 8$). Based on García de la Morena et al. (2015)

Southern Plateau and Andalusia), birds breeding in more eastern populations also wintered in more eastern quarters (Fig. 2a). These long-distance directional migrants represented 24% of the 71 birds monitored by García de la Morena et al. (2015).

Other Migration Patterns

García de La Morena et al. (2015) showed that Iberian little bustards might perform a second type of seasonal movement, not described in other parts of the species' range and depicted in Fig. 2b. It is a post-breeding summer movement undertaken by birds breeding in areas of very low summer productivity typical of the semi-arid Mediterranean climate of the southern half of Iberia. Birds may also completely leave these breeding areas and fly to northern and/or higher-altitude localities, where summer productivity can still provide feeding resources (it should be recalled that the Spanish Northern Plateau is notably higher than the Southern Plateau: 850 vs. 550 m.a.s.l. on average, respectively; see also García de la Morena et al. 2015). In fact, many 10 × 10 km Atlas squares in the northern half of Spain possess exclusive summer records of the species (García de la Morena 2015). However, some summering sites were not located in more northerly or higher areas but simply coincided with places offering enhanced summer primary production (e.g. areas with deeper, more productive soils, irrigated farmland, Silva et al. 2007, own unpubl. data).

Individuals from the same breeding locality may visit quite different summering sites, and distances travelled are shorter than those recorded for autumn migration, reaching a reported maximum of 162 km (but most often only several tens of km). Unlike typical autumn migration, birds do not usually gather in large flocks but spend the summer in small dispersed groups and often solitarily (own unpubl. data). Therefore, although this summer migration shares some features with the typical directional migration (i.e. geographically biased spread of movement directions, complete disappearance from some breeding areas), it closely resembles the so-called dispersive migration described in different seabird species (Newton 2010). Since birds undergo a partial moult during this phase (Cramp and Simmons 1980), it also somehow parallels the moult migration performed by many waterfowl (Newton 2010), although the large concentrations observed in those species do not occur in the little bustard. Birds performing summer migration were a majority in the sample monitored by García de la Morena et al. (2015), representing 75% of all individuals. In fact, the vast majority (ca. 77%) of 10 × 10 km Atlas squares in Spain with the presence of the species at some time of the year do not hold little bustards during summer, which means that birds leave these squares to summer elsewhere (García de la Morena 2015; see Silva et al. 2007 for Portugal). This indicates that the summer movement may actually be a dominant pattern in Iberia. However, summer migrants tagged were mainly males, which moved from lekking sites while breeding females were still rearing their chicks. Therefore, female movement patterns in these populations are poorly known and might differ from those reported for males.

Summer migrants may follow three different patterns of movement after the summer period (García de la Morena et al. 2015, Fig. 3). Some return directly to their breeding areas, where they stay for the rest of the year (Fig. 3b), while others stay through the winter at summering localities (Fig. 3c) or fly south to overwinter before returning the following spring (Fig. 3d, e). The former 'pure' summer migrants may start leaving breeding sites in May and begin their return in September. These individuals usually breed in localities where conditions are adequate for foraging and survival also in the winter, common in the Spanish Southern Plateau and Extremadura. Those staying at summering sites through the winter return to breeding sites during March. Among birds moving in autumn to a different overwintering site, called summer–winter migrants by García de la Morena et al. (2015), these authors identified two groups, those travelling to summering sites further away from the breeding grounds than the wintering areas (Fig. 3d), and those moving to relatively nearby summering sites (Fig. 3e), but overwintering in much more distant areas. Birds in this latter group usually belong to breeding populations where winters are cold, as it is usually the case in central Spain, but are close to higher-altitude areas or even irrigated farmland that can provide food resources during the summer drought period.

Sedentary Populations

Little bustard breeding localities where the species is present throughout the year (Fig. 3f), and thus can be considered sedentary, are spread across southern Iberia (Portuguese Alentejo, Extremadura, Spanish Southern Plateau, and Andalusia) and the Ebro valley (Silva and Pinto 2006; García de la Morena 2015), all under semi-arid meso- or thermo-Mediterranean climate conditions (García de la Morena et al. 2015). However, and at least in Spain, the number of 10 × 10 UTM squares monitored for national atlases where little bustards are strictly resident is a minority compared to those where they present a different movement status (García de la Morena 2015). These localities offer little bustards adequate habitat and resources throughout the year, and they either possess particularly benign climate conditions (e.g. thermo-Mediterranean areas with oceanic influence) or contain cultivated habitats exploitable by birds during both winter and summer, like irrigated alfalfa fields. In the study of García de la Morena et al. (2015), birds that never moved at a rate faster than 8 km/day were always considered to remain within the same area (Shimazaki et al. 2004) and thus classified as sedentary. These birds represented 11% of the total sample studied ($N = 8$). Nevertheless, it is important to highlight that, in all populations where sedentary individuals were tagged, at least some of them left the area in summer, thus undertaking dispersive summer migration (García de la Morena et al. 2015).

The little bustard is considered sedentary also in Sardinia and Mediterranean France (Snow and Perrins 1998). In Sardinia, published information about movements is completely lacking. In southern France, nearly 100 birds were tagged (VHF and GPS tags) over the course of two PhD studies (Wolff 2001; Devoucoux 2014), showing that patterns described for Spanish birds seem to hold also in this region. Post-breeding movements have been found for birds fitted with GPS in Costière de Nîmes (Devoucoux et al., unpubl.), as well as regular migration between Costières and La Crau (in both directions). However, the majority of birds are sedentary, staying in wintering locations within an 8 km radius. This non-migratory status has also been assumed for the virtually extinct populations of continental Italy and northern Africa, as well as for those relict ones in Turkey and the Middle East, including Iran (Snow and Perrins 1998). However, the latter are geographical and climatically heterogeneous regions, and it is possible that radio-tracking studies there might have eventually revealed a diversity of movement patterns similar to that found in Iberia.

Historical Hypotheses on the Evolution of Current Diversity of Little Bustard Migration Patterns

Little bustards seem well adapted to migration. Like all members of the Otididae family, they are suited to maintaining highly energy-demanding flapping flights thanks to their large heart relative to body size, well above the general allometric relationship described for birds (Bishop 1997). In addition, existing evidence from a translocation experiment, in which birds hatched from eggs collected in resident Spanish populations and released in post-breeding flocks of French migratory populations retained sedentary behaviour, indicating that migration is, at least partially, genetically controlled in the little bustard (Villers et al. 2010), and thus subject to natural selection. Therefore, the diversity of seasonal movement patterns described above for the little bustard, a species that occupies a vast range in the Palearctic and thus has faced different selection pressure regimes during its history, should not be a surprise.

However, the particular environmental and evolutionary changes through which such variation in movement patterns has developed remain almost entirely unknown, and explanatory hypotheses are largely speculative. In a phylogeographic study, García et al. (2011) used mitochondrial DNA markers to identify two historical periods of population expansion and genetic diversification in the little bustard: the cold periods just before and after the Last Glacial interstadial (127–111 Ky BP, see chapter "The little bustard and its family: an overview of relationships"). During such cold stadials, and throughout their Eurasian distribution range, little bustards must have left their northernmost breeding grounds to winter as far south as possible, where they probably joined local birds. When the climate warmed during the interglacial periods, forests expanded northwards, and steppe habitats must have been confined in certain continental climate areas of eastern Europe and Central Asia (e.g. Sümegi et al. 2013), as well as in some Mediterranean enclaves where saline or gypsum-rich soils, drought and continentality had preserved open vegetation habitats (e.g. saline basins in central Iberia; Suárez et al. 1992; Sainz-Ollero 2013). Such areas likely functioned as interglacial refugia for steppe-dwelling species, promoting the genetic isolation of populations (Hewitt 2001; García et al. 2011). During these phases, little bustard populations outside the Mediterranean Region (i.e. eastern Europe and Central Asia), which faced much harder winters, likely became totally migratory, using Mediterranean steppes as wintering grounds. Conversely, sedentary behaviour spread in Mediterranean populations. With the return of glacial conditions, all populations expanded and came into contact again. Such a scenario is consistent with the relatively higher haplotype diversity detected by García et al. (2011) in areas intermediate between central France, where little bustards are totally migratory, and southern Iberia, where sedentary birds are dominant. These areas correspond to populations breeding in the northern fringe of Iberia, from Galicia to Catalonia, where migration patterns are totally or partially migratory (see above). Indeed, García et al. (2011) infer that a genetic admixture occurred in those intermediate areas when birds from different interglacial refugia came into contact. This

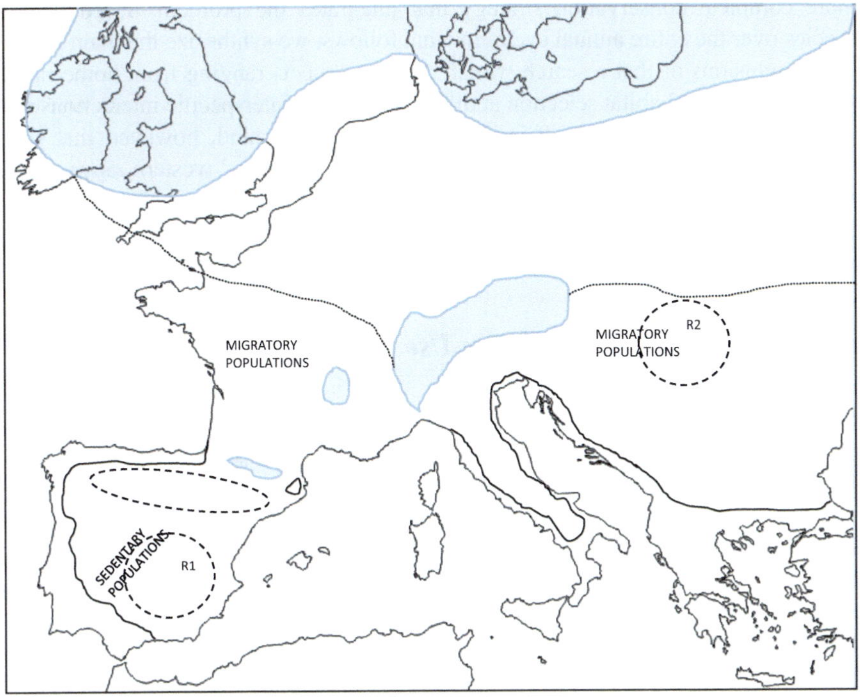

Fig. 4 Depiction of the hypothetical scenario explaining the evolution of migratory and sedentary little bustard populations in Europe during the Quaternary as suggested by García et al. (2011). Light blue area: regions covered by ice at the last Glacial Maximum (20,000–18,000 y BP). Pointed line: southern limit of permafrost. Solid line: limit of forested areas in Mediterranean Glacial refugia. The area between pointed and solid lines: open vegetation and steppes. Dashed circles: interstadial refugia (R1: Iberian refugium, R2: inferred north-eastern refugium). Dashed ellipsoid: genetic admixture area. Ice sheet, permafrost, and vegetation type covers are based on Tarbelet et al. (1998), and García-Antón et al. (2002) for the particular case of Iberia. See text for details

hypothetical past scenario, which is depicted in Fig. 4, could explain the diversity of movement strategies observed today in the western extreme of the species' range, although the role of present-day declines and recent extinctions cannot be ruled out in explaining the genetic impoverishment of some populations (e.g. central France, García et al. 2011). Moreover, the scenario must be completed with genetic data from central Europe, Italy, and northern Africa.

Post-Breeding Ecology

Although studies on flocking behaviour, space use and habitat needs of the little bustard out of the breeding season are still scarce, an important research effort has been made in recent years to improve knowledge of these topics and contribute to a more complete conservation strategy that integrates the problems faced by the species over the entire annual cycle. In what follows, we synthesize the main results and conclusions of that research, which concern aspects ranging from home range size variation and habitat selection at different scales to interspecific interactions and human-induced disturbance. It is important to keep in mind, however, that such information comes primarily from populations of the species' western range. As in many other aspects of their biology, the post-breeding ecology of eastern little bustards remains largely unknown.

Flocking Behaviour and Space Use

Outside the breeding season, little bustards are gregarious birds, gathering in flocks of variable size that include males and females, as well as juvenile individuals. Flocking starts in early summer (later as latitude increases) when males abandon their lekking sites to associate with one another and with non-breeding individuals such as first-year males and unsuccessful females. These birds may also completely leave their breeding grounds and migrate to their post-breeding summer areas or directly to their winter quarters (see Sect. "Other migration patterns"). At this time, most breeding females are still rearing their chicks in isolation (Tarjuelo et al. 2013), but as chicks grow and their size approaches that of adult females, families also gather in larger groups that may eventually be joined by adult males and non-breeding individuals (Cramp and Simmons 1980, own unpubl. data). These post-breeding summer flocks are usually small, ranging from a few birds to several tens, although they can eventually be much larger. In migratory populations, basically, all local individuals congregate in summer flocks before departing for the wintering grounds. For example groups of up to 200 individuals are regularly observed in Poitou-Charente, central-western France (V. Bretagnolle and A. Villers, unpublished data). In the 1950s and 1960s, such flocks frequently peaked at thousands of birds (Boutin and Métais 1995).

In wintering quarters that receive migrants from different breeding populations, the size of flocks and their level of aggregation increase as autumn progresses. For example in a Spanish wintering area of the mid-Tagus valley (see below), flocks attained their maximum size (200–300 individuals) in early December, when the number of existing flocks oscillated between one and two (own unpubl. data). Between January and February, these compact groups tended to split into smaller ones, while the total number of wintering birds decreased in March when little bustards started to return to breeding areas (own unpubl. data). Flock size also varied daily (own unpubl. data). The smallest flocks were observed early in the morning (ca. 100 individuals) when birds were still on the roosting sites, but their size increased steadily over the day to reach a maximum during the evening foraging and roosting periods (over 150 individuals). This circadian increase in flock size may be related to anti-predator strategies like enhanced vigilance or dilution of predation risk at roosts. In fact, the smaller morning flock size suggests group fragmentation overnight, perhaps caused by the attacks of nocturnal predators like red foxes (*Vulpes vulpes*). More detailed studies are needed to understand the origin of circadian variation in flock size.

These flocks tended to routinely commute between roosts and foraging sites (Wolff 2001, own unpubl. data). The former were usually ploughs or stubbles (irrigated cereals, maize) where birds probably seek high visibility to detect potential terrestrial predators, while foraging areas were basically located in large pivot-irrigated alfalfa fields, a preferred habitat where birds find both food and cover (see section "Overview"). In La Crau (southern France), roosts are often located in highly grazed grasslands, while foraging areas are often in rapeseed or mixed cereal and alfalfa crops. The size of these routine core areas (defined using 75 and 50% kernel area estimators) was much smaller than the total range of flock movements, measured as maximum convex polygons (MCP). For example over the course of the 2003–2004 wintering season, the mean size of core areas used by flocks in the mid-Tagus valley was estimated as 91 and 366 ha for 75 and 50% kernels, respectively, while the mean MCP size was 540 ha (own unpubl. data). The larger size of MCPs than estimated core areas usually reflects displacements caused by direct disturbance (generally human-induced) or changes in habitat availability (e.g. ploughing of preferred fields). Overall, the range of flock movements decreased from autumn (October–December) to winter (January–March) as the availability of alfalfa also decreased in the area (due to ploughing), although core areas notably increased in March (own unpubl. data), presumably associated with the start of pre-breeding migration. The fact that little bustard winter movements are mainly confined to a relatively small core range, even when the preferred habitats are largely available, suggests an influence of the species' social dynamics (conspecific attraction, site fidelity) on its habitat and space use. In this line of evidence, Cuscó et al. (2018) found that individual space use of GPS satellite-tracked female little bustards in the irrigated wintering areas of the Ebro valley was markedly constrained by spatial variables related to social and historical factors, which explained 47% of the variance in the probability of individual female occurrence.

Post-Breeding Habitat Requirements

Landscape Scale Preferences

Little bustards breeding in agricultural landscapes in western Europe continue to be linked mainly to such habitat after the breeding season. For example Suárez-Seoane et al. (2008) used landscape scale variables to spatially model and compare the potential breeding and winter distributions of little bustard males in central Spain and found high levels of niche overlap and habitat connectedness between seasons. Moreover, breeding habitat was a good predictor of winter distribution, although the converse relationship was weaker. This means that the preference of breeding birds for diversified agricultural landscapes (see chapter "Habitat selection and space use") generally holds also for the rest of the year. This has clear conservation implications, suggesting that preserving breeding sites closer to wintering grounds would allow the protection of a larger share of the total range. However, the species' habitat niche seems to present fewer constraints during the post-breeding period, thus increasing its width (Suárez-Seoane et al. 2008), which may reflect the absence of the more compelling habitat requirements related to reproduction and/or the need for birds to use a wider range of habitats to satisfy their food demands. Indeed, according to the study in question, breeding distribution was predicted only by habitat variables, while winter distribution was determined more by climate variables. Therefore, little bustard winter distribution in central Spain was determined not only by landscape composition variables such as the cover of dry arable crops and long-term fallows but also by mean rainfall and net radiation (Suárez-Seoane et al. 2008).

Non-breeding habitat must basically fulfil two key requirements with impact on survival: the need for food to overcome the adverse conditions associated with summer drought (in Mediterranean regions), winter cold, or both, and security from predators and other sources of disturbance (caused by humans, including poaching). For example little bustards may be more tolerant of the proximity of human infrastructures outside the breeding season than in spring (García de la Morena et al. 2007; Silva et al. 2007), if it implies better feeding or security conditions. They may also be less dependent on diverse landscapes with complementary habitats if a single habitat providing both food and security against predators is largely available. One such habitat may be stubbles (cereals, rapeseed, legumes) from late summer up to ploughing time (Silva et al. 2004, 2007; Faria and Silva 2010), in which birds can find sprouted weeds and crop plants (Bravo et al. 2017), along with shelter provided by standing straws (Martín et al. 2010; García de la Morena 2015). Another is fallows of various ages: Bravo et al. (2017) found that in wintering areas with a significant fallow cover, little bustards consumed a remarkable diversity of wild plants, particularly Cruciferae, Compositae, Leguminosae, and Papaveraceae, although cultivated legumes were also consumed if present. A third relevant habitat for wintering little bustards is alfalfa, particularly in irrigated landscapes (Cuscó et al. 2018, own unpubl. data), which provides anti-predator cover (García de la Morena 2015) as well as abundant protein-rich food in the form of leaves and shoots. It is worth noting here that, although alfalfa forms the bulk of the species' diet in irrigated wintering sites, wild plants like Cruciferae still have a significant role, accounting for around 25% of diet composition (Bravo et al. 2017).

Security from predators and human disturbance can also be provided by the topographic features of some wintering areas, where birds can benefit from the use of gentle hilltops for surveillance if they also offer an appropriate vegetation structure, as shown by Silva et al. (2004) in Alentejo. In synthesis, post-breeding areas must guarantee food resources and protection from disturbance and predation to support the little bustard, although whether they are fully suitable or not often depends on smaller-scale factors as already pointed out and described below.

Microhabitat Requirements

Little bustards must balance food availability with anti-predator safety when selecting suitable habitats for overwintering. The study by Silva et al. (2004) on winter habitat selection showed that wintering little bustards select food-rich habitats offering intermediate vegetation height that allows both visual surveillance and concealment. In the extensive dry farmland of Alentejo, such microhabitat structure is provided by grazed fallows and stubbles. However, in other post-breeding and wintering areas, other agricultural substrates can play that role. Such is the case of irrigated alfalfa fields in the mid-Tagus valley (central Spain) and Catalonia (García de la Morena 2015; Cuscó et al. 2018). Therefore, wintering little bustards select certain features of vegetation structure rather than habitat typologies, as has also been shown for the breeding season (see chapter "Habitat selection and space use"). For example García de la Morena (2015) shows that in the food-rich irrigated alfalfa fields of the Tagus valley, little bustards prioritize anti-predator cover, selecting higher and denser vegetation than the mean, as well as seeking good visibility by maximizing their distance to topographic horizons. Similar microhabitat features were selected in fallow fields of the dry farmland of the Spanish Southern Plateau, where sprouted weeds also provided abundant food for wintering birds (García de la Morena 2015). Therefore, very similar microhabitat selection patterns were found in two wintering sites dominated by quite different habitats and agricultural systems.

The little bustard in Iberia often shares wintering habitat with the smaller-sized pin-tailed sandgrouse (*Pterocles alchata*), frequently forming mixed flocks (Cramp and Simmons 1980). The studies by Martín et al. (2010) and García de la Morena (2015) have shown that little bustard microhabitat selection patterns do not differ between single-species and mixed-species flocks, while sandgrouse associated with bustards can access food resources in habitats they would otherwise avoid. For example, single-species sandgrouse flocks were very rarely seen in stubbles or alfalfa fields because vegetation there is too high to allow them anti-predator surveillance. However, mixed-species flocks were always found in those habitats, so that pin-tailed sandgrouse could exploit their abundant food resources while benefiting from little bustard surveillance effort and greater detection ability due to its larger size. This pattern was found in both dry cereal farmland and irrigated alfalfa fields (García de la Morena 2015). In this context, the little bustard would behave as the leader species (Sridhar et al. 2009), whereas the sandgrouse would be a follower species (García de la Morena 2015). Therefore, while this interspecific interaction is beneficial for the pin-tailed sandgrouse, it is rather neutral for the little bustard and could be considered a form of commensalism (see chapter "Interspecific relationships"). Nevertheless, certain advantages could also be expected for bustards, such as predator confusion or predation risk dilution due to enlarged group size (Terborgh 1990; Quinn and Cresswell 2005). Further research would be required to assess these possibilities.

Little Bustard Conservation in Winter

In this final section of the chapter, we review the main threats and conservation problems faced by the little bustard outside the breeding season, which can be grouped in two categories: direct human disturbance and landscape modification. The threats posed by these processes on little bustards are direct consequences of the species' winter behaviour and ecology described in the preceding sections and their interaction with humans and human land management. Therefore, we believe the reader will appreciate this overview of winter conservation issues after considering the species' post-breeding ecology, given the specific threats faced by the species during winter (more general threats and those specific of breeding populations are addressed in chapter "Threats affecting little bustards: human impacts").

Direct Human Disturbance and Shooting

While post-breeding little bustards may be able to minimize predation risk by raptors and terrestrial enemies through the landscape scale and microhabitat selection behaviour described above, they might still have to deal with other sources of disturbance, such as those associated with human activity in an increasingly humanized countryside. Human activities in farmland areas in autumn and winter are diverse, ranging from different agricultural labours (ploughing, agrochemical application, sowing, mowing) to leisure activities (walking with or without dogs, running, biking), including, of course, various forms of hunting. Humans engaged in these and other activities can be perceived as predators by animals, which often modify their behaviour accordingly.

Hunting is a major source of disturbance for wintering birds, both game and non-target species (Madsen and Fox 1995), although very few studies have paid attention to the effect of hunting on protected species. In one such study, Casas et al. (2009) showed that small game hunting altered the spatial distribution and behaviour of little bustards gathered in pre-migratory flocks in cereal farmland of central-western France. These authors compared flock activity on hunting days with that of days before and after hunting and found that flights occurred only on hunting days. Little bustards also increased the time devoted to vigilance on hunting days relative to days before and after and decreased the time spent resting. However, foraging time was not directly affected by hunting activity but varied with flock size, increasing with the number of individuals but decreasing in the largest flocks, presumably due to interference among birds (Sansom et al. 2008). Overall, these changes represent important alterations of the process of energy storage to prepare for their migration to Iberia, and even the lack of a negative response in foraging time could be partly due to a compensation effect to restore the energy lost in flights (Blanc et al. 2006). Also, as a consequence of hunting, little bustards concentrate in hunting-free areas set up as reserves for the game. Although these areas can adequately work as refuges for a threatened non-game species like the little bustard, they might not be large enough to fulfil all the requirements of birds throughout this preparatory migration period (Casas et al. 2009).

Besides these behavioural alterations, Tarjuelo et al. (2015) proved the existence of a stress response to increased human presence during weekends in a little bustard wintering area in the Spanish Southern Plateau. Not only does hunting take place mainly on weekends in this study area, but also other leisure activities then occur, such as walking with dogs, biking, and different agricultural activities. Tarjuelo et al. (2015) quantified corticosterone metabolites in little bustard faeces collected before, during and after weekends as a measure of physiological stress response to disturbance. The short-term release of glucocorticoids allows the reallocation of energy to improve survival possibilities (Sapolsky et al. 2000), although their continued secretion due to prolonged exposure to stressors reduces reproductive capacity and immunological competence (Romero et al. 2009). Apart from the behavioural changes observed in other studies (i.e. Casas et al. 2009), Tarjuelo et al. (2015) found a significantly higher physiological stress response during weekends, with increased levels of faecal corticosterone metabolites associated with an increase in human activities related to hunting. Therefore, the effect of prolonged hunting disturbance (from October to February plus the weeks of 'media veda' or half-season in countries like Spain, from mid-August to mid-September) on little bustards at wintering sites might adversely affect these populations (Tarjuelo et al. 2015), which still must face their return to breeding areas and the subsequent effort of reproduction.

Beyond behavioural and physiological stress responses, direct mortality caused by illegal hunting or poaching has been proved to be a major cause of human-induced mortality in the little bustard. Based on 139 radio, GPS or satellite-tracked little bustards, Marcelino et al. (2017) found that 32% of all known mortality records were due to poaching. Although those authors could not analyse mortality rates seasonally, it is reasonable to assume that most of that mortality occurs in the non-breeding season (summer, autumn, and winter), during hunting periods.

The impact of hunting during wintering and migration (legal and illegal), although not quantified, also seems to be very important in the eastern sub-range of the species. Iñigo and Barov (2010) list poaching on wintering grounds among the main global threats for the little bustard, with an estimated 30,000 migrant little bustards illegally hunted during migration, mainly in Azerbaijan. The killing of migrant birds has also been recorded in Lebanon in recent years (Ramadan-Jaradi et al. 2017), indicating high winter hunting pressure also in the Middle East (Collar et al. 2017). Such high hunting pressure is also reported for the above-mentioned wintering sites in Central Asia, where little bustards have traditionally been hunted by local people using various means of capture (E.A. Kreuzberg, A. Ten, V. Soldatov and R.S. Muratov, pers. comm.). In Iran, hunting pressure seems to have shaped the species' winter distribution since flocks are mainly found within borderland areas surveyed by the military and thus free of hunter presence (Yousefi et al. 2017). More recently, increasing use of falconry for hunting little bustards in Azerbaijan and other wintering areas has been detected (J. Burnside pers. comm.).

The impact of this pressure on population trends has not been assessed, but it may help explain the mentioned decrease in wintering numbers in some wintering sites from Central Asia.

Landscape Transformations

Little bustard wintering areas in the species' western sub-range have experienced significant land-use changes in the last two decades. In dry farmland areas, practices like early stubble ploughing or old fallow elimination clearly reduce their suitability as non-breeding habitat for little bustards (Silva et al. 2004; García de la Morena 2015). Nevertheless, the strongest transformations at the landscape level generally occur on irrigated farmland, where existing suitable crops may be totally replaced by alternative cultures depending on market-regulated demands. For example the pivot-irrigated alfalfa fields of the central Tagus valley in Spain, a traditional little bustard non-breeding site (from summer to late winter; García de la Morena 2015), have been largely replaced by other crops, such as irrigated olive groves, that are unsuitable as habitat for the species (Morales et al. 2015). In addition, sprinklers, which are clearly avoided by birds (pers. obs.), have replaced pivots in most remaining alfalfa fields. As a result, the number of birds wintering in the area has decreased from around 1000 in the late 1990s (García de la Morena et al. 2007) to ca. 50 in 2014 (own unpubl. data). Similar changes could potentially take place in other irrigated wintering sites, like those of the Ebro valley in Catalonia, where nearly the entire Catalonian population concentrates after breeding (Mañosa et al. 2015). It is, therefore, important that such non-breeding sites are shielded from the expansion of crops that are completely unsuitable for the little bustard such as fruit-tree orchards, nowadays increasing their cover in the Ebro valley.

It is also important to recall that irrigated landscapes involve high human influence and are prone to anthropogenic disturbances like those described above. In addition, irrigation facilities require high densities of electric powerlines, which increase the risk of bird collision (Silva et al. 2010), another significant cause of mortality in the little bustard (Marcelino et al. 2017), which is also a threat during migration (Voronova et al. 2012 in Collar et al. 2017). In any case, a significant proportion of the little bustard population in Western Europe summers and winters in irrigated farmland, where birds can spend up to 8 months of their year, so appropriate management of these areas is required, even if they are excluded from SPAs and other protection measures. In this context, the maintenance of preferred feeding habitats (i.e. alfalfa fields) with compatible irrigation systems (pivots, gravity irrigation), anti-poaching vigilance to keep disturbance to a minimum, creation of hunting reserves in areas used by bustards, and burying dangerous power lines represent desirable measures for these areas. Most of these recommendations also hold for non-breeding sites in dry farmland, where some of them are already included in current EU agri-environmental schemes (e.g. delayed stubble ploughing). Given the importance of leguminous plants in the diet of little bustards wintering in Iberia (Bravo et al. 2017), increasing the cover of dry leguminous crops (vetch, chickpea, or rain-fed alfalfa) would notably improve the quality of these landscapes as non-breeding habitat for the species.

Beyond Western Europe, very little is known about the fate of little bustard wintering areas in Azerbaijan, Iran, and other Asian regions (Table 1). However, given the large number of birds known to overwinter in some of them (see above and chapter "The little bustard around the world: distribution, global conservation status, threats and population trends"), as well as the recent agricultural changes (i.e. the return of intensive agriculture) reported in breeding grounds (Kamp et al. 2011), particular attention should be paid to these areas. Little bustard winter habitat in Azerbaijan is mainly grazed steppe, and relevant overgrazing problems favouring desertification have already been detected (Gauger 2007). The habitat occupied by the wintering flocks reported from Central Asia by E.A. Kreuzberg, A. Ten, V. Soldatov and R.S. Muratov (pers. comm.) is basically irrigated farmland (winter crops or cotton and cereal stubbles) on the banks and floodplains of main rivers, such as the Amu Darya in Uzbekistan, or the Javan and Sir Darya in Tajikistan. This habitat use clearly resembles that of western birds, which largely occupy irrigated crops in winter. As regards northern Iran, little bustards wintering there use mainly marsh steppe and rice stubbles (Sehhatisabet et al. 2012), although very little information about the potential threats associated with agricultural management in that region is available.

Acknowledgements Elena A. Kreuzberg, Anna Ten and Valentin Soldatov kindly provided recent and historical records of migrating and wintering little bustards in Uzbekistan and Kazakhstan. Dr. R. Sh. Muratov, from the Institute of Zoology and Parasitology of the Academy of Sciences of the Republic of Tajikistan, also kindly provided records from that country. We are thankful to all of them for this valuable information that otherwise would have remained out of our reach. Recent records provided by A. Ten and V. Soldatov were obtained with the help of the following additional observers: P Lampila and T. Eskelin. José Luis Tellería and Yves Hingrat kindly revised and commented on a previous version of this chapter. Finally, we are in debt with Nigel J. Collar, who thoroughly revised the English.

References

Alerstam T, Hedenström A, Åkesson S (2003) Long-distance migration: evolution and determinants. Oikos 103:247–260

Alonso JA, Martín CA, Morales MB, Alonso JC, Lane SJ (2001) Long distance seasonal movements of male Great Bustards *Otis tarda* in Central Spain. J Field Ornithol 72:504–508

Alonso JC, Morales MB, Alonso JA (2000) Partial migration, lek fidelity and nesting area fidelity in female great bustards Otis tarda. Condor 102:127–136

Bernis F (1966a) La migración en aves. Tratado teórico y práctico. Sociedad Española de Ornitología, Madrid

Bernis F (1966b) Aves migradoras ibéricas. Sociedad Española de Ornitología, Madrid

Berthold P (1993) Bird migration: a general survey. Oxford University Press, Oxford

Berthold P, Helbig AJ, Mohr G, Querner U (1992) Rapid microevolution of migratory behaviour in a wild bird species. Nature 360:668–670

Bishop CM (1997) Heart mass and the maximum cardiac output of birds and mammals: implications for estimating the maximum aerobic power input of flying animals. Philos Trans R Soc Lond B Biol Sci 352:447–456

Blanc R, Guillemain M, Mouronval J-B, Desmonts D, Fritz H (2006) Effects of non-consumptive leisure disturbances to wildlife. Rev Ecol Terre Vie 61:117–133

Boutin JM, Métais M (1995) L'outarde canepetière. Eveil Editeur, Saint-Yrieix

Bravo C, Cuscó F, Morales MB, Mañosa S (2017) Diet composition of a declining steppe bird, the Little bustard (*Tetrax tetrax*), in relation to farming practices. Avian Ecol Conserv 12(1):3. https://doi.org/10.5751/ACE-00938-120103

Casas F, Mougeot F, Viñuela J, Bretagnolle V (2009) Effects of hunting on the behaviour and spatial distribution of farmland birds: importance of hunting-free refuges in agricultural areas. Anim Conserv 12:346–354

Chapman BB, Brönmark C, Nilsson JA, Hansson LA (2011) The ecology and evolution of partial migration. Oikos 120:1764–1775

Collar NJ, Baral HS, Batbayar N, Bhardwaj GS, Brahma N, Burnside RJ, Choudhury AU, Combreau O, Dolman PM, Donald PF, Dutta S, Gadhavi D, Gore K, Goroshko OA, Hong C, Jathar GA, Jha RRS, Jhala YV, Koshkin MA, Lahkar BP, Liu G, Mahood SP, Morales MB, Narwade SS, Natsagdorj T, Nefedov AA, Silva JP, Thakuri JJ, Wang M, Zhang Y, Kessler AE (2017) Averting the extinction of bustards in Asia. Forktail 33:1–26

Combreau O, Riou S, Judas J, Lawrence M, Launay F (2011) Migratory pathways and connectivity in Asian houbara bustards: evidence from 15 years of satellite tracking. PLoS One 6:e20570

Cramp S, Simmons KEL (1980) Handbook of the birds of Europe the Middle East and North Africa: the birds of the Western Paleartic. In: Cramp S, Simmons KEL (eds) Hawks to Bustards. Oxford University Press, London

Cuscó F, Cardador L, Bota G, Morales MB, Mañosa S (2018) Interindividual consistency in habitat selection patterns and spatial range constraints of female little bustards during the non-breeding season. BMC Ecol 18:56

Delgado MP, Morales MB, García de la Morena EL (2010) Seguimiento de la invernada de sisones comunes franceses en España (2005–2009). Informe Final. Unpublished report. Sociedad Española de Ornitología, Madrid

Devoucoux P (2014) Conséquence et impacts prévisibles d'une perte d'habitat majeure sur une espèce menacée aux exigences écologiques complexes: dynamique de la population d'outardes canepetières des Costières de Nîmes et construction de la Ligne à Grand Vitesse CNM. Octubre 2014. PhD thesis, University of Poitiers, Poitiers

Faria N, Silva JP (2010) Habitat selection of the little bustard during the beginning of the agricultural year. Ardeola 57:363–373

García de la Morena EL (2015) Ecología y movimientos migratorios del sisón común (*Tetrax tetrax*) fuera del periodo reproductor. PhD Thesis, Autónoma University of Madrid, Madrid

García de la Morena EL, Morales MB, De Juana E, Suárez F (2007) Surveys of wintering little bustards (*Tetrax tetrax*) in central Spain: distribution and population estimates at regional scale. Bird Conserv Int 17:23–34

García de la Morena EL, Morales MB, Bota G, Silva JP, Ponjoan A, Suárez F, Mañosa S, De Juana E (2015) Migration patterns of Iberian Little Bustards *Tetrax tetrax*. Ardeola 62:95–112

García JT, Mañosa S, Morales MB, Ponjoan A, García de la Morena EL, Bota G, Bretagnolle V, Dávila JA (2011) Genetic consequences of interglacial isolation in a steppe bird. Mol Phylogenet Evol 61:671–676

García-Antón M, Maldonado-Ruiz J, Morla-Juaristi C, Sainz Ollero H (2002) Fitogeografía histórica de la península Ibérica. In: Pineda F, Casado MA, Montalvo J (eds) La Diversidad Bilógica de España. Pearson Educación, Madrid, pp 45–63

Gauger K (2007) Occurrence, ecology and conservation of wintering Little Bustards*Tetrax tetrax* in Azerbaijan. Arch Nat Landschaftsforschun 46:5–22

Gavrilov EI, Gavrilov AE (2005) The birds of Kazakhstan. Tethys Special Bulletin, Almaty

Heiss M (2013) The importance of Besh Barmag bottleneck (Azerbaijan) for Eurasian migrant birds. Acta Ornithol 48:151–164
Hewitt GM (2001) Speciation, hybrid zones and phylogeography—or seeing genes in space and time. Mol Ecol 10:537–549
Iñigo A, Barov B (2010) Action plan for the Little Bustard *Tetrax tetrax* in the European Union. SEO-BirdLife and BirdLife International for the European Commission, Madrid
Kamp J, Urazaliev R, Donald PF, Hölzel N (2011) Post-Soviet agricultural change predicts future declines after recent recovery in Eurasian steppe bird populations. Biol Conserv 144:2607–2614
Kessler AE, Batbayar N, Natsagdorj T, Batsuur' D, Smith AT (2013) Satellite telemetry reveals long-distance migration in the Asian great bustard *Otis tarda* dybowskii. J Avian Biol 44:311–320
Limiñana R, Soutullo A, López-López P, Urios V (2008) Pre-migratory movements of adult Montagu's harriers Circus pygargus. Ardea 96:81–90
Madsen J, Fox AD (1995) Impacts of hunting disturbance on waterbirds: a review. Wildl Biol 1: 193–207
Mañosa S, Bota G, Estrada J, Cuscó F (2015) Una oportunidad para el sisón en Cataluña. Quercus 356:24–35
Marcelino J, Moreira F, Mañosa S, Cuscó F, Morales MB, García de la Morena EL, Bota G, Palmeirim JM, Silva JP (2017) Tracking data of the Little Bustard *Tetrax tetrax* in Iberia shows high anthropogenic mortality. Bird Conserv Int. https://doi.org/10.1017/S095927091700051X
Martín CA, Casas F, Mougeot F, García JT, Viñuela J (2010) Positive interactions between vulnerable species in agrarian pseudo- steppes: hábitat use by pin-tailed sandgrouse depends on its association with the little bustard. Anim Conserv 13:383–389
Martin TE, Guerin R, Fages F, Martineau A, Hingrat Y (2018) Breeding populations of Great Bustard and Little Bustard in South Kazakhstan province, Republic of Kazakhstan. Sandgrouse 40:138–143
Morales MB, Alonso JC, Alonso JA, Martín E (2000) Migration patterns in male great bustards (*Otis tarda*). Auk 117:493–498
Morales MB, Suárez F, García de la Morena EL, De Juana E (2002) Movimientos estacionales y conservación de aves esteparias: el ejemplo del sisón. Quercus 193:493–498
Morales MB, Arroyo B, Traba J (2015) El declive del sisón en el centro de España. Quercus 356: 36–43
Newton I (1998) Population limitations in birds. Academic Press, London
Newton I (2010) Bird migration. Collins New Naturalist Library, London
Otero C (1985) Techniques for the capture of Little Bustards during the Autumn/Winter season. Bustard Studies 2:171–172
Palacín C (2007) Comportamiento Migratorio de la Avutarda Común en la Península Ibérica. PhD Thesis Complutense University of Madrid, Madrid
Palacín C, Alonso JC (2009) Probable population decline of the Little Bustard *Tetrax tetrax* in North-West Africa. Ostrich 80:165–170
Phillips-Wolley C (1881) Sport in the Crimea and Caucasus. Bentley and son, London
Pulido F (2007) The genetics and evolution of avian migration. Bioscience 57:165–174
Pulido F, Berthold P (2010) Current selection for lower migratory activity will drive the evolution of residency in a migratory bird population. Proc Natl Acad Sci U S A 107:7341–7346
Quinn JL, Cresswell W (2005) Escape response delays in wintering redshank, Tringa totanus, flocks: perceptual limits and economic decisions. Anim Behav 69:1285–1292
Ramadan-Jaradi G, Itani F, Serhal A (2017) Interesting bird records from Lebanon. Sandgrouse 39: 187–192
Romero LM, Dickens MJ, Cyr NE (2009) The Reactive Scope Model: a new model integrating homeostasis, allostasis, and stress. Horm Behav 55:375–389
Sainz-Ollero H (2013) Steppes across the world. An overview with emphasis on the Iberian Peninsula. In: Morales MB, Traba J (eds) Steppe ecosystems. Biological diversity, management and restoration. NOVA Publishers, New York, pp 1–25

Sampietro J, Rivas JL, Sanz J, Albero JC, Pelayo E, Gajón A, García de la Morena E (2013) Desplazamientos a larga distancia de machos de sisón común (*Tetrax tetrax*) reproductores en Aragón. Rocín Anuario Ornitol Aragón 8:35–52
Sansom A, Cresswell W, Minderman J, Lind J (2008) Vigilance benefits and competition costs in groups: do individual redshanks gain an overall foraging benefit? Anim Behav 75:1869–1875
Sapolsky RM, Romero LM, Munck AU (2000) How do glucocorticoids influence stress responses? Integrating permissive, suppressive, stimulatory, and preparative actions. Endocr Rev 21:55–89
Sardà-Palomera F, Puigcerver M, Brotons L, Rodríguez-Teijeiro JD (2012) Modelling seasonal changes in the distribution of common quail *Coturnix coturnix* in farmland landscapes using remote sensing. Ibis 154:703–713
Sehhatisabet ME, Abdi F, Ashoori A, Khaleghizadeh A, Khani A, Rabiei K, Shakiba M (2012) Preliminary assessment of distribution and population size of wintering Little Bustards *Tetrax tetrax* in Iran. Bird Conserv Int 22(279):28
Sherry TW, Holmes RT (1996) Winter habitat quality, population limitation and conservation of Neotropical-Neartic migrant birds. Ecology 77:36–48
Shimazaki H, Tamura M, Darman Y, Andronov V, Parilov MP, Nagendran M, Higuchi H (2004) Network analysis of potential migration routes for oriental white storks (*Ciconia boyciana*). Ecol Res 19:683–698
Shlyakhtin GV, Tabachishin VG, Khrustov AV, Zav'yalov EV (2004) Ecological segregation of bustards (Otididae) in the north of the lower Volga region: evolutionary and adaptive aspects. Russ J Ecol 35:247–253
Silva JP, Pinto P (2006) Final report of action 2 of the project LIFE02NAT/P/8476: conservation of the little bustard in Alentejo. Institute for Nature Conservation. Available via http://portal.icnb.pt/NR/rdonlyres/783CA43B-7C0E-4FA4-8286-F4EB120C4A49/0/Life_SISAO.pdf. Accessed 17 Jan 2011
Silva JP, Pinto M, Palmeirim JM (2004) Managing landscapes for the little bustard *Tetrax tetrax*. Lessons from the study of winter habitat selection. Biol Conserv 117:521–528
Silva JP, Faria N, Catry T (2007) Summer habitat selection and abundance of the threatened little bustard in Iberian agricultural landscapes. Biol Conserv 139:186–194
Silva JP, Santos M, Queirós L, Leitão D, Moreira F, Pinto M, Leqoc M, Cabral JA (2010) Estimating the influence of overhead transmission power lines and landscape context on the density of little bustard *Tetrax tetrax* breeding populations. Ecol Model 221:1954–1963
Sinclair ARE (1983) The function of distance movements in vertebrates. In: Swingland IR, Greenwood PR (eds) The ecology of animal movement. Clarendon Press, Oxford, pp 240–258
Snow DW, Perrins CM (1998) The birds of the Western Palearctic (concise edition) vol 1. Non-Passerines. Oxford University Press, Oxford
Sridhar H, Beauchamp G, Shanker K (2009) Why do birds participate in mixed-species foraging flocks? A large-scale synthesis. Anim Behav 78:337–347
Suárez F (2004) Aves y agricultura en España peninsular. Una revisión sobre el estado actual de conocimiento y una previsión de futuro. In: Tellería JL (ed) La ornitología hoy. Homenaje al profesor Francisco Bernis Madrazo. Editorial Complutense, Madrid, pp 223–265
Suárez F, Sainz H, Santos T, González Bernáldez F (1992) Las estepas ibéricas. MOPT, Madrid
Suárez-Seoane S, García de la Morena EL, Morales MB, Osborne PE, De Juana E (2008) Maximum entropy niche-based modelling of seasonal changes in little bustard (*Tetrax tetrax*) distribution. Ecol Model 219:17–29
Sümegi P, Szilágyi G, Gulyás S, Jakab G, Molnár A (2013) The late quaternary paleoecology and environmental history of the Hortobágy, an unique mosaic alkaline steppe from the heart of the Carpathian basin, Central Europe. In: Morales MB, Traba J (eds) Steppe ecosystems. Biological diversity, management and restoration. NOVA Publishers, New York, pp 165–193
Tarbelet P, Fumagalli L, Wust-Saucy AG, Cosson JF (1998) Comparative phylogeography and postglacial colonization routes in Europe. Mol Ecol 7:453–464

Tarjuelo R, Delgado MP, Bota G, Morales MB, Traba J, Ponjoan A, Hervás I, Mañosa S (2013) Not only habitat but also sex: factors affecting spatial distribution of Little Bustard *Tetrax tetrax* families. Acta Ornithol 48:119–128

Tarjuelo R, Barja I, Morales MB, Traba J, Benítez-López A, Casas F, Arroyo B, Delgado MP, Mougeot F (2015) Effects of human activity on physiological and behavioral responses of an endangered steppe bird. Behav Ecol 26:828–838

Terborgh J (1990) Mixed flocks and polyspecific associations: costs and benefits of mixed groups to birds and monkeys. Am J Primatol 21:87–100

Terraube J, Mougeot F, Cornulier T, Verma A, Gavrilov A, Arroyo B (2012) Broad wintering range and intercontinental migratory divide within a core population of the near-threatened pallid harrier. Divers Distrib 18:401–409

Terrill SB, Able KP (1988) Bird migration terminology. Auk 105:205–206

Vickery JA, Ewing SR, Smith KW, Pain DJ, Bairlein F, Skorpilov J, Gregory RD (2014) The decline of Afro-Palaearctic migrants and an assessment of potential causes. Ibis 156:1–22

Villers A, Millon A, Jiguet F, Lett JM, Attie C, Morales MB, Bretagnolle V (2010) Migration of wild and captive-bred Little Bustards *Tetrax tetrax*: releasing birds from Spain threatens attempts to conserve declining French populations. Ibis 152:254–261

Wolff A (2001) Changements agricoles et conservation de la grand avifaune de plaine: Etude des rélations espèce-habitats à differentes échelles chez l'Outarde canepetière. Dissertation. Université Montpellier II. Montpellier

Yousefi M, Kafash A, Malakoutikhah S, Ashoori A, Khani A, Mehdizade Y, Ataei F, Ilanloo SS, Rezaie HR, Silva JP (2017) Distance to international border shapes the distribution pattern of the growing Little Bustard *Tetrax tetrax* winter population in northern Iran. Bird Conserv Int. https://doi.org/10.1017/S0959270917000181

Behavioural Ecology of the Little Bustard: Sexual Selection and Mating Systems

Vincent Bretagnolle, Santi Mañosa, François Mougeot, and Manuel B. Morales

Introduction

Some aspects of little bustard behaviour are already dealt with in chapter "Natural History of the Little Bustard: Morphology, Biometry, Diet, Sexual Dimorphism, and Social and Breeding Behaviour", such as those mainly relating to general behaviour (display, calls) and social behaviour. In this chapter, we focus on sexual behaviour, and more specifically on the lekking and mating behaviours, which are so remarkable in bustards in general and in the little bustard in particular (Morales et al. 2001). We thus cover mainly the behavioural ecology of the species (i.e. how behaviour relates to ecology and evolution). We particularly emphasize the species' lekking behaviour and its variation and the evolutionary forces that drive this particular mating strategy and

V. Bretagnolle (✉)
Centre d'Etudes Biologiques de Chizé, CNRS & La Rochelle Université, Beauvoir sur Niort, France
e-mail: vincent.bretagnolle@cebc.cnrs.fr

S. Mañosa
Facultat de Biologia, Departament de Biologia Evolutiva, Ecologia i Ciències Ambientals, Institut de Recerca de la Biodiversitat (IRBio), Universitat de Barcelona, Barcelona, Catalonia, Spain
e-mail: smanosa@ub.edu

F. Mougeot
Institute for Game and Wildlife Research (IREC, CSIC-UCLM-JCCM), Ciudad Real, Spain
e-mail: Francois.Mougeot@uclm.es

M. B. Morales
Terrestrial Ecology Group, Departamento de Ecología, Universidad Autónoma de Madrid, Madrid, Spain

Centro de Investigación en Biodiversidad y Cambio Global (CIBC-UAM), Universidad Autónoma de Madrid, Madrid, Spain
e-mail: manuel.morales@uam.es

V. Bretagnolle et al. (eds.), *Little Bustard: Ecology and Conservation*, Wildlife Research Monographs 5, https://doi.org/10.1007/978-3-030-84902-3_8

analyse it within a sexual selection theory framework. At the end of the chapter, we also explore other, less well-studied aspects of little bustard behavioural ecology.

Sexual selection is a mode of natural selection in which members of one sex choose mates of the other sex to mate with (intersexual selection), while they compete with members of the same sex for access to the opposite sex (intrasexual selection). These two forms of selection mean that some individuals have better reproductive success than others within a population. Therefore sexual selection results either from unequal access of one sex (generally males) to individuals of the opposite sex through competition or from one sex generating selection on the opposite sex through mate choice, with females being the choosy sex in polygynous species (Andersson 1994). Females may mate randomly (i.e. no choice) or select amongst many mates based on a comparative assessment using one or several relative preference criteria (Janetos 1980). A better understanding of mating strategies depends on the identification of characters associated with male and female mating success. Traits involved in female choice are expected to be more variable among than within males, allowing reliable cues of male identity and quality to be available for prospecting females (Gibson and Bradbury 1985). Such traits include plumage coloration and extravagance in ornaments or display performance and are usually condition-dependent, i.e. males with better body condition develop traits preferred by females (Carranza and de Trucios 1993; Andersson 1994; Sardà-Palomera et al. 2011; Morales et al. 2003). In the little bustard, mating success is probably highly skewed among males (see below, and Jiguet and Bretagnolle 2006), and thus females likely exert strong selection pressure on male traits.

Leks

What Are Leks?

Leks are clusters of displaying males that females visit solely for the purpose of mating (Höglund et al. 1995). In birds characterized by a lekking mating system, males provide females with no other resources than their gametes. Male reproductive success (number of offspring sired) shows the high variance in lekking species (Fiske et al. 1998). In typical (also known as classical) leks, females have the opportunity to assess simultaneously multiple potential mates and are supposed to choose freely among them (Bradbury and Gibson 1983). Lekking species are generally sexually dimorphic as a result of intense sexual selection (Payne 1984; Höglund 1989; Oakes 1992). Several studies have identified male traits that correlate with male mating success in lekking species, such as displays and calls (Gibson 1996), plumage (Gibson and Bradbury 1985; Petrie et al. 1991) or even territorial behaviour (Höglund and Robertson 1990). In lek mating systems, females show a strong discriminating capacity to select mates, sometimes reaching high levels of consensus (e.g. Payne 1984; Gibson and Bradbury 1985), unlike resource-based mating systems, in which the benefits for females of mating with a male defending a good territory are more important than mate choice (Thornhill 1976).

In birds, lekking is rare and has been described only in a handful of avian families (Höglund and Alatalo 1995; Jiguet et al. 2000), including grouse (e.g. black grouse *Tetrao tetrix*), scolopacids (ruff *Philomachus pugnax*, great snipe *Gallinago media*), birds of paradise, cotingas, and manakins (see review in Höglund and Alatalo 1995). Bradbury (1981) suggested that lekking species should fulfil four criteria: (1) no male parental care (males only contribute their gametes); (2) males aggregate at specific sites for display (the lek arenas); (3) the only resource females find at a lek are the males themselves; and (4) females can select their mate(s), although this latter criterion has been debated (see Höglund and Alatalo 1995). Some bird species present lek-like mating systems that do not completely fulfil Bradbury's conditions, although they are still considered lekking. One of the main categories of such non-classical leks is the so-called "exploded" or "dispersed" lek (Gilliard 1969; Emlen and Oring 1977; Höglund and Alatalo 1995). In exploded leks, males are separated by larger distances than in classical ones. Aggregation still exists, although it might not be detectable until males are carefully mapped over a large area (Bradbury 1981), and spatial statistics is carried out in order to detect non-uniform and aggregated distribution. Thus, in exploded leks, males can hold large territories within which females may forage or even nest (Ligon 1999), therefore failing to fulfil Bradbury's third criterion. As reviewed in chapter "Natural History of the Little Bustard: Morphology, Biometry, Diet, Sexual Dimorphism, and Social and Breeding Behaviour", bustards (Otididae) are one of the avian families in which exploded leks seem to be most widespread (Jiguet et al. 2000; Morales et al. 2001; Hingrat et al. 2008).

When resources critical for breeding are patchily distributed so that males aggregate on patches that females need to use to breed, the mating system becomes what has been called a resource-based lek (Alexander 1975). Finally, another interesting form of male breeding aggregation is represented by the so-called "hidden leks," in which territories of socially monogamous males are clumped in order to favour the pursuit of extra-pair copulations (regardless of territory quality) by their owners (Wagner 1998), as described in several passerines (Tarof et al. 2005). Many behavioural and evolutionary predictions of classical leks also hold for hidden leks.

Evolutionary Forces Shaping Leks

Two aspects of lek evolution deserve attention: the lek paradox and the aggregation (clustering) of males. Both led to several theories and arguments that we will briefly summarize here. The lek paradox (Borgia 1979) arises from the intensity of female mate selection on males' secondary sexual characters and the choice for "good genes:" if females consistently choose males with the most exaggerated secondary sexual trait as a cue for the genes of the highest quality, the unanimous female choice for these males would lead to a rapid loss of variation in male genetic quality and, subsequently, females would gain nothing from their choice (Kirkpatrick and Ryan 1991). Pomiankowski and Møller (1995) proposed a solution to this paradox:

genetic variability of secondary sexual traits is higher than that of non-sexual ones. That would explain the absence of fixation of advantageous male traits despite the extreme consensus of female choice found in leks. Their conclusion is, therefore, that no lek paradox exists. A proposed mechanism to maintain genetic variability that has received support on both empirical and theoretical grounds was proposed by Hamilton and Zuk (1982): coevolutionary cycles with short-lived parasites or pathogens continuously select for new variation for resistance and thus for fitness. Since parasites continuously modify their infective method, there is always selection pressure to show a healthy condition (Boyce 1990; Balenger and Zuk 2014). However, the issue is still open, and other solutions include a female preference for outbred sires (Neff and Pitcher 2008), condition-dependent traits, or that individual body condition shows high genetic variance, therefore maintaining secondary sexual trait variability (Miller and Moore 2007; see Chargé et al. 2010 for a study in the houbara bustard *Chlamydotis undulata*).

Male clustering in leks has been explained by four basic theoretical models: hotspot, female preference, hotshot, and black hole (Höglund and Alatalo 1995; Ligon 1999). According to the hotspot model (Bradbury and Gibson 1983), males cluster in areas where female density is highest. In the female preference model (Bradbury and Gibson 1983; Beehler and Foster 1988), females prefer to mate with males in larger leks in which the costs associated with searching or assessing potential mates are reduced. The hotshot model proposes that females prefer to mate with a high-quality male (hotshots) and that suboptimal males aggregate around such males to parasitize their attractiveness (Beehler and Foster 1988); hotshots are therefore attractive to females, but also to males. Finally, the black hole model proposes that females avoid sexual harassment by certain males and leave their territories (Clutton-Brock et al. 1992), and leks arise because the probability that males retain females is higher if males are clustered than if they are solitary (Stillman et al. 1993). Few studies have tested, empirically or experimentally, these different evolutionary models, but studies conducted on little bustards have provided novel insights into the processes shaping male clustering in leks (Jiguet et al. 2000; Jiguet and Bretagnolle 2006, 2014; Morales et al. 2014). Exploded leks, in which males defend territories with resources, further raise the question of whether territory quality per se (size, food resources) might also be selected by females (Jiguet et al. 2002; Traba et al. 2008).

The Little Bustard Mating System

Little bustards are characterized by an exploded lek mating system (Schulz 1986; Jiguet et al. 2000, 2002; Morales et al. 2001; Lett et al. 2000). Early reports described the species as socially monogamous, i.e. forming pairs tied by permanent bonds lasting until the chick-rearing period (Dementiev and Gladkov 1966), with males contributing some sort of parental care (probably general protection of the brood). However, nest attendance and chick-rearing are conducted only by the females:

males try to attract and mate with as many females as they can without providing any sort of parental care (Schulz 1986). They establish their dispersed territories at the beginning of the breeding season, and they defend these throughout the mating period (Schulz 1986; Jiguet et al. 2000). However, varying degrees of within-season territorial turnover and satellite behaviour have also been described (Jiguet and Bretagnolle 2001; Ponjoan et al. 2012, see below). The size of these territories can be quite variable, ranging 0.7–66.6 ha (mean $\pm$ sd $=$ 18.7 $\pm$ 16.2 ha) in western France, although the most frequent values were around 5 ha (Jiguet et al. 2000). In Catalonia (north-eastern Spain), Ponjoan et al. (2012) report core breeding male ranges of 17 $\pm$ 17 ha (mean $\pm$ sd), with the most frequent values below 20 ha. Delgado et al. (2010) calculated display areas of 1.73 $\pm$ 2.18 ha and 1.49 $\pm$ 1.12 ha in two consecutive years, although these figures cannot be compared with the previous two due to methodological differences. In any case, these reports indicate mating territories large enough to encompass food and other resources valuable for females. Nevertheless, all the criteria that Bradbury (1981) used to define a lek mating system seem to be fulfilled in the case of little bustards, except for the third criterion ("the only resource females find at a lek are the males themselves"). Consequently, the critical point in such non-classical leks is not the level of male aggregation or the presence of resources per se, but whether or not resources play a determinant role in shaping the male distribution and female mate choice, i.e. the degree to which males regulate female access to those resources in order to obtain copulations (Bradbury 1985). If some kind of resources were key to male mating success, a resource-based polygyny system would be a better description of the species' mating system.

Jiguet et al. (2000) tested whether male little bustards in western France defend and/or provide resources in their territories and whether these influence female choice. They determined the degree of male clustering and tested the independence of male spatial distribution with respect to that of suitable display habitat as a way to discard the possibility of landmark aggregations. Jiguet et al. (2000) found that resources were mainly selected according to males' requirements (food and display traits) rather than females' requirements (food and shelter for nests, provided by permanent crops). Furthermore, females did not apparently use the amount of resources in the defended territory as a criterion for mate choice, as male attractiveness and resources within the territory were unrelated. Consequently, they concluded that lekking behaviour was truly a characteristic of the species.

Male Spacing Behaviour: Why Males Aggregate in Leks

According to Jiguet et al. (2000), displaying male little bustards attract each other, and larger leks might be more attractive to females than smaller ones, as predicted by the female preference model of lek evolution (Bradbury et al. 1986). However, Jiguet et al. (2002) suggested that aggregations occur as males concentrate in areas where females are more likely to be encountered, for example due to the presence of resources (*hotspot model*). Later on, Jiguet and Bretagnolle (2006) used an experimental approach and little bustard decoys (see Plate 1) and showed that males were not attracted to female decoys. The results of this experiment supported the *hotshot* model of lek formation: male aggregations appear around a hotshot (preferred) male, with less attractive males concentrating around the hotshot in order to obtain matings. Indeed, the experiments showed that both males and females were attracted to decoys featuring particularly attractive male traits (Jiguet and Bretagnolle 2006). The *female preference* hypothesis, which states that females prefer clustered males, has also received mixed support. The local density of females has been found to increase with the density of displaying males (Wolff et al. 2002; Devoucoux et al. 2018), and decoy experiments conducted by Jiguet and Bretagnolle (2006) showed that females were more attracted to a particular lek size of four male decoys. Jiguet and Bretagnolle (2014) later showed that individual male attractiveness did not increase with (natural) lek size. The *black hole* model states that, in a situation where females exhibit no male choice and are highly mobile, for example as a result of male aggressiveness and harassment, the optimal male spacing involves reduced nearest-neighbour distances (Höglund and Alatalo 1995). This model was only weakly supported by the experiments carried out on little bustards: although male harassment clearly occurs in this species, females do exhibit some mate preferences (Jiguet and Bretagnolle 2006). In fact, each of the proposed mechanisms contributes to explaining the lekking behaviour of the little bustard, but at different spatial scales. At larger scales, female mobility may induce male aggregation, as suggested by the black hole hypothesis. At smaller scales, female preference for particularly attractive males or lek sizes may induce male clustering, which would further be maintained and structured as a result of the hotshot mechanism.

Plate 1 Two examples of male decoys attracting real Little Bustard males, the second one even showing a male displaying next to the decoy (lower photo)

During displays, male little bustards are aggregated (see Jiguet et al. 2000; Ponjoan et al. 2012), but they are also highly territorial and aggressive to each other, each one defending a more or less restricted area within the lek from the intrusion of other males (i.e. a territory). The intensity of the agonistic response to experimentally placed male decoys has been shown to be density-dependent (Morales et al. 2014), which clearly indicates that males are defending some kind of resource from other males. However, it is not clear if this behaviour is related to the defence of the feeding resources or habitat themselves or if this is to prevent the access of other males to visiting females. Resources clearly play a role in the little bustard mating system, although, in principle, they are not used as a criterion in female choice as expected in exploded leks (Höglund and Alatalo 1995; Ligon 1999). Little bustard males prefer food-rich territories, perhaps first as a way to afford the costs of the mating activity (Jiguet et al. 2000, 2002; Traba et al. 2008; Faria et al. 2012). Food-rich male territories may also allow females to forage within these territories, at least while visiting leks (Jiguet et al. 2000), or perhaps even to nest (Morales et al. 2013), although there is no evidence that females use territory resources as a criterion in mate choice or that resources are controlled by males to improve their mating success. Moreover, Jiguet et al. (2000, 2002) showed that this was not the case in populations of central-western France.

The Flexible Exploded Lek of the Little Bustard

In species with exploded leks, males defend territories and are only loosely aggregated (Gilliard 1969; Höglund and Alatalo 1995). In some species, males can hold large territories containing resources potentially exploitable by breeding females (Höglund and Alatalo 1995; Ligon 1999), although female use of male territory resources is unclear in the little bustard (Jiguet et al. 2000; Morales et al. 2001). Recent evidence suggests that the lekking behaviour of little bustards varies geographically, depending on environmental conditions (landscape characteristics) and endogenous factors (little bustard density). Where habitat suitability is strongly impaired due to agricultural intensification, exploded leks can shift to resource-based mating systems as a result of female concentration in high-quality patches and subsequently of displaying male concentration within the few remaining patches of suitable habitat. This occurred in a rapidly declining population of little bustards in Deux-Sèvres, western France, where the existence of resource-based leks was reported (Jiguet and Bretagnolle 2006, 2014). Following a further decline in bird density, the mating system could even shift from resource-based leks to resource-based polygyny, in which females are attracted to the best territories rather than to the males displaying the best attributes. Delgado et al. (2010) and Morales et al. (2014) suggested that changes from exploded leks to so-called resource-based leks or polygyny could also be occurring in the increasingly intensified Spanish cereal steppes as well, although the results of Ponjoan et al. (2012) in Lleida (north-eastern Spain) do not support this possibility. However, in more or less intensified farmland areas, which are nowadays the most frequent landscapes where little bustards occur in the species' western range, favourable habitats for males and/or females are progressively clumped and limiting. If males concentrate in the few suitable habitat patches for display but distribute uniformly within those patches, this could lead to "landmark aggregations" of males (a situation that has not yet been described in the species). If, on the contrary, females concentrate in a few good patches for breeding or feeding, and males consequently concentrate in such places (where males inevitably will compete—and may also fight—for accessing and controlling the females), then female-defence polygyny may be favoured. Both are superficially similar to classical leks, in terms of their degree of male aggregation, but result from strong habitat concentration and subsequent male and female clumping.

Decoy experiments conducted with little bustards have revealed that male territory defence intensity is density-dependent in exploded leks, pointing to flexibility in territorial behaviour. Morales et al. (2014) have shown that, after a simulated territorial intrusion (decoy presentation), males responded more intensively and took less time to return to their display sites when male and female abundances were higher. Such flexibility in territorial behaviour may be considered as evidence supporting density-dependent models of lek formation, such as the female preference and hotshot models (Bradbury 1981; Beehler and Foster 1988; Höglund and Alatalo 1995).

Overall, current levels of habitat fragmentation and degradation may constrain little bustard mating systems, such that exploded leks, resource-based leks and even solitary polygyny successively occur when population density declines and/or resources diminish (Jiguet et al. 2000; Morales et al. 2001; Delgado et al. 2010). In very intensive agricultural habitats with very low bustard density, the mating system could even shift to resource-based monogamy, although without male parental care (Jiguet et al. 2000; Morales et al. 2001). Classical monogamy involving male parental care has never been described in the little bustard (but see Dementiev and Gladkov 1966). Conversely, in populations with very high density (e.g. southern France: Devoucoux et al. 2018), classical lek systems may occur, in which males do not defend territories and are mobile (e.g. change their display sites from day to day). Such variability has been observed in other lekking species, like great bustard *Otis tarda* (Carranza et al. 1989) or fallow deer (Clutton-Brock and Parker 1992). In Fig. 1, we propose a general framework linking little bustard mating system flexibility to variations in habitat quality as well as bird density, which co-vary to some extent. This framework is based on both correlational and experimental evidence (Jiguet et al. 2000; Jiguet and Bretagnolle 2006, 2014; Morales et al. 2001, 2014).

Relatively little is known regarding lek formation in little bustards. A recent study conducted in Portugal used high-resolution GPS tracking to study male movements at lekking sites (Silva et al. 2017). Within a breeding season, only one-third of little bustard males remained constantly territorial, while others exhibited a floating behaviour, similar to that described by Ponjoan et al. (2012), or shifted between being a floater and territorial. From 1 year to the next, males showed territorial site fidelity, and floating males reduced their movement to establish a territory. Both processes may explain the temporal persistence of leks within stable habitats (Silva et al. 2017).

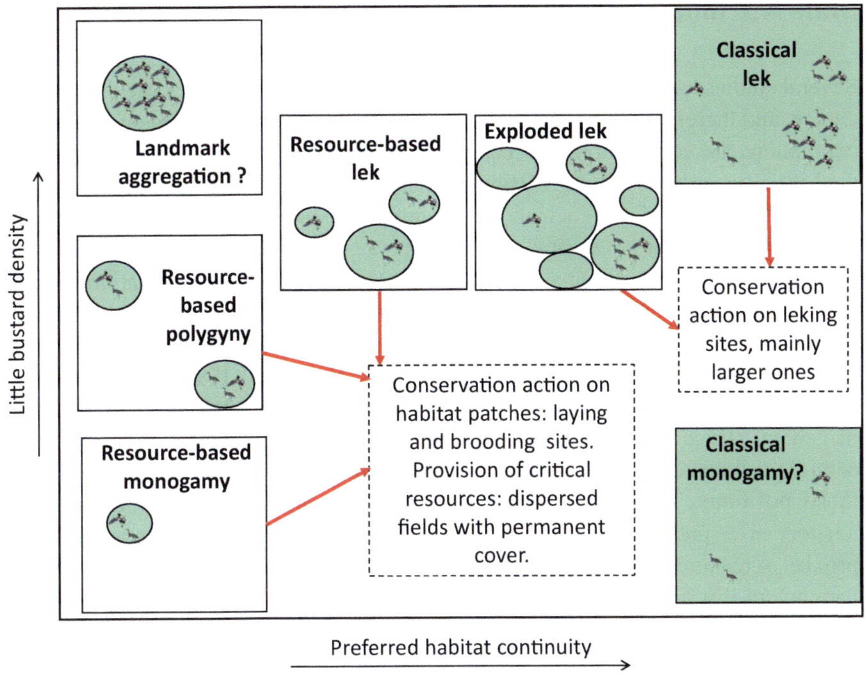

Fig. 1 A theoretical framework describing little bustard mating system flexibility in relation to density and degree of continuity of preferred habitat. Question marks indicate mating strategies theoretically possible within the framework but not actually demonstrated in the little bustard. Red arrows connect each mating strategy with its corresponding most suitable conservation approach (dashed rectangles). The mating strategies represented are described in the text. Here we mention their main traits to aid figure interpretation: (1) Classical lek: males aggregate closely to attract females regardless of preferred habitat distribution (here assumed continuous); high variation in female attraction among males. (2) Exploded lek: males aggregate loosely defending larger territories in fragmented preferred habitat. Females have the opportunity to use resources in habitat patches but do not use them as mate choice criterion. Considerable variation still exists in female attraction success among males. (3) Resource-based lek: males aggregate loosely defending larger territories in further fragmented preferred habitat, which becomes scarce. Females use resources in habitat patches as mate choice criterion. Variation in female attraction success among males decreases. (4) Resource-based polygyny: Males disperse (non-aggregated) over further fragmented and scarce preferred habitat. Females use resources in habitat patches as mate choice criterion, breeding in habitat patches. Some variation in female attraction success among males exists. (5) Resource-based monogamy: Solitary males occupy the few remaining preferred habitat patches. Females are also so scarce that they mate with the few solitary males but with no stable male–female bond and no male parental investment. No variation in female attraction success among males exists. (6) Classical monogamy: Females form a stable bond with solitary males, but irrespective of preferred habitat distribution (here assumed continuous). Some degree of male parental investment is expected. (7) Landmark aggregations: males congregate on remaining preferred habitat patches, although not necessarily in a spatially aggregated manner within the patch, which is also visited by females for mating

Male Attributes and Attractiveness

Several studies have attempted to identify which male traits females use during mate choice, and therefore which male attributes could predict their mating success. Since copulations are very seldom observed in little bustards (chapter "Natural History of the Little Bustard: Morphology, Biometry, Diet, Sexual Dimorphism, and Social and Breeding Behaviour"), Jiguet and Bretagnolle (2014) compared males that were visited by females or not during observations (an attractive male was defined as one that attracted at least one female). Lek attendance (measured as the duration of stay), wing flash rate (but not snort call rate), and body condition were significantly greater for attractive than non-attractive males. Lek attendance is regarded as the main factor involved in mating success on leks (Fiske et al. 1998; Hill 1991; Rintamäki et al. 1995), and early breeding is usually associated with higher breeding success in birds in general (Price et al. 1988). Males arriving earlier may also be able to choose central positions on the lek, although it is not known whether they actually fight for those positions. Invertebrate availability in grasslands and in sunflower fields, territory size, proportions of grass and sunflower cover were also investigated as proxies of resources defended by males (i.e. male territory quality), but none of these parameters discriminated between attractive and non-attractive males (Jiguet and Bretagnolle 2014). Moreover, and perhaps more surprisingly, no studied plumage traits (except for the symmetry of the neck collar, see below) differed significantly between attractive and non-attractive males, nor did any of the acoustic call parameters analysed. Other parameters that varied to a lesser extent with male attractiveness were the symmetry of the neck collar pattern and the performance of the jump display (for a subset of males only). The black and white neck collar pattern is displayed by males during courtship when males erect their neck feathers laterally so that the "V-shape" pattern is made very obvious to a female standing in front of a male (see chapter "Natural history of the little bustard: morphology, biometry, diet, sexual dimorphism, and social and breeding behaviour"). The symmetry of the neck collar may therefore be easy to assess for a female. Fluctuating asymmetry has been suggested as a female choice criterion in the peacock *Pavo cristatus* (Swaddle and Cuthill 1994), in which the symmetry of a male's train increases with the number of eye-spots (Manning and Hartley 1991) and influences male mating success (Petrie and Halliday 1994). In a sample of 13 little bustard males (central Spain), males with a greater neck collar symmetry also showed a higher frequency of jump displays (M. B. Morales, unpubl. data). Using a sample of males for which all phenotypic traits were recorded simultaneously, Jiguet and Bretagnolle (2014) also found evidence for correlations among traits involved in male little bustard attractiveness.

Females in leks are expected to choose among males using several uncorrelated traits (Andersson 1994), as this should decrease the probability of choosing mates based on a dishonest signal (Pomiankowski and Møller 1993; Johnstone 1995). Given this dependence on multiple criteria, Jiguet and Bretagnolle (2014) proposed a scenario in which male snort call rate may be used during the earliest stages (on arrival from migration) for intrasexual competition and female attraction. Snort call may primarily be used in male–male contests (Jiguet and Bretagnolle 2001), and males with lower snort call rates may attract more females because they have been able to defend their territory without fighting, and those territories may appear as safer places (Alonso et al. 2012). Once the female visits the male territory, other male traits could be used by the female to assess its quality, especially the symmetry in ornaments and the wing flash or jump display rates, the latter two displays occurring especially in the presence of a female (Jiguet and Bretagnolle 2001). The proportion of males displaying symmetric collar patterns was found to increase with lek size in the population studied by Jiguet and Bretagnolle (2014) in central-western France, being highest in leks of size four. It is noteworthy that in experimental work with decoys in that particular study area, females preferred a lek size of four males.

In other bustard species, males use calls for intrasexual competition and to attract females. For instance, male houbara bustards perform low-frequency booming calls that not only attract females from very long distances but also convey information about the quality of the caller, informing other males about its competitive ability (Cornec et al. 2014, 2015, 2017). Great bustard males position themselves relative to the sun during displays so that their white plumage is maximally detectable by distant females (Olea et al. 2010).

Female Mating Behaviour

Leks may evolve without female choice, e.g. if the female has no other choice than mating with the dominant male (Clutton-Brock et al. 1992), such as in the black hole model. Alternatively, male territory quality may play some role: a radio-tagging study in Lleida, north-eastern Spain (Ponjoan et al. 2012) showed that older (and probably higher status) males defended smaller and higher quality territories (in terms of habitat composition) than younger males, which spent much less time on the lek and were classified as satellites (see also Moseykin 1992). This is in agreement with the results of Traba et al. (2008), who found that displaying males settled in territories with a higher abundance of coleopterans than expected from a random distribution (see also Faria et al. 2012). Such evidence suggests that habitat quality and resources within male territories are variable and influence male settlement but says nothing about their role in female choice. Therefore, whether females use resource information remains an open question, although females tend to nest close to or within male territories in France (Lett et al. 2000; Jiguet et al. 2002) and central Spain (Morales et al. 2013). Little bustard broods also tend to be found closer to male territories than expected by chance in central (Ciudad Real) and north-eastern (Lleida) Spain (Tarjuelo et al. 2013). Such associations may, however, simply be the result of a common habitat selection pattern by males and females at the landscape level. In Ciudad Real, these associations occur even within large patches of optimal homogeneous habitat, so females may obtain some benefits from nesting next to male territories. These observations support a resource-based scenario, even if there is no evidence yet for female mate choice to be influenced by the amount of resources contained within male territories (Jiguet et al. 2000, 2002).

The question of whether or not the female choice is active is of critical importance, although hard to demonstrate. In most lekking species, the female choice seems to be based on an array of criteria (Gibson et al. 1991; Höglund and Alatalo 1995), but experimental evidence for active female choice exists only in a few species (Clutton-Brock et al. 1989; Gibson 1989; Höglund et al. 1990; Petrie and Halliday 1994; Sæther et al. 2000). Experimental results using little bustard decoys with different neck coloration patterns established that female attraction depends on the symmetric pattern of plumage ornaments (Jiguet and Bretagnolle 2006), a result confirmed with observations of wild birds (Jiguet and Bretagnolle 2014). Observations also indicate that females seem to prefer males with higher wing flash display rates (Jiguet and Bretagnolle 2014). There is theoretical support for the prediction that less fit males display less symmetrical ornaments (Møller 1990) and evidence consistent with this prediction has been found in the closely related houbara bustard (Chargé et al. 2010). In the red-legged partridge *A. rufa*, a condition during moult was shown experimentally to affect the expression of a neck plumage pattern (Perez-Rodriguez et al. 2013). The symmetry of the neck collar pattern displayed by male little bustards could similarly be affected by male body condition during moult and thereby reliably indicate male condition and quality. Indeed, the development of morphological sex traits (e.g. plumage patterns and symmetry) and display performance (e.g. ability to display at a regular rate without having to feed) of lekking great bustards have been shown to be condition-dependent (Carranza and de Trucios 1993; Morales et al. 2003).

Measuring mating success by copulation rate, which is easily done in classical lek species where males display and females mate on the arena, is less easy in exploded lek species in which males are dispersed and thus cannot be monitored simultaneously. Moreover, copulations are rarely seen in bustards: in the great bustard, during observations of 17 radio-tagged males in Villafáfila, north-western Spain, only five copulations were recorded, two of them performed by the same individual (Morales 2000). The scarcity of copulation records is even more important for the little bustard, where only one copulation was observed during a 4-year study in western France (Jiguet 2001). One possible explanation for this could be that copulations occur mainly at night (although light conditions should allow females to use visual cues for mate choice, for example on full-moon nights), which deserves further research. Male mating success has not been assessed directly through the proportion of offspring sired by each male using DNA fingerprinting of hatchlings, a seemingly impossible task in the little bustard. Only one study has used microsatellite markers and paternity analyses to study reproductive skew in a bustard species. Lesobre et al. (2010) studied houbara bustards in Morocco and, contrary to expectations, found no evidence for reproductive skew. The results rather suggest a successful off-lek male mating strategy, with no apparent reproductive benefit of lekking, and provide evidence for polyandry and sperm competition. There is a need for more studies assessing paternity and reproductive skew in other bustard species, and in a wide range of contexts, to fully understand mating system variation.

Conservation Implications of Lekking Behaviour

Basic ecological knowledge has proved critical in establishing efficient conservation and management strategies (Caro 1997). Studying a species' mating system and evaluating mechanisms of male and female distributions, as well as the processes determining mate choice and the role of resources, indeed provides useful conservation cues. The study of little bustard mating systems has helped define effective conservation measures for the species in highlighting that males and females have different requirements, and thus that conservation measures directed to improve reproductive success might be different from those aimed at favouring males abundance. In addition, the fact that females are less attracted to small leks indicates that declines may be faster than expected from usual population dynamics since breeding areas will be deserted when male density falls below a given threshold value, a kind of Allee effect (see chapter "Little bustard population dynamics").

Studies on the closely related great bustard have also emphasized the importance of understanding mating systems for conservation. Both males and females show a strong conspecific attraction and a high inter-annual fidelity to lekking and nesting sites (Alonso et al. 2000; Morales et al. 2000). As a result, individuals do not colonize vacant patches of suitable habitat but instead, concentrate in already occupied areas. Therefore, when a lek disappears due to habitat change or human disturbance, dispersing great bustards do not form new leks in suitable empty habitat (Lane et al. 2001) but join the closest and largest already existing leks (Alonso et al. 2004), as predicted by theoretical models of conspecific attraction (Reed and Dobson 1993). Such a process of population concentration increases their vulnerability to local catastrophes, as well as the risk of genetic diversity loss, eventually leading to the extinction of smaller and more peripheral leks (Alonso et al. 2004). Even growing populations may not necessarily imply healthy meta-populations in such circumstances (Alonso et al. 2004; Morales et al. 2001). Although such processes are not yet documented in the little bustard, all the patterns described for the great bustard do exist in the former species. Indeed, there is anecdotal evidence that similar processes occur in little bustards, where solitary displaying males are extremely rare, and where apparently suitable habitats, or places which were previously occupied by leks, are deserted even when little bustards occur nearby (V. Bretagnolle, pers. obs.).

Other, Less Studied Aspects of Little Bustard Behavioural Ecology

Although most behavioural research on the little bustard has focused on the species' mating system within the framework of sexual selection theory, other aspects have also been investigated, such as non-breeding flocking behaviour and interaction with human disturbances (chapter "Migration, movements and non-breeding ecology"), or behaviour-mediated interactions with other species (chapter "Interspecific relationships"). Other aspects, very little studied, and also rooted in sexual selection theory, including sperm competition and parent–offspring and sibling conflicts (Andersson 1994; Birkhead 2000; Birkhead and Møller 1992). These remain unstudied mostly because of important methodological challenges. Nevertheless, the existing knowledge of the species' breeding biology, life history and behaviour, along with findings in related species, allows some reasonable speculation about what future studies may tell, should challenges be overcome.

Sperm Competition

As mentioned in chapter "Natural history of the little bustard: morphology, biometry, diet, sexual dimorphism, and social and breeding behaviour", the scarce and rather anecdotal evidence regarding sperm competition is contradictory. The apparently long time elapsing between copulation and egg-laying would allow for potential cryptic female choice through sperm competition (Birkhead and Møller 1992). However, the very low rate of observed copulation seems to contradict the possibility of intense sperm competition. Inspection of female reproductive tracts allows us to infer the level of sperm competition in a species from the number of spirals and pockets present, whose function is storing sperm from different males and stimulating their competition within the tract (e.g. Brennan et al. 2007). Unfortunately, no such study has been done in the little bustard, although evidence for sperm competition has been provided for the houbara bustard (Chargé et al. 2010; Vuarin et al. 2018, 2019). An alternative paternity guard to frequent copulation is, however, mate guarding (Birkhead and Møller 1992). The latter does not seem to be a frequent behaviour in male little bustards, since males have to attend their lek position, given that male attractiveness relates to male lek attendance. Nevertheless, female-pursuit flights by males are quite common and could be reasonably interpreted as an escorting flight behaviour aimed at preventing access of competing males to visiting females, as commonly found in ducks.

In addition, sperm competition has been suggested to be high in the closely related houbara bustard, in which genetic analyses proved the existence of multiple paternity in a majority (60%) of clutches and thus the occurrence of polyandry in the species (Lesobre et al. 2010). Nevertheless, and although a detailed genetic fingerprinting of wild little bustard chicks seems unfeasible, the importance of multiple paternity (and thus the potential for sperm competition) could be assessed through the use of DNA micro-satellites sampled from a large enough number of captured families and putative male sires (Beebee and Rowe 2004). This type of information would help in completing our current view of the evolution of little bustard mating strategies and behaviour.

Personality Traits

Personality has so far never been investigated in any bustard species. Unpublished information on little bustards, however, strongly suggests that males behave very differently from each other and do so consistently: behavioural differences among individuals are, to some extent, repeatable. Decoy presentation to male little bustards in Deux-Sèvres, France, showed that some males flee, some approach the decoy, crouch and remain on the ground, and others behave agonistically and call (snort call). Similar differences are also obvious when undertaking playback experiments with snort calls and recording male responses (V. Bretagnolle, unpubl. data). These behavioural characteristics seem to be consistent, at least over one breeding season. In addition, males vary strongly in the way they perform their mating displays (see chapter "Natural history of the little bustard: morphology, biometry, diet, sexual dimorphism, and social and breeding behaviour" for a description of display variation). Again, it seems that males differ in the type of display they perform and consistently perform certain types of display. Jump displays, for instance, seem restricted to a few males per breeding season (typically, no more than 15–20% of all males), and the males that perform jump displays do so repeatedly within a breeding season. All these observations suggest that male little bustards do show different personalities, though one may argue that display categories may differ according to male age. In fact, age has been identified as a determinant factor in male display and mating performance in the houbara bustard (Preston et al. 2011, 2015).

Females also seem to exhibit different personalities. We have been able to observe these different personalities when capturing female little bustards with funnel traps (Ponjoan et al. 2010). Some are very confident and do not hesitate to enter the funnel. Others are very distrustful and reluctant to approach the trap, consistently flying away as we tried to shepherd them into it. However, previous experiences, age, or genetic traits may also shape these differences in behaviour, and further research is needed on this topic.

Parent (Female)–Offspring Interactions

Little bustard families are inconspicuous and try to keep concealed in relatively dense vegetation while chicks are smaller than their mothers. When fledglings attain their mother's size, families gradually integrate into post-breeding flocks, joining other families as well as adult males and non-breeding individuals, so that distinguishing between chicks and mothers is extremely difficult. Thus, documenting mother–offspring or inter-sibling interactions is problematic, and no information on these aspects is so far available. The larger size and less secretive female-and-chick behaviour of other bustard species like the great bustard have allowed the study of maternal care and competition between siblings, as well as the timing and drivers of juvenile dispersal. In the great bustard, mother–offspring bonds last beyond the formation of post-breeding flocks, so that young and female tend to remain close to each other within the flock, and mothers may even continue to feed them through the non-breeding period, although at a steadily declining rate (Martín and Martin 1997). Whether and to what extent these bonds are maintained in little bustard post-breeding flocks is largely unknown, but careful observations of such flocks in Deux-Sèvres, France, suggest that family interactions do not occur (fledglings indeed tend to gather together in such flocks, without adults; V. Bretagnolle unpublished data).

The observation of female great bustard chick-feeding behaviour has also revealed the existence of size-based competition between siblings for access to food items (Alonso et al. 2018). This may also occur in the little bustard since there is variation in chick size within families. Although all chicks hatch within a few hours, within-brood competition for food creates some size differences. The currently ongoing captive breeding programmes might consider incorporating adequately designed behavioural studies in their activities in order to tackle these and other knowledge gaps, and this might also provide insight for conservation measures. In the great bustard, young males become independent earlier than young females, possibly responding to a pressure to integrate into the hierarchy of male groups as soon as they acquire the size and condition to do so. Young males usually join distant leks, while females tend to remain around natal ones to benefit from acquired knowledge on the local environment, so they maximize the time they remain with their mother (Alonso et al. 1998). The use of increasingly lighter GPS devices might allow investigating temporal and spatial dispersal patterns in juvenile little bustards and would provide insights into juvenile behaviour during the pre-dispersal and natal dispersal phases.

Concluding Remarks on Little Bustard Behavioural Ecology

The little bustard mating system is described in most localities as an exploded lek strategy but can be quite flexible, and a variety of scenarios, ranging from resource-based monogamy to classical leks, has been suggested historically by different authors. All (non-exclusive) theoretical models of lek formation proposed are consistent, to some degree, with the little bustard's lekking behaviour. However, two of them seem to fit particularly well at different spatial scales: the hotshot and the black hole models. On larger scales, female mobility and harassment by individual males may induce male aggregation as described by the black hole mechanism. At smaller scales, female preference for particularly attractive (hotshot) males or certain lek sizes may induce an initial clustering, which would be further maintained and structured as a result of the hotshot mechanism.

Overall, the female choice seems to be based on multiple criteria (male snort call rate, neck collar symmetry and displays). Habitat quality and resources within male territories are variable, and evidence suggests that these determine male settlement, although existing evidence does not support a decisive role of resources in mate choice. Evidence on sperm competition in little bustard is scarce and contradictory. The apparently long time elapsing between matings and laying would allow for potential cryptic female choice through sperm competition, but the apparently low rate of copulation suggests that sperm competition may not be intense in little bustards. Decoy experiments and observations suggest that male and female little bustard behaviours (displays, behaviour during captures) are variable but consistent within individuals and may define individual personalities. These aspects of little bustard behaviour, and their relevance for conservation, require further research.

Acknowledgements We are thankful to Juan Moreno Klemming and Michel Saint-Jalme for their valuable comments and suggestion on a previous version of this chapter. Particular thanks are due to Nigel J. Collar for kindly reviewing the English.

References

Alexander RD (1975) Natural selection and specialized chorusing behaviour in acoustic insects. In: Pimentel D (ed) Insects, science and society. Academic Press, New York, pp 35–77

Alonso JC, Alonso JA, Martín E, Morales MB (1998) Proximate and ultimate causes of natal dispersal in the great bustard, *Otis tarda*. Behav Ecol 9:243–252

Alonso JC, Morales MB, Alonso JA (2000) Partial migration, lek fidelity and nesting area fidelity in female great bustards *Otis tarda*. Condor 102:127–136

Alonso JC, Martín CA, Alonso JA, Palacín C, Magaña M, Lane SJ (2004) Distribution dynamics of a great bustard metapopulation throughout a decade: influence of conspecific attraction and recruitment. Biodivers Conserv 13:1659–1674

Alonso JC, Alvarez-Martinez JM, Palacín C (2012) Leks in ground-displaying birds : hotspots or safe places ? Behav Ecol 23:491–501

Alonso JC, Martín E, Morales MB, Alonso JA (2018) Sibling competition and not maternal allocation drives differential offspring feeding in a sexually size-dimorphic bird. Anim Behav 137:35–44

Andersson M (1994) Sexual selection. Monographs in behavior and ecology. Princeton University Press, Princeton, NJ

Balenger SL, Zuk M (2014) Testing the Hamilton–Zuk hypothesis: past, present, and future. Integr Comp Biol 54:601–613

Beebee TJC, Rowe G (2004) An introduction to molecular ecology. Oxford University Press, New York

Beehler BM, Foster MS (1988) Hotshots, hotspots, and female preferences in the organisation of lek mating systems. Am Nat 131:203–219

Birkhead TR (2000) Promiscuity. An evolutionary history of sperm competition. Harvard University Press, Cambridge, MA

Birkhead TR, Møller AP (1992) Sperm competition in birds: evolutionary causes and consequences. Academic Press, London

Borgia G (1979) Sexual selection and the evolution of mating systems. In: Blum MS, Blum NA (eds) Sexual selection and reproductive competition in insects. Academic Press, New York, pp 19–80

Boyce MS (1990) The red queen visits sage grouse leks. Am Zool 30:263–270

Bradbury JW (1981) The evolution of leks. In: Alexander RD, Tinkle D (eds) Natural selection and social behaviour. Chiron Press, New York, pp 138–169

Bradbury JW (1985) Contrasts between insects and vertebrates in the evolution of male display. Fortschr Zool 31:273–289

Bradbury JW, Gibson RM (1983) Leks and mate choice. In: Bateson PPG (ed) Mate choice. Cambridge University Press, Cambridge, pp 107–138
Bradbury JW, Vehrencamp SL, Gibson RM (1986) Leks and the unanimity of female choice. In: Greenwood PJ, Harvey PH, Slatkin M (eds) Evolution: essays in honor of John Maynard-Smith. Cambridge University Press, Cambridge, pp 301–314
Brennan PLR, Prum RO, McCracken KG, Sorenson MD, Wilson RE, Birkhead TR (2007) Coevolution of male and female genital morphology in waterfowl. PLoSOne 2(5):e418. https://doi.org/10.1371/journal.pone.0000418
Carranza J, Hidalgo de Trucios SJ, Ena V (1989) Mating system flexibility in the great bustard: a comparative study. Bird Study 36:192–198
Caro T (1997) Behavioural ecology and conservation biology. Oxford University Press, Oxford
Carranza J, de Trucios SJH (1993) Condition-dependence and sex traits in the male great bustard. Ethology. https://doi.org/10.1111/j.1439-0310.1993.tb00559.x
Chargé R, Saint-Jalme M, Lacroix F, Cadet A, Sorci G (2010) Male health status, signalled by courtship display, reveals ejaculate quality and hatching success in a lekking species. J Anim Ecol 79:843–850
Clutton-Brock TH, Parker GA (1992) Potential reproductive rates and the operation of sexual selection. Q Rev Biol 67:437–456
Clutton-Brock TH, Hiraiwa-Hasegawa M, Robertson A (1989) Mate choice on fallow deer leks. Nature 340:463–465
Clutton-Brock TH, Price O, MacColl A (1992) Mate retention, harassment and the evolution of ungulate leks. Behav Ecol 3:234–242
Cornec C, Hingrat Y, Rybak F (2014) Individual signature in a lekking species: visual and acoustic courtship parameters may help discriminating conspecifics in the Houbara Bustard. Ethology 120:726–737
Cornec C, Hingrat Y, Robert A, Rybak F (2015) The meaning of boom calls in a lekking bird: identity or quality information? Anim Behav 109:249–264
Cornec C, Hingrat Y, Aubin T, Rybak F (2017) Booming far: the long-range vocal strategy of a lekking bird. R Soc Open Sci 4:170594
Delgado MP, Traba J, García de la Morena EL, Morales MB (2010) Habitat selection and density-dependent relationships in spatial occupancy by male Little Bustards *Tetrax tetrax*. Ardea 98: 185–194
Dementiev G, Gladkov NA (1966) Birds of the Soviet Union. Israel Program for Scientific Translations, Jerusalem
Devoucoux P, Besnard A, Bretagnolle V (2018) Sex-dependent habitat selection in a high-density Little Bustard *Tetrax tetrax* population in southern France, and the implications for conservation. Ibis. https://doi.org/10.1111/ibi.12606
Emlen ST, Oring LW (1977) Ecology, sexual selection and the evolution of mating systems. Science 197:215–223
Faria N, Rabaça J, Morales MB (2012) Linking plant composition and arthropod abundance to establish little bustard breeding requirements in pastureland dominated landscapes. Biodivers Conserv 21:2109–2125
Fiske PP, Rintamäki T, Karvonen E (1998) Mating success in lekking males: a meta-analysis. Behav Ecol 9:328–338
Gibson RM (1989) Field playback of male display attracts females in lek-breeding sage grouse. Behav Ecol Sociobiol 24:439–443
Gibson RM, Bradbury JW, Verhencamp SL (1991) Mate choice in lekking sage grouse revisited: the roles of vocal display, female site fidelity, and copying. Behav Ecol 2:165–180
Gibson RM (1996) Female choice in sage grouse: the roles of attraction and active comparison. Behav Ecol Sociobiol 39:55–59
Gibson RM, Bradbury JW (1985) Sexual selection in lekking sage grouse: phenotypic correlates of male mating success. Behav Ecol Sociobiol 18:117–123
Gilliard ET (1969) Birds of paradise. Natural History Press, New York

Hamilton WD, Zuk M (1982) Heritable true fitness and bright birds: a role of parasites. Science 218: 384–387
Hill WL (1991) Correlates of male mating success in the ruff *Philomachus pugnax*, a lekking shorebird. Behav Ecol Sociobiol 29:367–372
Hingrat Y, Saint Jalme M, Chalah T, Orhant N, Lacroix F (2008) Environmental and social constraints on breeding site selection. Does the exploded-lek and hotspot model apply to the Houbara bustard *Chlamydotis undulata undulata*? J Avian Biol 39:393–404
Höglund J (1989) Size and plumage dimorphism in lek-breeding birds: a comparative analysis. Am Nat 134:72–87
Höglund J, Alatalo RV (1995) Leks. Princeton University Press, Princeton, NJ
Höglund J, Robertson JGM (1990) Spacing of leks in relation to female home ranges, habitat requirements and male attractiveness in the great snipe (*Gallinago media*). Behav Ecol Sociobiol 26:173–180
Höglund J, Eriksson M, Lindell LE (1990) Females of the lek-breeding great snipe, *Gallinago media*, prefer males with white tails. Anim Behav 40:23–32
Höglund J, Alatalo RV, Gibson RM, Lundberg A (1995) Mate-choice copying in black grouse. Anim Behav 49:1627–1633
Janetos AC (1980) Strategies of female mate choice: a theoretical analysis. Behav Ecol Sociobiol 7: 107–112
Jiguet F (2001) Défense des ressources, choix du partenaire et mécanismes de formation des leks chez l'outarde canepetière (*Tetrax tetrax*), une espèce menacée des plaines céréalières. PhD Thesis, University of Paris 6, Paris
Jiguet F, Bretagnolle V (2001) Courtship behaviour in a lekking species: individual variations and mating tactics in male Little Bustard. Behav Process 55:107–118
Jiguet F, Bretagnolle V (2006) Manipulating lek size and composition using decoys: an experimental investigation of lek evolution models. Am Nat 168:758–768
Jiguet F, Bretagnolle V (2014) Sexy males and choosy females on exploded leks: correlates of male attractiveness in the Little Bustard. Behav Process. https://doi.org/10.1016/j.beproc.2014.01.008
Jiguet F, Arroyo B, Bretagnolle V (2000) Lek mating system: a case study in the Little Bustard *Tetrax tetrax*. Behav Process 51:63–82
Jiguet F, Jaulin S, Arroyo B (2002) Resource defence on exploded leks: do male little bustards, *T. tetrax*, control resources for females? Anim Behav 63:899–905
Johnstone RA (1995) Honest advertisement of multiple qualities using multiple signals. J Theor Biol 177:87–94
Kirkpatrick M, Ryan MJ (1991) The evolution of mating preferences and the paradox of the lek. Nature 350:33–38
Lane SJ, Alonso JC, Martín CA (2001) Habitat preferences of great bustard *Otis tarda* flocks in the arable steppes of Central Spain: are potentially suitable area unoccupied? J Appl Ecol 38:193–203
Lesobre L, Lacroix F, Le Nuz E, Hingrat Y, Chalah T, Saint Jalme M (2010) Absence of male reproductive skew, along with high frequency of polyandry and conspecific brood parasitism in the lekking Houbara bustard *Chlamydotis undulata undulata*. J Avian Biol 41:117–127
Lett JM, Guyomarch JC, Boutin JM (2000) Occupation de l'espace par les femelles reproductrices chez l'outarde canepetière (*Tetrax tetrax*). Cahiers d'Ethol 20:1–20
Ligon JD (1999) The evolution of avian breeding systems. Oxford University Press, Oxford
Manning JT, Hartley MA (1991) Symmetry and ornamentation are correlated in the peacock's train. Anim Behav 42:1020–1021
Martín LJ, Martín I (1997) Selección de habitat y gregarismo de la Avutarda en La Moraña. El Cervunal 5:109–175
Miller CW, Moore AJ (2007) A potential resolution to the lek paradox through indirect genetic effects. Proc R Soc B Biol Sci 274:1279–1286

Møller AP (1990) Fluctuating asymmetry in male sexual ornaments may reliably reveal male quality. Anim Behav 40:1185–1187
Morales MB (2000) Ecología Reproductiva y Movimientos Estacionales en la Avutarda (*Otis tarda*). PhD Thesis Complutense University of Madrid, Madrid
Morales MB, Alonso JC, Alonso JA, Martín E (2000) Migration patterns in male great bustards (Otis tarda). Auk 117:493–498
Morales MB, Jiguet F, Arroyo B (2001) Exploded leks: what bustards can teach us. Ardeola 48:85–98
Morales MB, Alonso JC, Martin A, Martin E, Alonso JA (2003) Male sexual display and attractiveness in the great bustard *Otis tarda*: the role of body condition. J Ethol 21:51–56
Morales MB, Traba J, Delgado MP, García de la Morena EL (2013) The use of fallows by nesting little bustard *Tetrax tetrax* females: implications for conservation in mosaic cereal farmland. Ardeola 60:85–97
Morales MB, Casas F, García de la Morena EL, Ponjoan A, Calabuig C, Martínez-Padilla J, García JT, Mañosa S, Viñuela J, Bota G (2014) Density dependence and habitat quality modulate the intensity of display territory defense in an exploded lekking species. Behav Ecol Sociobiol 68: 1493–1504
Moseykin VN (1992) Ecology and protection of the Little Bustard in the Saratov region. Bustard Stud 5:78–91
Neff BD, Pitcher TE (2008) Mate choice for non-additive genetic benefits: a resolution to the lek paradox. J Theor Biol 254:147–155
Oakes EJ (1992) Lekking and the evolution of sexual dimorphism in birds: comparative approaches. Am Nat 140:665–684
Olea PP, Casas F, Redpath S, Vinuela J (2010) Bottoms up: great bustards use the sun to maximise signal efficacy. Behav Ecol Sociobiol 64:927–937
Payne RB (1984) Sexual selection, lek and arena behavior, and sexual size dimorphism in birds. Ornithol Monogr 33:1–53
Perez-Rodriguez L, Jovani R, Mougeot F (2013) Fractal geometry of a complex plumage trait reveals bird's quality. Proc R Soc B Biol Sci 280(1755):20122783
Petrie M, Halliday T (1994) Experimental and natural changes in the peacock's (*Pavo cristatus*) train can affect mating success. Behav Ecol Sociobiol 35:213–217
Petrie M, Halliday T, Saunders C (1991) Peahens prefer peacocks with elaborate trains. Anim Behav 41:323–331
Pomiankowski A, Møller AP (1993) Why have birds got multiple sexual ornaments? Behav Ecol Sociobiol 32:167–176
Pomiankowski A, Møller AP (1995) A resolution of the lek paradox. Proc R Soc Lond 360:31–39
Ponjoan A, Bota G, Mañosa S (2010) Trapping techniques for little bustards *Tetrax tetrax* according to age, sex and season. Bird Study 57:252.255
Ponjoan A, Bota G, Mañosa S (2012) Ranging behavior of little bustard males, *Tetrax tetrax*, in the lekking grounds. Behav Process 91:35–40
Preston BT, Saint Jalme M, Hingrat Y, Lacroix F, Sorci G (2011) Sexually extravagant males age more rapidly. Ecol Lett 14:1017–1024
Preston BT, Saint-Jalme M, Hingrat Y, Lacroix F, Sorci G (2015) The sperm of aging males retards their offspring's development. Nat Commun 6. https://doi.org/10.1038/ncomms7146
Price T, Kirkpatrick M, Arnold SJ (1988) Directional selection and the evolution of breeding date in birds. Science 240:798–799
Reed MJ, Dobson AP (1993) Behavioural constraints and conservation biology: conspecific attraction and recruitment. TREE 8:253–256
Rintamäki PT, Alatalo RV, Höglund J, Lundberg A (1995) Male territoriality and female choice on black grouse leks. Anim Behav 49:759–767
Sæther SA, Fiske P, Kålås JA, Gjul JM (2000) Females of the lekking great snipe do not prefer males with whiter tails. Anim Behav 59:273–280

Sardà-Palomera F, Puigcerver M, Vinyoles D, Rodríguez-Teijeiro JD (2011) Exploring male and female preferences, male body condition, and pair bonds in the evolution of male sexual aggregation: the case of the common quail (*Coturnix coturnix*). Can J Zool 89:325–333

Schulz H (1986) Agonistisches Verhalten, Territorialverhalten und Balz der Zwergtrappe (*Tetrax tetrax*). J Ornithol 127:125–204

Silva JP, Moreira F, Palmeirim JM (2017) Spatial and temporal dynamics of lekking behaviour revealed by high-resolution GPS tracking. Anim Behav 129:197–204

Stillman RA, Clutton-Brock TH, Sutherland WJ (1993) Black holes, mate retention, and the evolution of ungulate leks. Behav Ecol 4:1–6

Swaddle JP, Cuthill IC (1994) Female zebra finches prefer males with symmetric chest plumage. Proc R Soc Lond B 258:267–271

Tarjuelo R, Delgado MP, Bota G, Morales MB, Traba J, Ponjoan A, Hervás I, Mañosa S (2013) Not only habitat but also sex: factors affecting spatial distribution of Little Bustard *Tetrax tetrax* families. Acta Ornithol 48:119–128

Tarof SA, Ratcliffe LM, Kasumovic MM, Boag PT (2005) Are least flycatcher (*Empidonax minimus*) clusters hidden leks? Behav Ecol 16:207–217

Thornhill R (1976) Sexual selection and paternal investment in insects. Am Nat 110:153–163

Traba J, Morales MB, de la Morena ELG, Delgado MP, Kristin A (2008) Selection of breeding territory by little bustard (*Tetrax tetrax*) males in Central Spain: the role of arthropod availability. Ecol Res 23:615–622

Vuarin P, Bouchard A, Lesobre L, Levêque G, Chalah T, Saint Jalme M, Lacroix F, Hingrat Y, Sorci G (2018) No evidence for prezygotic postcopulatory avoidance of kin despite high inbreeding depression. Mol Ecol 27:5252–5262

Vuarin P, Bouchard A, Lesobre L, Levêque G, Chalah T, Saint Jalme M, Lacroix F, Hingrat Y, Sorci G (2019) Post-copulatory sexual selection allows females to alleviate the fitness costs incurred when mating with senescing males. Proc R Soc B 286:20191675

Wagner RH (1998) Hidden leks: sexual selection and the clustering of avian territories. In: Parker PG, Burley N (eds) Female and male extra-pair mating tactics in birds. American Ornithologists' Union, Washington DC, pp 123–146

Wolff A, Dieuleveult T, Martin JL, Bretagnolle V (2002) Landscape context and little bustard abundance in a fragmented steppe: implications for reserve management in mosaic landscapes. Biol Conserv 107:211–220

Interspecific Relationships: Predation, Competition or Coexistence, Parasites

Rocío Tarjuelo, Manuel B. Morales, and Juan Traba

Introduction

The distribution of species at different temporal and spatial scales is partly governed by the physical characteristics of the environment (Pearson and Dawson 2003; Chen et al. 2011). However, organisms do not live in isolation. They interact with individuals of other species in many different ways, for instance, impeding the use of resources or facilitating the establishment of new species by modifying environmental conditions (Martin and Martin 2001; Ellis et al. 2011). Biotic interactions are ubiquitous and therefore are a crucial force shaping population dynamics and community structure (Ellis et al. 2011; Martínez-Padilla et al. 2014). The effects of biotic interactions are not only restricted at local scales or ecological times (Pearson and Dawson 2003); indeed they can scale up and influence the presence or abundance of species at broader scales and induce evolutionary change (Grant and Grant 2006; Heikkinen et al. 2007). Despite the great influence of biotic interactions at shaping species distributions, this topic remains largely unexplored for the little bustard compared to other ecological aspects of the species, such as its habitat preferences.

R. Tarjuelo (✉)
Instituto de Investigación en Recursos Cinegéticos, IREC (CSIC, UCLM), Ciudad Real, Spain

Sustainable Forest Management Institute, University of Valladolid-INIA. ETS Ingenierías Agrarias, University of Valladolid, Palencia, Spain
e-mail: rocio.tarjuelo@uva.es

M. B. Morales · J. Traba
Terrestrial Ecology Group, Departamento de Ecología, Universidad Autónoma de Madrid, Madrid, Spain

Centro de Investigación en Biodiversidad y Cambio Global (CIBC-UAM), Universidad Autónoma de Madrid, Madrid, Spain
e-mail: manuel.morales@uam.es; juan.traba@uam.es

V. Bretagnolle et al. (eds.), *Little Bustard: Ecology and Conservation*, Wildlife Research Monographs 5, https://doi.org/10.1007/978-3-030-84902-3_9

The present chapter describes the existing knowledge about the interactions established between the little bustard and other species and the potential implications for its conservation. It begins by introducing the scarce available information on predation and parasitism, two types of biotic interaction that may negatively affect the little bustard's population dynamics. We summarize the potential predators and parasites studies to date and how they can impact the species at individual and population levels. Next, we focus on competitive interactions, in particular with the co-occurring great bustard (*Otis tarda*), the effects on density-dependent habitat selection and ecological niche, and how competition can affect the conservation of the little bustard. Finally, we provide details on the formation of mixed flocks with a sympatric steppe bird species and the potential anti-predator benefits obtained by this association.

Predation

The prominent role of predator-prey interactions in population, community, and evolutionary ecology has been extensively described in the literature. The predator obtains energy by consuming all or part of another organism, the prey (Begon et al. 2006). Predation exerts an evident and direct negative effect on the prey (i.e. mortality), but there are many other indirect effects associated with predation risk—such as energy investment in anti-predator behaviours, changes in habitat use, reduced foraging in favour of vigilance, movements towards safe but low-quality areas—that may reduce growth, reproduction, or even survival (Creel and Christianson 2008). The impacts of predation go beyond population levels and scale up to community structure. For instance, a predator can regulate competitive interactions between different prey (Chase et al. 2002), and particularly intriguing is the role of top predators on trophic cascades that control the abundance patterns across levels in a food web (Pace et al. 1999; Ripple et al. 2014). However, little is known about the role of predation in the little bustard's population dynamics. This section summarizes the information available in the literature.

Predation Rates and Types

Most predatory events documented in the literature correspond with rough analyses of marks on isolated carcasses or observations of unsuccessful predatory attacks. However, very recent studies directly addressing the effects of predators on little bustard populations have yielded novel insights. Predation appears as a major cause of natural mortality in the little bustard (Schulz 1985; Marcelino et al. 2017). Marcelino et al. (2017) estimated that predation causes around 8% of adult annual deaths in Iberian populations. They recorded data from 151 radio-tracked birds over more than a decade throughout the southwest, central, and northeast regions of the

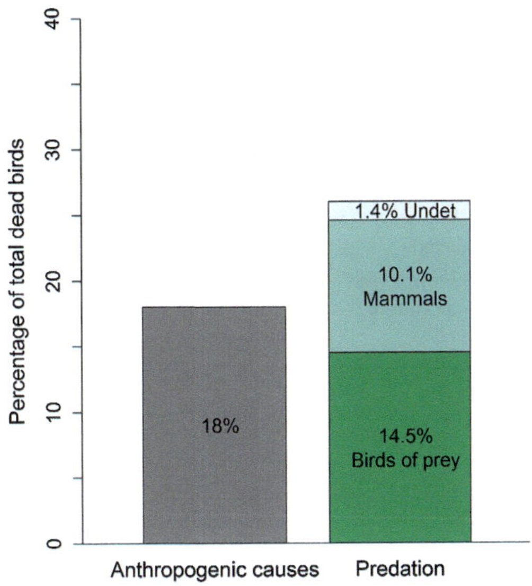

Fig. 1 The relative incidence of different predators on the mortality of little bustard adults across several Iberian populations [*Undet* refers to predation events for which the predator type could not be determined; after data provided by Marcelino et al. (2017)]

Iberian Peninsula. From a total of 82 dead birds, 26% of radio-tracked adults were killed by predators against 18% of birds that died due to anthropogenic causes, such as collision with powerlines or illegal hunting (Fig. 1).

Beyond the sporadic records described in the literature (see Table 1 for a review on predation events published), they showed for the first time that the main predators of the little bustard in Iberian populations are birds of prey, secondarily mammals (55.6% and 38.9% of predation events respectively) (Marcelino et al. 2017). However, these numbers are likely to vary depending on the particular predator community coexisting with little bustards. For instance, 90% of little bustard carcasses observed in Azerbaijan seem to be predated by mammals (Gauger 2007). Among them, the red fox (*Vulpes vulpes*) appears as a key predator for little bustards. Verestchagin (1940) found evidence of little bustard remains in the stomach content of eight foxes shot in the Shirvan steppe (Azerbaijan). The role of red foxes as predators affecting the dynamics of little bustard populations has not yet been studied in Western Europe, but recent evidence suggests that this predator may influence the distribution of the species. Hunting estates in central Spain where fox control (legal culling of red foxes) is applied more intensively have a higher abundance of little and great bustards than hunting estates where fox control is less intense (Estrada et al. 2015), although additional studies confirming these positive trends are required to clarify how predation by red foxes influences population dynamics of the little bustard.

Differential predation risk to males, females, and juveniles may critically impact the population dynamics of a species (Newton 1998). This could be particularly relevant in the case of the little bustard, as many of their populations in Western Europe are currently biased toward males (Inchausti and Bretagnolle 2005; Morales et al. 2005a, 2015; but see Silva et al. 2014). If predation pressure is higher for

Table 1 Predatory events on little bustards recorded in the literature or provided by observers

Species	Observation	References
Eagle owl (*Bubo bubo*)	Adult remains in the nest of an eagle owl in Portugal (found by the photographer M. D. England)	Ferguson-Lees (1967)
Spanish imperial eagle (*Aquila adalberti*)	Predated adult found in the nest of an imperial eagle in Doñana (Spain)	Valverde (1967)
Booted eagle (*Hieraatus pennatus*)	Juvenile remains in nest in Central Spain	Garzón (1974)
Egyptian vulture (*Neophron percnopterus*)	Adult remains in nest in Central Spain (probably scavenging)	Garzón (1974)
Eagle owl		Hellmich (1984)
Spanish imperial eagle	Predatory attacks and used as a food resource for chicks in Central Spain	Veiga (1985)
Montagu's harrier (*Circus pygargus*)	Remains of eggs found in Montagu's harrier's pellets and nests.	Corbacho et al. (2005)
Hen harrier (*Circus cyaneus*)	Unsuccessful hunting attempts on little bustards	Gauger (2007)
Peregrine falcon (*Falco peregrinus*)	Hunting attempts	Gauger (2007)
Eastern imperial eagle (*Aquila heliaca*)	Hunting attempts	Gauger (2007)
Spanish imperial eagle	Successful predatory attack in Central Spain	J. Martínez-Padilla (pers. obs.) (2003)
Red fox (*Vulpes vulpes*)	Remains of little bustards in red foxes' stomachs	Verestchagin (1940)

females than for males, then predation would create an unbalanced sex-ratio and could jeopardize the species' recovery. Unfortunately, no studies have analysed this important issue for the conservation of the species, and additional research is urgently required to elucidate the factors underlying the extremely low abundance of little bustard females in Western Europe (see chapter "Little bustard population dynamics").

Low productivity values, like low female abundance, seem to be rather the norm in most Iberian populations (Lapiedra et al. 2011; Tarjuelo et al. 2013). Factors associated with agricultural intensification are among the major causes underlying productivity declines (Lapiedra et al. 2011; Bretagnolle et al. 2011), but the potential effects of predation on nest and chick survival are largely unstudied. Predation seems to be responsible for a minimum of 42.8% of nest failures in a population located in the Lleida Plains, Catalonia (Lapiedra et al. 2011). Predation rate on clutches was, however, below 8% in Western France ($N = 67$ nests; Bretagnolle et al. 2018), and there was little predation on chicks in the same study. Indeed, studies in Western France provided only anecdotal evidence of active predation of chicks by mammals. However, on post-gathering flocks (August–October), predation of fledglings (in particular released birds; see chapter "Captive breeding, handling and care, and the impact of releases on wild populations") by domestic cats (*Felis silvestris catus*) and foxes seems to be rather high (e.g. about half of the fledglings were predated by mammals in 2017; V. Bretagnolle, pers. obs.). Another study conducted by Corbacho et al. (2005) also indicates that Montagu's harrier (*Circus pygargus*) predates little bustard nests, consuming the eggs (9.7% of total biomass found in food remains and pellets). Opportunistic mammals, like red foxes, feral dogs (*Canis lupus familiaris*) and cats, may exert high predation pressure on the nests of this ground nesting species, as in other steppe birds (Suárez et al. 1993; De Borbón et al. 1999b; Calero-Riestra et al. 2013).

Anti-predator Behaviours

The evolution of anti-predator behaviours and other phenotypic adaptations allow individuals to cope with potential predators (Begon et al. 2006). The little bustard has a cryptic plumage that, as in many other steppe species, reduces its conspicuousness in order to avoid predators in an open landscape with few secure shelters (De Juana 2005). The cryptic coloration consists in vermiculated sandy-brown upperparts, white underparts and black tips in the wings, a pattern that resembles the colour of those substrates occupied by the species (Cramp and Simmons 1980). The little bustard also displays behaviours against predators, although this is an issue poorly explored from an ecological and evolutionary perspective. Little bustards are gregarious during winter, with adults and juveniles congregating in large flocks that dilute individual predation risk (García de la Morena 2015). During the breeding season, however, adults are solitary and may avoid a predator simply by flying up or running into areas of dense vegetation (Cramp and Simmons 1980). Nesting females show different anti-predator behaviours to divert predators' attention from nests or chicks. For instance, they move a short distance from the nest before flying away and show distraction behaviours or a special shock display, similar to that of the great bustard, in which the bird ducks, raising and fanning its tail, shaking its wings, and pointing the bill at the source of the threat (Cramp and Simmons 1980). Females with chicks also make rushes with the head lowered at potential predators (Cramp and Simmons 1980). Other anti-predator behaviours displayed by little bustards from very early stages include jumps, mid-air twists, and zigzag runs (Cramp and Simmons 1980).

Predators also shape species' habitat selection (Morris 2003; Creel et al. 2005) (for a detailed description of which, see chapter "Habitat selection and space use"). The preference of little bustards for flat areas (Tarjuelo et al. 2014) and for semi-permanent vegetation types of a certain height, such as fallows, pastures, and short shrubs (e.g. Morales et al. 2005b; Delgado et al. 2010; Faria et al. 2012), may have evolved to facilitate the detection and avoidance of predators. The vegetation height of the little bustard's most preferred habitats enables birds to survey their surroundings and detect the approach of aerial or terrestrial predators. Likewise, the selection of flat areas offers higher visibility to monitor larger proportions of the terrain.

Interestingly, the segregation between breeding males and females at the microhabitat scale may reflect a different trade-off between concealment and food acquisition, with males and females displaying different strategies against predation risk (Morales et al. 2008). During the breeding season, males perform their sexual display within their territories to signal individual quality to potential mates and other competing males. The reproductive success of males in a species without male parental care depends on the number of successful mating events attained (see, for example Kruijt and de Vos 1988), which is directly linked to the quality of the sexual display (Morales et al. 2003; Jiguet and Bretagnolle 2014). This energy-consuming activity may push little bustard males to establish their territories in sites not only containing a great abundance of food resources (Morales et al. 2008; Traba et al. 2008) but also ensuring their visibility over considerable distances, at the cost of increased predation risk (Morales et al. 2008). In contrast, female reproductive success relies on offspring survival (Clutton-Brock 1988). Little bustard females trade-off the search for food and the need for cover against predators (Morales et al. 2008; Traba et al. 2015). They use patches of dense high vegetation that constitute good nesting sites but allow to evaluate the quality of potential mates (Morales et al. 2008, 2013; Lapiedra et al. 2011; Traba et al. 2015).

Little Bustard as a Predator

Little bustard adults are mainly herbivorous, but they also incorporate arthropods into their diet during the breeding season (Cramp and Simmons 1980), although in a very low proportion, around 2% of faeces content (Jiguet 2002). The proportion of arthropods in diet, however, seems to be higher when using metabarcoding analyses (Cabodevilla 2021). Most of the arthropods taken by nesting females are used to feed their chicks, which are also able to capture their own prey from their fifth day (Cramp and Simmons 1980). Chicks only consume this food resource during the first 3 weeks (Cramp and Simmons 1980; Jiguet 2002). According to Jiguet (2002), Coleoptera is the order most represented in the little bustard's diet, of both adults and juveniles (more than 56% of the total number of arthropods present in faeces). Dermaptera and Orthoptera are also common prey (around 20% and 10% of the total number of arthropods found in faeces, respectively). During the summer, the proportion of Orthoptera found in faeces increases (13% of the total number of arthropods), coinciding with the peak of the abundance of this arthropod group in regions of southwestern Europe (Badenhausser et al. 2009). There are very few studies investigating chick diet: however, based on focal observations, as well as some studies carried out in captive breeding flocks, there is strong evidence that chicks feed almost exclusively on insects, particularly grasshoppers (genus *Cortypus*, *Eucortypus*, and *Calliptamus*). In captivity, chicks can feed on 200 grasshoppers per day, the grasshopper size depending on chick age (larger items at later age; V. Bretagnolle, pers. obs.). Species of Hymenoptera are also present in little bustard faeces (8% of total arthropods found in faeces during winter), but they remain almost intact, indicating that they are a poorly digestible food resource for the little bustard, and hence of extremely low quality. Jiguet (2002) also found that little bustard eats arachnids (Opiliones and Araneidae), Lepidoptera, Heteroptera and Diptera, but these groups represented less than 5% of the arthropod number detected in faeces. The sporadic consumption of small vertebrates (small frogs, field voles) has also been reported (Cramp and Simmons 1980; but see below).

Parasites

Parasitism also plays a key role in host population and community dynamics. As in predation, the parasite obtains benefits because it extracts energetic resources from its host or uses it as a habitat for survival and reproduction (Begon et al. 2006). When parasite abundance exceeds the load supported by the host, the host may suffer a loss of body condition that can alter its behaviour, reduce opportunities for reproduction or offspring survival, and eventually result in death. These effects may scale up and greatly affect population dynamics and even community processes by, for instance, changing the functional role of the host within the community (McCallum and Dobson 1995; Poulin 1999; Hudson et al. 2006). Despite the diverse impacts of parasites on population regulation, our knowledge of parasitic diseases affecting the little bustard and how they impact individual conditions and population-level processes is almost non-existent. Only two published studies document infection events in little bustards by two different parasite species (López-Neyra 1944; Villanúa et al. 2007).

The first parasite ever described for the little bustard is the cestode *Hispaniolepis villosa* (López-Neyra 1944). This parasite has been found in the intestine of little bustards but also in the guts of great bustards (López-Neyra 1944). The second parasite reported in the literature is *Eucoleus contortus*, a capillarid nematode that typically parasitizes regions of the gastrointestinal tract in bird species (Yabsley 2008). This nematode was reported for the first time in the corpse of a little bustard male that died during his capture in winter in central Spain (Villanúa et al. 2007). The autopsy revealed five adults—three males and two females—of this parasite within the bird's crop. This infected bird was close to a hunting estate on which 8% of widely released farm-reared red-legged partridges (*Alectoris rufa*) carried *E. contortus*. This capillarid nematode was detected in partridges only on hunting estates where farm-bred juveniles were released (Villanúa et al. 2008), and seems to be almost exclusive to farm-reared individuals (Millán et al. 2004). Villanúa and collaborators suggested that the most plausible cause of this condition in the little bustard was a heterospecific transmission from the red-legged partridge (Villanúa et al. 2007), a very likely scenario given the broad range of hosts of *E. contortus*, which has been found in at least nine orders of birds (Yabsley 2008).

These findings are clearly relevant to the ecology and conservation of the declining little bustard because this parasite may affect the host's body condition (Bosch et al. 2000), induce severe pathologies (Cruz et al. 2016), and increase predation risk for its host (Millán et al. 2002). The infected little bustard presented a 28% poorer body condition, measured as weight over cube tarsus length, than the other 15 birds captured during the same winter (Villanúa et al. 2007). However, direct links between parasite presence and weakened body condition remain to be demonstrated.

The case of *E. contortus* transmission to a wild bird focuses attention on the absence of information on the parasitic community of this and other threatened species. Further understanding of the parasite burdens of little bustard populations and how they affect survival and reproductive success are needed in order to gain a clear understanding of the biotic threats impacting little bustard populations.

Interspecific Competition

In natural communities, ecologically or phylogenetically closely related species may engage in competition if they share limited resources (Begon et al. 2006). Interspecific competition may affect the abundance of the species involved, their spatial and temporal distributions, and drive evolutionary change (Ziv et al. 1993; Adams 2004; Cooper et al. 2007). The community of birds co-occurring with the little bustard in natural or anthropogenic landscapes such as extensive cereal croplands in Europe is diverse and belongs to several families and genera (De Juana 2005). Some species are not confined to steppe-like environments, while others are tightly linked to these kinds of landscapes, as so-called "steppe birds." Different species from these families have evolved particular adaptations to survive in steppe-like landscapes, including cryptic plumage, ground nesting, and short incubation and fledgling periods (De Juana 2005). Steppe-like landscapes are characterized by their low productivity and the lack of secure sites for nesting, so that organisms inhabiting them are likely to experience sparse availability of critical resources like food or safe nesting sites. Therefore, the competition must have occurred in the past between steppe bird species sharing similar ecological requirements, and especially between closely related ones, playing a major role in biodiversity maintenance in these bird communities by, for instance, favouring the differentiation of their breeding phenologies, habitat preferences, or diet (Morales et al. 2006; Benítez-López et al. 2014, 2017; Tarjuelo et al. 2017a, b).

Under current processes of global change, many populations may lose their ecological equilibrium due to habitat alteration or destruction (Auer and Martin 2013) and thereby affect the stable coexistence with competitors. Farmlands in Western Europe, where the little bustard maintains a high proportion (chapter "Natural history of the little bustard: morphology, biometry, diet, sexual dimorphism, and social and breeding behaviour") of its global breeding population, have undergone important changes in the structure and dynamics of habitats due to agricultural intensification since the middle of the last century (Robinson and Sutherland 2002; Benton et al. 2003), resulting in an overall decrease in food resources, nesting sites, or good quality territories for little bustard males

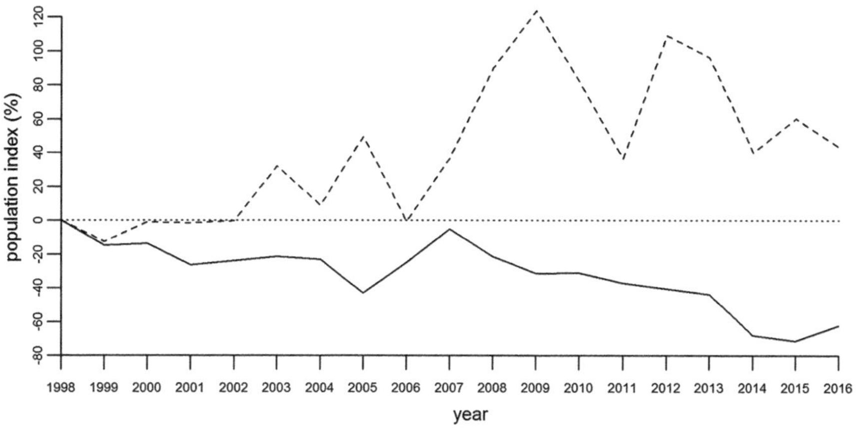

Fig. 2 Temporal variation in population index (%) for the little (solid line) and great bustards (dashed line) calculated with TRIM (Pannekoek and Van Strien 2005) from data from the Spanish common bird monitoring programme (SACRE) conducted during the breeding season from 1998 to 2016. Dotted line indicates no variation in population numbers from the first year of surveys in 1998 [Based on data provided by SEO/BirdLife (2012) and updates]

(Bretagnolle et al. 2011; Morales et al. 2013) (see chapter "Habitat selection and space use"). This reduced availability of shared resources may intensify competitive interactions with other steppe bird species, which can be resolved by density-dependent habitat selection (Morris 2003).

Although two studies have explicitly explored the effects of intraspecific competition on habitat selection by little bustards (Delgado et al. 2010; Morales et al. 2014), the role of interspecific competition on population regulation of this species remains mostly unknown. To date, only three studies have directly investigated the existence of competition between the little bustard and a closely related species, the great bustard (Shlyakhtin et al. 2004; Tarjuelo et al. 2017a, b). Both species are members of the Otididae family and coexist in many regions across their geographic distribution (Del Hoyo and Collar 2014). Both species inhabit mainly open grasslands and extensive cereal croplands in Western Europe, which account for a large proportion of their global breeding population (Cramp and Simmons 1980) and most conservation efforts to aid in their recovery (see chapter "Little bustard and humans: conservation and management"). Over the twentieth century, populations of little bustard and great bustard have declined abruptly, and nowadays are classified as "Near Threatened" and "Vulnerable" respectively, according to IUCN criteria (IUCN 2016). Monitoring programmes in Spain suggest a recovery in the great bustard's populations over recent years (results from SACRE program, SEO/BirdLife-MAPAMA, Spain), probably owing to targeted conservation policies, whereas the little bustard maintains its declining trends (see chapter "The little bustard around the world: distribution, global conservation status, threats and population trends"; Fig. 2).

Although both species occupy the same landscapes at the regional scale, studies of habitat selection during the breeding season indicate different preferences at the landscape scale (Lane et al. 1999; Morales et al. 2006; Delgado et al. 2010). The little bustard shows a more specialized habitat selection, using preferentially semi-permanent habitats with a low height and dense vegetation structure like fallows of different ages, legume crops, pastures, and short shrubs (Salamolard and Moreau 1999; Morales et al. 2005b; García et al. 2007). In contrast, the great bustard behaves as a generalist with no clear preferences for particular agrarian substrates (Lane et al. 2001; Morales et al. 2006; Tarjuelo et al. 2017b). Although great bustards are frequently observed using mostly cereals, their habitat use depends more on the particular landscape configuration and the availability of different habitats within the landscape (López-Jamar et al. 2011; Tarjuelo et al. 2014). However, this segregation of their habitat preferences does not prevent current competitive interactions between the two species. Two recent studies address the existence and effects of competition between little bustard males and great bustards in dry cereal croplands during the breeding season by evaluating changes in their habitat use.

Theories of habitat selection assume that stable coexistence occurs when competitors spatially segregate in their preferred habitats (Rosenzweig 1981; Morris 1988). However, competitors can coexist by adjusting the use of their preferred habitat in a density-dependent manner (Morris 1999). Tarjuelo et al. (2017b) addressed the existence of interspecific competition between little bustard males and great bustards by using habitat isodars, the set of densities in occupied habitats such that an individual's expected fitness is equal in each habitat (Morris 1987). These authors found that the little and the great bustard compete in an asymmetric manner: the little bustard's habitat use is altered by the abundance of great bustards, whereas great bustards do not change their habitat use in the presence of the little bustard. Interspecific competition is not widespread and only occurs in cereals, a suboptimal habitat for the little bustard (see chapter "Habitat selection and space use"), and apparently by both exploitation and interference. As the density of great bustards increases in this habitat, little bustards progressively abandon it and concentrate in fallows, natural vegetation, and legume crops, their preferred habitats (Fig. 3). The presence of great bustard leks or great bustard nesting females may reduce opportunities for little bustard males to establish territories, not only because the space is already occupied but also because the chance of encounters decreases the attractiveness of cereals as male display sites (Tarjuelo et al. 2017b).

Another study further explored the potential effects of intra- and interspecific competition (with great bustard) on the little bustard's ecological niche (Tarjuelo et al. 2017a). They found that the degree of niche overlap with great bustard did not correlate with the density of competitors (Tarjuelo et al. 2017a). This result apparently contradicts classic competition theory that predicts a decrease in niche overlap as the intensity of competition rises (Pianka 1974). The authors argued that this result could be due to unresolved competition between the species.

Tarjuelo et al. (2017a) also investigated how the habitat niche of little bustard males varies between sympatry and allopatry with great bustard and whether niche shifts are density-dependent. The little bustard's niche breadth tended to be wider in

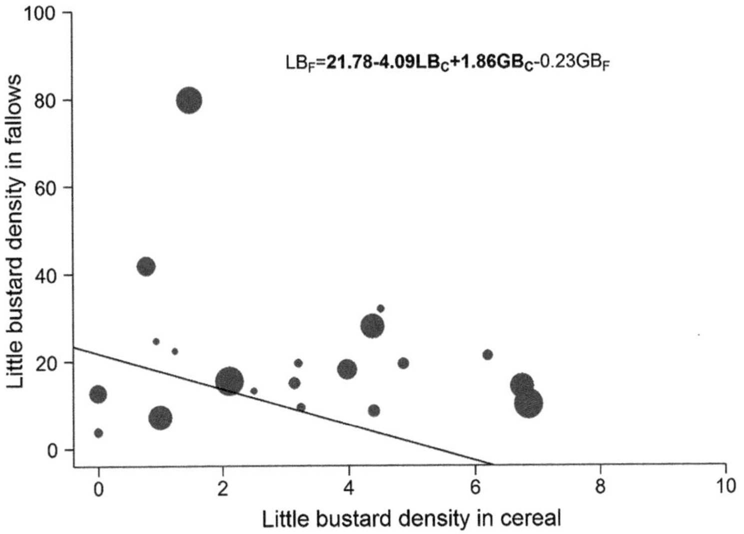

Fig. 3 Multispecies isodar for male little bustards considering fallows as the preferred habitat and cereals as the secondary habitat choice. The isodar is the set of join densities in two habitats such that individuals' expected fitness is equal in each (Morris 1987). The regression model is displayed with significant ($p < 0.05$) coefficients in bold type. The model shows a positive effect of great bustard density in cereals on the little bustard's habitat selection (point size reflects the density of the great bustards in this habitat). As great bustard density increases in cereals, little bustards use a greater proportion of fallows. The negative slope of the isodar points to additional competitive effects with great bustard due to interference (modified from Tarjuelo et al. 2017b)

sympatry than in allopatry, although this result was only marginally significant. In addition, the little bustard niche also shifted toward increased use of natural vegetation when in sympatry. Interspecific competition may restrict the use of those particular habitats where competition occurs (i.e. cereal crops), displacing little bustards toward one of their most preferred habitats, natural vegetation (e.g. Delgado et al. 2010), which could be close to saturation levels. However, the increased use of the species' preferred habitat may intensify processes of intraspecific competition. In order to mitigate increased levels of intraspecific competition in the preferred habitat, some birds may occupy low-quality habitats that otherwise would be avoided.

Although the little bustard's niche breadth tended to be narrower in allopatry than in sympatry, the authors also found that niche breadth varies in a density-dependent manner with the density of great bustards (Tarjuelo et al. 2017a). Niche breadth decreases with the increased density of great bustards in regions where both species occur. Little bustards seem to reduce the proportional use of shared habitats where they compete for resources with great bustards because the increased presence of the latter progressively limits their capacity to exploit resources.

These patterns of density-dependent habitat selection are clearly relevant for the conservation of little bustards, given that one of its key habitats, fallows, has been greatly reduced due to agricultural intensification (see chapters "Habitat selection and space use" and "Threats affecting little bustards: human impacts") and the populations of its competitor, the great bustard, have been steadily increasing in recent years (SEO/BirdLife 2012; Fig. 2). The lack of little bustard breeding territories of high quality caused by the decrease of fallow fields may intensify intraspecific competition, which can be mitigated by increased use of cereals (Tarjuelo et al. 2017b). However, in regions where great bustard density is also high, interspecific competition may prevent the exploitation of cereal habitats by little bustards, forcing them to occupy the scarce fallow fields and suboptimal habitats like ploughed lands (Tarjuelo et al. 2017a, b). Little bustard's reproductive success may thus be compromised, which could eventually translate into further population declines.

These studies provide insight into the effects of interspecific competition on little bustard males, but nothing is known about how competition, either intra- or interspecific, could affect females' ecology. The secretive behaviour of females imposes a major constraint on robust scientific research, but it is crucial to face this challenge to improve conservation strategies that can reverse the current negative trends, particularly if we take into account the importance of female reproductive success for the population viability (Morales et al. 2005a).

Competition between these two bustard species may also occur at a finer scale. Both are mainly herbivorous but consume arthropods during the breeding season, especially their chicks (Cramp and Simmons 1980; Lane et al. 1999; Jiguet 2002; Bravo et al. 2012). Shlyakhtin et al. (2004) evaluated dietary segregation between little and great bustards by analysing the content of faecal samples collected in regions where both species exist, finding only slight differences in their diet. Although the proportion of plants in faeces was greater in the little bustard (90%) than in the great bustard (75%) during the mating period, the lack of statistical significance prevented the authors from identifying a clear trophic segregation. Interestingly, the two species differed in the type of insect they capture. The great bustard preys on bigger insects than the little bustard and the presence of vertebrates such as lizards and voles in the diet has been confirmed in the great bustard but not in the little bustard (Lane et al. 1999; Shlyakhtin et al. 2004). Future studies should clarify whether interspecific competition for food occurs and the potential role of diet segregation mediating the coexistence of these bustard species.

Another coexisting species that deserves further attention in relation to competition with the little bustard is the red-legged partridge. This latter is widely distributed in the Iberian Peninsula, largely overlapping with little bustard distribution (Del Hoyo and Collar 2014) and reaching high densities in the extensive cereal farmlands also occupied by the little bustard (Carrascal and Palomino 2008). However, we still do not know whether interspecific competition may be operating between them, but it is certainly a real possibility due to their similar body size and use of similar habitats and food resources in Mediterranean farmlands (Morales and Traba 2009; Traba et al. 2015).

Other Interactions

Not all is conflict in nature, and positive interactions in which all interacting parts obtain benefits are widespread and crucial for the well-being of communities (Bruno et al. 2003). Reduced predation risk, use of cues, or creation of favourable environments (exploitation of resources not previously available) are only a few examples of how organisms can benefit by interacting with other species. To date, two studies conducted by Martín et al. (2010) and García de la Morena (2015) directly addressed the existence of positive interactions between the little bustard and the also threatened pin-tailed sandgrouse (*Pterocles alchata*). The pin-tailed sandgrouse is a steppe bird that inhabits farmlands and natural steppes and frequently co-occurs with the little bustard. Mixed flocks formed by little bustards and pin-tailed sandgrouse are common in cereal landscapes in the Iberian Peninsula and can concentrate several hundreds of birds of both species (De Borbón et al. 1999a; Martín et al. 2010). The formation of mixed flocks can produce several benefits related to foraging efficiency and decreased predation risk (Greenberg 2000; Sridhar et al. 2009). Mixed flocks of little bustard and pin-tailed sandgrouse can be found from October to March, but their proportions, as well as their size, vary greatly between seasons (Martín et al. 2010). At the end of the breeding season, little bustards aggregate in flocks formed by males, females, and juveniles (García de la Morena 2015). However, it is during the autumn and winter months when the association of little bustards and pin-tailed sandgrouse is more frequent in dry cereal landscapes of central Spain (Martín et al. 2010). Wintering mixed flocks usually involve larger numbers than monospecific flocks (Martín et al. 2010; García de la Morena 2015). Moreover, mixed flocks involve higher numbers of birds of each species than single-species flocks. By gathering in mixed flocks, both little bustards and pin-tailed sandgrouse may obtain anti-predator benefits because either more individuals are vigilant against potential predators (Lima 1995) or individual predation risk diminishes (Hamilton 1971).

During winter, little bustards and pin-tailed sandgrouse in mixed flocks behave as members of one group, coordinating activities like feeding or flying (Martín et al. 2010). Habitat selection differs in these mixed flocks but only for the sandgrouse, which adopts the habitat use by little bustards and thus diversifies the spectrum of habitats they can exploit, in particular those characterized by high vegetation (Martín et al. 2010; García de la Morena 2015). Vegetation height, body size, and predator-avoidance strategy are factors highly related to the predation risk facing species living in open landscapes (Whittingham and Evans 2004; Butler et al. 2005). Habitats with high vegetation may provide more food but hinder the early detection of predators. On the other hand, they provide shelter for species capable of efficiently scanning the surroundings (Butler et al. 2005). Pin-tailed sandgrouse in mixed flocks are able to occupy high vegetation they would otherwise avoid because they can take advantage of the superior abilities of longer-necked, longer-legged little bustards to detect aerial and terrestrial predators. For instance, pin-tailed sandgrouse preferentially forage in stubbles when associating with little bustards, whereas they forego this option in single-species flocks (Martín et al. 2010). The preferential use of this habitat by pin-tailed sandgrouse is apparently determined by the little bustard's habitat use during winter, which involves substrates of relatively tall vegetation such as fallows, alfalfa, and stubble, which may provide shelter against predators (Leitão and Costa 2001; Silva et al. 2004; García de la Morena 2015). Similarly, the use of irrigated crops by pin-tailed sandgrouse almost exclusively occurs in mixed flocks (García de la Morena 2015).

The formation of mixed flocks by little bustards and pin-tailed sandgrouse seems to provide asymmetrical benefits. These studies suggest that pin-tailed sandgrouse encounter clear anti-predator and foraging benefits when associating with little bustards. The benefits for the little bustard remain less clear, but they might be related to reduced individual predation risk. Indeed, Tarjuelo et al. (2015) found that after anthropogenic disturbance events, the proportion of little bustards being vigilant decreases as flock size increases. Future studies should explore the role of positive interactions with the pin-tailed sandgrouse and other species inhabiting steppe landscapes. Nevertheless, increasing group size may also impose foraging costs when competition for food exceeds anti-predator benefits (Clark and Mangel 1986; Sansom et al. 2008). The different diets of little bustards and pin-tailed sandgrouse—the former feeding mainly on green plants during winter and the latter mostly on seeds—may prevent competition for food resources and facilitate the association between them.

Concluding Remarks

In summary, the role of biotic interactions in regulating little bustard population dynamics remains one of the most unexplored issues in the ecology of this species. Recent works shed some light on this topic, but future studies are needed to determine the relative importance of different biotic interactions on population dynamics to improve the design of conservation strategies for this declining species. Competition has recently received increasing attention, and studies conclude that little and great bustards compete asymmetrically. The little bustard changes its habitat use in a density-dependent manner when in sympatry with great bustards. Great bustards force the much smaller little bustards to abandon cereals, a secondary habitat occupied to mitigate intraspecific competition, in favour of the species' preferred habitats as fallows, natural vegetation, and legume crops. However, the extent of these habitats has been reduced in recent years (Traba and Morales 2019), and thus increased intraspecific competition may jeopardize the recovery of little bustard populations. Predation seems to be the main cause of natural mortality for the species, with raptors and opportunistic mammals being the most important predators, at least for little bustard adults. Of particular interest is the role of red foxes, a key predator on Asian steppes that deserves further attention in hunting regions inhabited by little bustards because of the potential effects that may cause hunting management practices such as predator control. Parasites may also play a role in little bustard mortality. The presence of *Eucoleus contortus* in a corpse of a little bustard male with a poor body condition indicates the possibility that new pathogens are being introduced into little bustard populations due to the release of farm-reared red-legged partridges for hunting activities. Although positive interactions are ubiquitous in nature, these remain to be confirmed between the little bustard and co-occurring species. The formation of mixed flocks with the pin-tailed sandgrouse could be understood as a mutualism in which the pin-tailed sandgrouse obtains the major advantage (access to a food resource), whereas the little bustard appears only to benefit by reduced individual predation risk.

Acknowledgements We are in debt with Jesús Martínez-Padilla, Ana Benítez and Vincent Bretagnolle for their valuable comments on previous versions of this chapter, and with Nigel Collar for kindly reviewing the English.

References

Adams DC (2004) Character displacement via aggressive interference in Appalachian salamanders. Ecology 85:2664–2670

Auer SK, Martin TE (2013) Climate change has indirect effects on resource use and overlap among coexisting bird species with negative consequences for their reproductive success. Glob Chang Biol 19:411–419

Badenhausser I, Amouroux P, Lerin J, Bretagnolle V (2009) Acridid (Orthoptera: Acrididae) abundance in Western European grasslands: sampling methodology and temporal fluctuations. J Appl Entomol 133:720–732

Begon M, Townsend CR, Harper JL (2006) Ecology: from individuals to ecosystems. Blackwell, Oxford

Benítez-López A, Viñuela J, Suárez F, Hervás I, García JT (2014) Niche-habitat mechanisms and biotic interactions explain the coexistence and abundance of congeneric sandgrouse species. Oecologia 176:193–206

Benítez-López A, Viñuela J, Mougeot F, García JT (2017) A multi-scale approach for identifying conservation needs of two threatened sympatric steppe birds. Biodivers Conserv 26:63–83

Benton TG, Vickery JA, Wilson JD (2003) Farmland biodiversity: is habitat heterogeneity the key ? TREE 18:182–188. https://doi.org/10.1016/S0169-5347(03)00011-9

Bosch M, Torres J, Figuerola J (2000) A helminth community in breeding yellow-legged gulls (*Larus Cachinnans*): pattern of association and its effect on host fitness. Can J Zool 78:777–786

Bravo C, Ponce C, Palacín C, Alonso JC (2012) Diet of young great bustards *Otis Tarda* in Spain: sexual and seasonal differences. Bird Study 59:243–251

Bretagnolle V, Villers A, Denonfoux L, Cornulier T, Inchausti P, Badenhausser I (2011) Rapid recovery of a depleted population of little bustards *Tetrax Tetrax* following provision of alfalfa through an agri-environment scheme. Ibis 153:4–13

Bretagnolle V, Denonfoux L, Villers A (2018) Are farming and birds irreconcilable? A 21-year study of bustard nesting ecology in intensive agroecosystems. Biol Conserv 228:27–35

Bruno JF, Stachowicz JJ, Bertness MD (2003) Inclusion of facilitation into ecological theory. TREE 18:119–125

Butler SJ, Bradbury RB, Whittingham MJ (2005) Stubble height affects the use of stubble fields by farmland birds. J Appl Ecol 42:469–476

Cabodevilla X (2021) Multidisciplinary approach to elucidate aspects involved in the decline of farmland birds. PhD thesis. Universidad del Pais Vasco

Calero-Riestra M, García JT, Herranz J, Suárez F (2013) Breeding output and nest predation patterns in steppe-associated Mediterranean birds: the case of the tawny pipit *Anthus Campestris*. J Ornithol 154:289–298

Carrascal LM, Palomino D (2008) Las aves comunes reproductoras En España. Población en 2004-2006. SEP/Birdlife, Madrid

Chase JM, Abrams PA, Grover JP, Diehl S, Chesson P, Holt RD, Richards SA, Nisbet RM, Case TJ (2002) The interaction between predation and competition: a review and synthesis. Ecol Lett 5: 302–315

Chen I, Hill JK, Ohlemüller R, Roy DB, Thomas CD (2011) Rapid range shifts of species associated with high levels of climate warming. Science 333:1024–1026

Clark CW, Mangel M (1986) The evolutionary advantages of group foraging. Theor Popul Biol 30: 45–75

Clutton-Brock TH (1988) Reproductive success. Studies of individual variation in contrasting breeding systems. In: Clutton-Brock TH (ed) Reproductive success. Chicago University Press, Chicago, pp 472–486

Cooper CB, Hochachka WM, Dhondt AA (2007) Contrasting natural experiments confirm competition between house finches and house sparrows. Ecology 88:864–870

Corbacho C, Morán R, Villegas MA (2005) La alimentación del aguilucho cenizo *Circus pygargus* en relación a los usos del suelo en áreas pseudoestepáricas de Extremadura (SO Península Ibérica). Ardeola 52:3–19

Cramp S, Simmons KEL (1980) The birds of the Western Paleartic, vol II. Oxford University Press, Oxford

Creel S, Christianson D (2008) Relationships between direct predation and risk effects. TREE 23: 194–201

Creel S, Winnie J Jr, Maxwell B, Hamlin K, Creel M (2005) Elk alter habitat selection as an antipredator response to wolves. Ecology 86:3387–3397

Cruz CSF, Fredo G, Casagrande R, Oliveira L, Rolim V, Marques S, Pavarini S, Driemeier D (2016) *Eucoleus contortus* parasitism in captive-bred valley quail *Callipepla californica* (Shaw, 1798): disease and control. Zool Gart 85:152–159

De Borbón MN, Barros C, De Juana E (1999a) El gregarismo en las gangas ibérica y ortega. In: Herranz J, Suárez F (eds) La ganga ibérica (*Pterocles alchata*) y la ganga ortega (*Pterocles orientalis*) en España. Distribución, abundancia, biología y conservación, Ministerio de Medio Ambiente. Organismo Autónomo de Parques Nacionales, Madrid, pp 195–214

De Borbón MN, Barros C, Guadalfajara R, De Juana E, Herranz J (1999b) Parámetros reproductivos de la ganga ibérica (*Pterocles alchata*) y la ganga ortega (*Pterocles orientalis*). In: Herranz J, Suárez F (eds) La ganga ibérica (*Pterocles alchata*) y la ganga ortega (*Pterocles orientalis*) en España. Distribución, abundancia, biología y conservación, Ministerio de Medio Ambiente. Organismo Autónomo de Parques Nacionales, Madrid, pp 239–259

De Juana E (2005) Steppe birds: a characterisation. In: Bota G, Morales MB, Mañosa S, Caprodon J (eds) Ecology and conservation of steppe-land b, irds. Lynx Edicions & Centre Tecnològic Forestal de Catalunya, Barcelona, pp 25–48

Del Hoyo J, Collar NJ (2014) HBW and Birdlife international illustrated checklist of the birds of the world. Volumen 1: non-passerines. Lynx Edicions & Birdlife International, Barcelona

Delgado MP, Traba J, García de la Morena E, Morales MB (2010) Habitat selection and density-dependent relationships in spatial occupancy by male little bustards *Tetrax tetrax*. Ardea 98: 185–194

Ellis BK, Stanford JA, Goodman D, Stafford CP, Gustafson DL, Beauchamp DA, Chess DW, Craft JA, Deleray MA, Hansen BS (2011) Long-term effects of a trophic cascade in a large lake ecosystem. PNAS 108:1070–1075

Estrada A, Delibes-Mateo M, Caro J, Viñuela J, Díaz-Fernández S, Casas F, Arroyo B (2015) Does small-game management benefit steppe birds of conservation concern? A field study in Central Spain. Anim Conserv 18:567–575

Faria N, Rabaça JE, Morales MB (2012) The importance of grazing regime in the provision of breeding habitat for grassland birds: the case of the endangered little bustard (*Tetrax Tetrax*). J Nat Conserv 20:211–218
Ferguson-Lees IJ (1967) Studies of less familiar birds 142: little bustard. Br Birds 60:80–84
García de la Morena E (2015) Ecología y movimientos migratorios del sisón común (*Tetrax tetrax*) fuera del periodo reproductor. PhD dissertation. Universidad Autónoma de Madrid, Madrid
García J, Suárez-Seoane S, Miguélez D, Osborne PE, Zumalacárregui C (2007) Spatial analysis of habitat quality in a fragmented population of little bustard (*Tetrax tetrax*): implications for conservation. Biol Conserv 137:45–56
Garzón J (1974) Contribución al estudio del status, alimentación y protección de las Falconiformes en España Central. Ardeola 19:279–330
Gauger K (2007) Occurrence, ecology and conservation of wintering little bustards *Tetrax tetrax* in Azerbaijan. Arch Nat Landschafts 46:5–27
Grant PR, Grant BR (2006) Evolution of character displacement in Darwin's finches. Science 313: 224–226
Greenberg R (2000) Birds of many feathers: the formation and structure of mixed-species flocks of forest birds. In: Boinski S, Gerber PA (eds) On the move: how and why animals travel in groups. Chicago University Press, Chicago, pp 521–558
Hamilton WD (1971) Geometry for the selfish herd. J Theor Biol 31:295–311
Heikkinen RK, Luoto M, Virkkala R, Pearson RG, Körber JH (2007) Biotic interactions improve prediction of boreal bird distributions at macro-scales. Global Ecol Biogeogr 16:754–763
Hellmich J (1984) Über geflügelte beutegreifer der zwergtrappe (*Tetrax tetrax*) unter besonderer berücksichtigung des uhus (*Bubo bubo*). Beitr Vogelkunde 30:169–177
Hudson PJ, Dobson AP, Lafferty KD (2006) Is a healthy ecosystem one that is rich in parasites? TREE 21:381–385
Inchausti P, Bretagnolle V (2005) Predicting short-term extinction risk for the declining little bustard (*Tetrax tetrax*) in intensive agricultural habitats. Biol Conserv 122:375–384
IUCN (2016) The IUCN red list of threatened species. Version 2016-2. http://www.iucnredlist.org. Downloaded on 03 November 2016
Jiguet F (2002) Arthropods in diet of little bustards *Tetrax Tetrax* during the breeding season in Western France. Bird Study 49:105–109
Jiguet F, Bretagnolle V (2014) Sexy males and choosy females on exploded leks: correlates of male attractiveness in the little bustard. Behav Proc 103:246–255
Kruijt JP, de Vos GJ (1988) Individual variation in reproductives success in male black grouse, *Tetrao tetrix* L. In: Clutton-Brock TH (ed) Reproductive success. Studies of individual variation in contrasting breeding systems. Chicago University Press, Chicago, pp 279–290
Lane SJ, Alonso JC, Alonso JA, Naveso MA (1999) Seasonal changes in diet and diet selection of great bustards (*Otis tarda*) in Northwest Spain. J Zool 247:201–214
Lane SJ, Alonso JC, Martín CA (2001) Habitat preferences of great bustard *Otis tarda* flocks in the arable steppes of Central Spain: are potentially suitable areas unoccupied? J Appl Ecol 38:193–203
Lapiedra O, Ponjoan A, Gamero A, Bota G, Mañosa S (2011) Brood ranging behaviour and breeding success of the threatened little bustard in an intensified cereal farmland area. Biol Conserv 144:2882–2890
Leitão D, Costa LT (2001) First approach to the study of the non-breeding abundance and habitat use by the little bustard *Tetrax tetrax* in the lower Tejo grassland (South Portugal). Airo 11:37–43
Lima SL (1995) Back to the basics of anti-predatory vigilance: the group-size effect. Anim Behav 49:11–20
López-Jamar J, Casas F, Díaz M, Morales MB (2011) Local differences in habitat selection by great bustards *Otis tarda* in changing agricultural landscapes: implications for farmland bird conservation. Bird Conserv Int 21:328–341

López-Neyra CR (1944) Compendio helmintología Ibérica (continuación). Parte II. Capítulo III. Rev Ibér Parasitol 4:403–491

Marcelino J, Moreira F, Mañosa S, Cuscó F, Morales MB, García de la Morena E, Bota G, Palmeirim JM, Silva JP (2017) Tracking data of the little bustard *Tetrax tetrax* in Iberia shows high anthropogenic mortality. Bird Conserv Int. https://doi.org/10.1017/S095927091700051X

Martin PR, Martin TE (2001) Ecological and fitness consequences of species coexistence: a removal experiment with wood warblers. Ecology 82:189–206

Martín CA, Casas F, Mougeot F, García JT, Viñuela J (2010) Positive interactions between vulnerable species in agrarian pseudo-steppes: habitat use by pin-tailed sandgrouse depends on its association with the little bustard. Anim Conserv 13:383–389

Martínez-Padilla J, Redpath SM, Zeineddine M, Mougeot F (2014) Insights into population ecology from long-term studies of red grouse *Lagopus lagopus scoticus*. J Anim Ecol 83:85–98

McCallum H, Dobson A (1995) Detecting disease and parasite threats to endangered species and ecosystems. TREE 10:190–194

Millán J, Gortázar C, Tizzani P, Buenestado FJ (2002) Do helminths increase the vulnerability of released pheasants to fox predation? J Helminthol 76:225–229

Millán J, Gortázar C, Villafuerte R (2004) A comparison of the helminth faunas of wild and farm-reared red-legged partridges. J Wildlife Manag 68:701–707

Morales MB, Traba J (2009) Compromisos adaptativos en la selección de hábitat de aves esteparias. In: Dopazo H, Navarro A (eds) Adaptación y Evolución: 150 años después del Origen de las Especies. SESBE, Madrid, pp 304–313

Morales MB, Alonso JC, Martín CA, Martín E, Alonso JA (2003) Male sexual display and attractiveness in the great bustard *Otis tarda*: the role of body condition. J Ethol 21:51–56

Morales MB, Bretagnolle V, Arroyo B (2005a) Viability of the endangered little bustard *Tetrax tetrax* population of Western France. Biodivers Conserv 14:3135–3150

Morales MB, García JT, Arroyo B (2005b) Can landscape composition changes predict spatial and annual variation of little bustard male abundance? Anim Conserv 8:167–174

Morales MB, Suárez F, García de la Morena E (2006) Résponses des oiseaux de steppe aux différents niveaux de mise en culture et d'intensification du paysage agricole: Analyse comparative de leurs effects sur la densité de population et la sélection de l'habitat chez l'outarde canepetière *Tetrax tetrax* et l' outarde barbue *Otis tarda*. Rev Ecol (Terre Vie) 61:261–269

Morales MB, Traba J, Carriles E, Delgado MP, García de la Morena E (2008) Sexual differences in microhabitat selection of breeding little bustards *Tetrax tetrax*: ecological segregation based on vegetation structure. Acta Oecol 34:345–353

Morales MB, Traba J, Delgado MP, García de la Morena E (2013) The use of fallows by nesting little bustard *Tetrax tetrax* females: implications for conservation in mosaic cereal farmland. Ardeola 60:85–97

Morales MB, Casas F, García de la Morena E, Ponjoan A, Calabuig G, Martínez-Padilla J, García JT, Mañosa S, Viñuela J, Bota G (2014) Density dependence and habitat quality modulate the intensity of display territory defence in an exploded lekking species. Behav Ecol Sociobiol 68:1493–1504

Morales MB, Traba J, Arroyo B (2015) El declive del sisón en el centro de España. Quercus 356:36–43

Morris DW (1987) Spatial scale and the cost of density-dependent habitat selection. Evol Ecol 1:379–388

Morris DW (1988) Habitat-dependent population regulation and community structure. Evol Ecol 2:253–269

Morris DW (1999) Has the ghost of competition passed? Evol Ecol Res 1:3–20

Morris DW (2003) Toward an ecological synthesis: a case for habitat selection. Oecologia 136:1–13

Newton I (1998) Population limitations in birds. Academic Press, London

Pace ML, Cole JJ, Carpenter SR, Kitchell JF (1999) Trophic cascades revealed in diverse ecosystems. TREE 14:483–488

Pannekoek J, Van Strien A (2005) TRIM 3 manual (Trends & Indices for monitoring data). Statistics Netherlands, Voorburg
Pearson RG, Dawson TP (2003) Predicting the impacts of climate change on the distribution of species: are bioclimate envelope models useful? Global Ecol Biogeogr 12:361–371
Pianka ER (1974) Niche overlap and diffuse competition. PNAS 71:2141–2145
Poulin R (1999) The functional importance of parasites in animal communities: many roles at many levels? Int J Parasitol 29:903–914
Ripple WJ, Estes JA, Beschta RL, Wilmers CC, Ritchie EG, Hebblewhite M, Berger J et al (2014) Status and ecological effects of the world's largest carnivores. Science 343:1241484
Robinson RA, Sutherland WJ (2002) Post-war changes in arable farming and biodiversity in Great Britain. J Appl Ecol 39:157–176
Rosenzweig ML (1981) A theory of habitat selection. Ecology 62:327–335
Salamolard M, Moreau C (1999) Habitat selection by little bustard *Tetrax tetrax* in a cultivated area of France. Bird Study 46:25–33
Sansom A, Cresswell W, Minderman J, Lind J (2008) Vigilance benefits and competition costs in groups: do individual redshanks gain an overall foraging benefit? Animal Behav 75:1869–1875
Schulz H (1985) Grundlagenforschung Zur Biologie Der Swergtrappe *Tetrax tetrax*. Staatlichen Naturhistorischen Museum, Braunschweig
SEO/Birdlife (2012) Resultados del programa de seguimiento de aves comunes en primavera. SEO/Birdlife, Madrid
Shlyakhtin GV, Tabachishin VG, Khrustov AV, Zav'yalov EV (2004) Ecological segregation of bustards (Otididae) in the north of the lower Volga region: evolutionary and adaptive aspects. Russ J Ecol 35:247–253
Silva JP, Pinto M, Palmeirim JM (2004) Managing landscapes for the little bustard *Tetrax tetrax*: lessons from the study of winter habitat selection. Biol Conserv 117:521–528
Silva JP, Estanque B, Moreira F, Palmeirim JM (2014) Population density and use of grasslands by female little bustards during lek attendance, nesting and brood-rearing. J Ornithol 155:53–63
Sridhar H, Beauchamp G, Shanker K (2009) Why do birds participate in mixed-species foraging flocks? A large-scale synthesis. Anim Behav 78:337–347
Suárez F, Yanes M, Herranz J (1993) Nature-reserves and the conservation of Iberian shrubsteppe passerines—the paradox of nest predation. Biol Conserv 64:77–81
Tarjuelo R, Delgado MP, Bota G, Morales MB, Traba J, Ponjoan A, Hervás I, Mañosa S (2013) Not only habitat but also sex: factors affecting spatial distribution of little bustard *Tetrax tetrax* families. Acta Ornithol 48:119–128
Tarjuelo R, Morales MB, Traba J, Delgado MP (2014) Are species coexistence areas a good option for conservation management? Applications from fine scale modelling in two steppe birds. PLoS One 9(1):e87847
Tarjuelo R, Barja I, Morales MB, Traba J, Benítez-López A, Casas F, Arroyo B, Delgado MP, Mougeot F (2015) Effects of human activity on physiological and behavioral responses of an endangered steppe bird. Behav Ecol 26:828–838
Tarjuelo R, Morales MB, Arroyo B, Mañosa S, Bota G, Casas F, Traba J (2017a) Intraspecific and interspecific competition induces density-dependent habitat niche shifts in an endangered steppe bird. Ecol Evol 7:9720–9730
Tarjuelo R, Traba J, Morales MB, Morris DW (2017b) Isodars unveil asymmetric effects on habitat use caused by competition between two endangered species. Oikos 126:73–81
Traba J, Morales MB (2019) The decline of farmland birds in Spain is strongly associated to the loss of fallow-land. Sci Rep 9:9473
Traba J, Morales MB, García de la Morena E, Delgado MP, Krištin A (2008) Selection of breeding territory by little bustard (*Tetrax tetrax*) males in Central Spain: the role of arthropod availability. Ecol Res 23:615–622
Traba J, Morales MB, Carmona CP, Delgado MP (2015) Resource partitioning and niche segregation in a steppe bird assemblage. Community Ecol 16:178–188

Valverde JA (1967) Estructura de una comunidad mediterránea de vertebrados terrestres. Monografías de Ciencia Moderna, vol 76. Madrid

Veiga JP (1985) Ecología de las rapaces de un ecosistema mediterráneo de montaña. Aproximación a su estructura comunitaria. PhD dissertation. Universidad Complutense de Madrid

Verestchagin NK (1940) On little bustards and great bustards wintering in Eastern Transcaucasia. Rep Azerbaijan Branch Acad Sci 5:57–65

Villanúa D, Casas F, Viñuela J, Gortázar C, García de la Morena E, Morales MB (2007) First occurrence of *Eucoleus contortus* in a little bustard *Tetrax tetrax*: negative effect of red-legged partridge *Alectoris rufa* releases on steppe bird conservation? Ibis 149:405–406

Villanúa D, Pérez-Rodríguez L, Casas F, Alzaga V, Acevedo P, Viñuela J, Gortázar C (2008) Sanitary risks of red-legged partridge releases : introduction of parasites. Eur J Wildlife Res 54: 199–204

Whittingham MJ, Evans KL (2004) The effects of habitat structure on predation risk of birds in agricultural landscapes. Ibis 146:210–220

Yabsley MJ (2008) Capillarid nematodes. In: Atkinson CT, Thomas NJ, Bruce Hunter D (eds) Parasitic diseases of wild birds. Wiley-Blackwell, Oxford, pp 463–497

Ziv Y, Abramsky Z, Kotler BP, Subach A (1993) Interference competition and temporal and habitat partitioning in two gerbil species. Oikos 66:237–246

Captive Breeding, Handling and Care, and the Impact of Releases on Wild Populations

Carole Attié, Alicia Munoz, Oriane Chevasson, and Vincent Bretagnolle

Introduction

Many bustard species in general, and the western-Europe population of little bustards in particular, are currently endangered and threatened with extinction. Among the several conservation strategies that may be put in place to reverse documented declines, captive breeding, reintroduction, or the reinforcement of local populations, also called "supportive breeding" (Chargé et al. 2014), have been repeatedly tried in various bustard species, including the little bustard. In this chapter, we review the reasons why, as well as the way, captive breeding has been or might be successfully carried out in this species.

C. Attié
LPO France, Rochefort Cedex, France

Coursan, France
e-mail: carole.attie@orange.fr

A. Munoz · O. Chevasson
ZOODYSSEE, Elevage conservatoire de l'outarde canepetière, Virollet, Villiers-en-Bois, France
e-mail: Alicia.MUNOZ@deux-sevres.fr; oriane.chevasson@deux-sevres.fr

V. Bretagnolle (✉)
Centre d'Etudes Biologiques de Chizé, CNRS & La Rochelle Université, Beauvoir sur Niort, France
e-mail: vincent.bretagnolle@cebc.cnrs.fr

V. Bretagnolle et al. (eds.), *Little Bustard: Ecology and Conservation*, Wildlife Research Monographs 5, https://doi.org/10.1007/978-3-030-84902-3_10

Reasons for Little Bustard Captive Breeding: The Case of France

The little bustard was historically present in most agricultural plains in France, but its population faced a major crash between the 1970s (with c.10,000 males: Yeatman 1976) and the 1990s (c.1300 males in 2000, i.e. an 87% decline: Jolivet and Bretagnolle 2002). A particularly severe decline occurred in central-western France (Poitou-Charentes region), from 6000 males in 1985 to only 250–300 males by 2000 (see chapter "The little bustard around the world: distribution, global conservation status, threats and population trends"). A conservation plan was prepared for the species in central-western France by LPO (French BirdLife partner), CNRS and Ministry of Environment (DREAL Poitou-Charentes), and evaluated several alternative management interventions. Regulatory measures (e.g. nature reserves) for the protection of habitats are ill-suited for cultivated plains, and a land-sparing strategy was rejected (Bretagnolle et al. 2011). Therefore, agri-environmental measures (AES), as a land-sharing strategy, were considered as the main if not the only tool for implementing environmental restoration measures in favour of farmland birds.

However, AES measures are not only slow to apply because of their implementation mechanism, but they are also contractual in nature and cannot be imposed on farmers (Bretagnolle et al. 2018). Their application requires negotiation with farmers, and measures are only contracted for short time periods, usually, 5 years, and they often change markedly between Common Agricultural Policy (CAP) reforms. The little bustard in the centre-west lowlands of France is now largely restricted to 10 Special Protection Areas (157,000 ha) where farmers can subscribe to agri-environment measures. Such measures have proved effective in stopping the decline: numbers have been more or less maintained between 2004 and 2018 thanks to the implementation of the Natura 2000 network with c.10,000 ha of agri-environment measures (Bretagnolle et al. 2011, 2018). But the decoupling of CAP subsidies (2006), the abandonment of compulsory set-aside in order to facilitate the deployment of energy crops (2007), the high prices of cereals (e.g. in 2009) and the last CAP reform (2014) strongly limited the effectiveness of these measures. Thus, despite four European and National Action Plans with two successive LIFE projects in western France, the decline continues (300 males in 2009, 244 males in 2014; Gendre et al. 2018), and little bustard populations still face a high risk of extinction (Inchausti and Bretagnolle 2005). Another strategy was thus tested in the second LIFE project (RENF TETRAX n ° LIFE04 NAT/FR/000091, 2004–2009): reinforcing the little bustard's migratory population by supportive breeding to compensate for the lack of agri-environment measures, with the constitution of a conservatory breeding stock (2009), the building of a breeding centre, and the release of captive-raised fledglings into the wild.

Before any captive breeding programme is initiated, in particular when endangered species or populations are involved, a careful examination of alternative strategies is needed (see Dolman et al. 2015 for an evaluation of a possible captive breeding programme for the Indian great bustard *A. nigriceps*). A feasibility study on the breeding of bustards in captivity was thus led in 2002 by LPO, CNRS, and the Museum National d'Histoire Naturelle (MNHN), at the request of the French Ministry of Environment. The purpose of this feasibility study was to show that reinforcing populations of little bustards, in parallel with environmental management measures, would be necessary to avoid the extinction of the species in French farmlands. Population reinforcement was therefore not meant to replace or be done at the expense of other conservation actions, but rather as a complement to accompany them temporarily, as the total area under agri-environmental contracts was insufficient to halt the decline and was obviously not improving the species' conservation status. Simulation studies (Bretagnolle and Inchausti 2005; Inchausti and Bretagnolle 2005; Bretagnolle et al. unpubl. report) showed that without reinforcement this 'flagship' species, the particular target of the NATURA 2000 designations, was paradoxically likely to disappear at the precise moment when these designations came into force. The feasibility report indicated that the major cause of the decline was a sharp decrease in female productivity due to the destruction of nests during agricultural work and the high mortality of chicks due to lack of food (productivity per female fluctuated between 0.3 and 0.5 chicks per year, i.e. less than half the productivity required for population stability: Morales et al. 2005). Increasing productivity was not achieved sufficiently by AES (e.g. Bretagnolle et al. 2018) except locally in a few cases, so artificially increasing the production of chicks within these populations through captive breeding and release of fledglings appeared to be a promising solution. The study proposed both to take eggs from the wild, rearing the chicks in order to supplement populations directly and to captive-breed adult birds and release their offspring for the same purpose. This action was only feasible if pursued within a policy of conservation, part of the national restoration plan, and if adopted by the Ministry of Environment, following the opinion of the National Council for the Protection of Nature (CNPN).

Captive Breeding of Bustards

Great Bustard

Between 1973 and 1981, in former East Germany, the systematic destruction of great bustard nests during agricultural work motivated egg collecting from the wild in order to rear chicks in captivity for later release. By 1982, nearly 500 eggs had been artificially incubated, producing 190 juvenile birds which were released but with unknown fate (Martin et al. 1996). Up to 2017, 1070 chicks had been reared at the Buckow Nature Conservation Centre and released during 1980–2017, while adults in captivity produced additional chicks that were later released (Langgemach 2008). Hatching success of fertilized eggs improved from 64.7% in the 1980s to 79.9% in 2010–2017, as did the percentage of released birds from hatched eggs, 53.6% in the 1980s and 85.4% for 2010–2017. In addition, measures such as the prohibition of aircraft flights over core areas (2300 ha bought or leased in two SPAs) significantly reduced disturbance and improved egg fecundity in the wild, from 74% in the 1980s to 93% by 2017. Correspondingly, the wild population increased from 55 to 283 individuals during 1995–2017 (https://www.tierpark-berlin.de/en/nature-and-species-conservation/worldwide/great-bustard).

Similarly, in the 1980s in the former USSR, a number of eggs were taken from the wild, e.g. 81 chicks were born in captivity at the Severtsov Institut (from the Russian National Academy of Science) and subsequently released, but unfortunately again without post-release monitoring (Martin et al. 1996). In total, between 1973 and 1994, over a thousand great bustard eggs were collected in conjunction with agricultural work. The hatching rate was 57%, whereas the survival rate was 24% (for a summary, see Martin et al. 1996).

In the UK, the great bustard became extinct in the nineteenth century, but in 2003 a reintroduction programme was established, with the Severtsov Institute providing chicks between 2004 and 2012. In total, 428 eggs were collected in Russia, resulting in 308 chicks (72% hatching success), of which 167 were released in the UK (Burnside et al. 2012). After 2013 eggs were collected in Spain, with 33 juveniles being released in 2014 (hatching success of 44/56) and 29 released in 2015. The last bird was released in 2019. Despite changes in rearing and release strategies, initial post-release survival remained consistently low, with only 11.3% of 167 bustards surviving from release to 1 year post-release between 2004 and 2014 (Ashbrook et al. 2016). The first reproduction in the wild occurred in 2007 on Salisbury Plain. The objective of the project is to reach 100 adult bustards in the wild; in 2018, the UK population of great bustard was around 75 individuals on Salisbury Plain, and by 2020, the total population is said to have reached 100 birds, with second-generation breeding. However, survival of young birds after release remains the most difficult issue, with failure mainly attributable to the lack of maternal guidance and protection: in the wild, chicks associate with their mother for 275–330 days, while the captive-reared chicks are all released at 90 days (Martin et al. 1996). The future of this reintroduced population, therefore, appeared uncertain (Ashbrook et al. 2016), but in 2020, the adult population passed the threshold of self-sustainability, with about 20 nests.

Captive breeding in great bustard first took place in East Berlin Zoo in 1970 with four chicks (Ardamatskaya 1992). Twenty years later, Gödöllö University of Agriculture (Hungary), in collaboration with the Budapest Zoo, tried natural reproduction in captivity. After several years of unsuccessful attempts, the use of artificial insemination allowed four females to produce broods from 7.2 eggs on average (Modlinger et al. 1998). In Spain, the Breeding Centre for Steppic Birds (a private centre in Jerez de la Frontera) has successfully bred captive great bustards, also using artificial insemination, as did, in 2017, the Centre for Wildlife Recovery and Environmental Education 'Los Hornos' in Sierra de Fuentes (Cáceres, Spain) which obtained a chick from a pair reared in captivity. Overall, captive breeding in great bustard is challenging, and reinforcement, supportive breeding or reintroduction of populations in the wild has provided mixed results (Dolman et al. 2015). For instance, the Pannonian population of great bustard seems to survive through in situ conservation efforts without the help from either captive breeding or supportive breeding (Raab et al. 2010).

Houbara Bustard

In response to environmental degradation and excessive hunting pressure, by the beginning of the 1980s, the resident houbara bustard population in Saudi Arabia was on the verge of extinction. In 1986, captive breeding was launched at the National Wildlife Research Centre (NWRC) in Taif (Saint Jalme et al. 1996). The original stock consisted of eggs collected in Baluchistan (Pakistan) and Algeria (Saint-Jalme and van Heezik 1996). In 1995, another captive breeding centre, the Emirates Centre for Wildlife Propagation (ECWP) started in Morocco (Chargé et al. 2014). The founder stock came from wild-caught birds provided by the NWRC from collecting expeditions in Algeria in 1986 and 1987, their descendants born in Taif and repatriated to Morocco, and eggs taken from Morocco (Chargé et al. 2014). At these two centres, the hatching success of wild bird eggs was higher than 85%. An artificial insemination programme was used to fertilize captive females, and as early as 1993, 285 chicks were produced with fertility rates as high as 90% (e.g. van Heezik et al. 2002). At ECWP, the first releases started in 2000 (Lesobre et al. 2010), and after 6 years of the programme the production of chicks was around 1000 birds per year, which rapidly increased wild populations. Currently, c.10,000 chicks are produced yearly. These birds are primarily produced for hunting purposes and to support an off-take of around 2000 birds per year (Bacon et al. 2019). The release of thousands of fledglings each year in Morocco has not, however, translated into a major increase in the wild population (L. Bacon, unpubl. PhD thesis).

In Saudi Arabia (though recognized as a different species now), the first bustards resulting from the rearing of eggs collected in Baluchistan were reintroduced successfully into a reserve in 1993 and 1994, and the first record of reproduction in the wild was in 1995 (Gelinaud et al. 1997). However, the success of releases heavily depends on founders learning anti-predator behaviour, whereas minimizing contact with humans had no effect on survival after release (van Heezik et al. 1999). Many other breeding centres have now been established, e.g. in 2007 by the Emirates Bird Breeding Centre for Conservation (EBBCC) in Uzbekistan, and in the UAE, where a resident breeding population once existed. There, since 2004, the National Avian Research Centre (NARC) had released over 13,000 captive-bred houbara by the end of 2016 in a network of protected areas (Azar et al. 2018). In the latter case, it was found that group size at release and food availability in release sites affected survival probability (Azar et al. 2016), although this may not be the case on other life-history traits, such as fecundity parameters in released females (Bacon et al. 2017). Whether resident houbara populations can be sustained by released birds from captive breeding or whether releases can mitigate or compensate for hunting mortality has been assessed (Burnside et al. 2017; Azar et al. 2018; Dolman et al. 2018). Such studies, however, concluded that supportive breeding to reinforce exploited populations over the long term is a high-risk strategy, so long as source-sink dynamics operate between hunted and non-hunted areas (Hardouin et al. 2015).

Little Bustard

Captive breeding in the little bustard has been achieved in at least two places before it started under the second French LIFE project. In Spain, in Jerez de la Frontera, a small farm held at least two pairs and a group (1 male and 4 females) in covered aviaries. Birds were habituated to captivity, and one of the two males displayed. The two females produced on average eight eggs, and three chicks were reared in total (pers. comm. of the owner). The four females kept in a group with the male never laid eggs. The second case is the Centre for Biological Studies of Chizé (CEBC-CNRS) in France, which successfully reared several chicks from abandoned, destroyed nests during the first French LIFE Programme (1997–2001). In total 31 eggs from the Poitou-Charentes region, but also some from neighbouring regions, were incubated at CEBC; all chicks raised were released at the post-breeding gathering site of a wild bustard population in Poitou-Charentes; 78% of eggs brought back to CEBC yielded fledglings. Based on this latter trial experience, and the feasibility study, a decision was made to start a captive breeding programme in Poitou-Charentes in 2002–2003, and a second LIFE project was submitted to Europe by LPO in 2004.

Buildings and Infrastructures for Little Bustard Captive Breeding

Legal Aspects of Breeding Centres

A national exemption was established on 4 June 2010, by the Ministry of Environment, allowing the collection, transport, and keeping of bustards throughout the Poitou-Charentes region. This was regularly updated and is still valid in 2020. All birds kept in the various breeding centres have been the subject of Intra Community Certificates (CIC) issued individually (i.e. for each bird or chick to be kept in the captive breeding stock) by the competent authorities from the DREAL Poitou-Charentes. Since 7 September 2011, by order of the Ministry of Environment, LPO, and now the Zoodyssée Team were authorized to proceed with the release of 50–100 young bustards per year for the period 2010–2016, extended in 2017 to 2022. An “Arrêté Prefectoral” provides the list of all sites where birds can be released, as decided within the framework of the National Action Plan, and an ad hoc scientific committee that was created in 2018 and which is directed by the DREAL.

Facilities at the Successive Breeding Centres

During the LIFE project (2005–2009), a first breeding centre was established (the Sainte-Blandine breeding centre). It belonged to and was maintained by LPO. After the end of the LIFE project, in 2009, a new structure was used, the breeding centre for the preservation of farmland birds (CEPOP hereafter), a property of the Conseil Général des Deux Sèvres (CD79), based at Zoodyssée Centre, and financially and logistically supported by the CD79, the DREAL Poitou-Charentes, and the Ministry of Environment. CEPOP was still under the responsibility of LPO, however, but in 2014 LPO transferred this responsibility to CD79. Since 2006, another breeding centre was used in the "La Haute Touche" Park, managed by MNHN. In 2012 the very small breeding centre at CEBC-CNRS was closed, and the females were transferred to CEPOP. Since 2014, therefore, only two breeding centres have been maintained in France, Zoodyssée (former CEPOP) and La Haute Touche (MNHN).

In the first breeding centre (established 2005, Sainte-Blandine), facilities consisted of 11 aviaries 20 × 6 m, an office, a breeding room for the chicks, an insect farming room and a sheltered aviary 4 × 5 m. Transferred to Zoodyssée in 2010, the breeding centre was housed in a temporary structure (until 2014) comprising three rooms, a kitchen, an air-conditioned incubation room and office, a breeding room, and a land of c.1 ha. At the same time, outdoor structures evolved from 11 to 23 aviaries of 20 × 6 m, 7 aviaries 18 × 6 m and 2 aviaries 10 × 6 m. This was completed by a small aviary of 2 × 5 m for the chicks. In 2015, the centre moved again to its current site, a plot of 1 ha with a permanent building dedicated to the incubating and rearing of small chicks. It consists of an incubation and hatching room of 5 × 2.5 m, two breeding rooms for chicks (5 × 5 m and 4 × 5 m), a farmhouse kitchen (4 × 5 m), a storage room for food and small equipment (5 × 2.4 m), an office and desk facilities, and finally 17 outdoor aviaries for the breeders and the young, of which 8 aviaries were split into two smaller ones (half-size). Attached to the building is an outdoor aviary arranged for young chicks (2 × 5 m, semi-covered with a hard structure and natural ground). In 2015, a hangar was built to house an aviary (20 × 6 m), connected to 6 outdoors aviaries of 12 × 6 m, this being established to protect the birds during winter, even under snow. Since 2017, therefore, 17 aviaries 12 × 6 m have been created, 8 being split into two smaller ones, one aviary with a roof, and two isolated aviaries for late chick rearing. In 2019, the floor of the outdoor aviary for young chicks was covered with a geotextile with a new substrate each year to improve sanitary conditions and preserve non-concrete soil for chicks.

Eggs and Incubation

Origin of Eggs

Eggs can either be taken from the wild or come from captive breeding stock. Each of these strategies has advantages and disadvantages. In 2002, when the decision was made to have a breeding centre, an evaluation of whether eggs could be taken from Spain was tested under a pilot study. Spain was chosen because, a priori, the number of eggs available was potentially high, egg collection was facilitated by high little bustard breeding density, and the impact on the source population would be low given the Spanish population size. On the other hand, the complexity of the international administrative procedures and the legislation in this area constituted a serious obstacle. In addition, given that the Spanish population of little bustards is largely sedentary (see chapter “Migration, movements and non-breeding ecology”), there were risks in terms of migration and genetics since the reinforcement project specified that the migratory behaviour of the western France population should be absolutely preserved, as it is a biological trait specific to this population and likely under selective pressure (Villers et al. 2010). Birds from Spain might not perform winter migration or could be genetically too different from French bustards. A trial was thus conducted in 2003, with seven eggs brought from Spain, incubated, hatched, and raised in France. Five fledglings were released and fitted with GPS. The result was that these birds actually did not migrate (Villers et al. 2010), so this solution was discounted. In 2005 and 2006, five more eggs were collected, but chicks were not released but were kept in the breeding stock (see the section below).

An alternative source population had to be considered targeting the Poitou-Charentes region itself. The possibilities to collect eggs were indeed fewer and the impact on the population potentially greater, but there was no risk in terms of migration and genetics since it was the source population. There were two options for the removal of eggs. The first was to collect whole clutches in order to induce replacement laying, while the second was to collect only a part of the clutch to allow the female to continue incubation. The latter option was preferred since very few females were shown to fledge four or even three chicks successfully, most families being one or sometimes two fledglings per female (Bretagnolle et al. 2011, 2018). It would then be possible to take one or two eggs without affecting the status of wild populations.

A third possibility consisted in having a breeding stock of females laying their eggs in captivity. Although not mutually exclusive, the constitution of the captive stock (a priority) would be to some extent at risk of compromise by the releasing of captive-reared fledglings to the wild (for a detailed examination of the trade-off between strategies, see Dolman et al. 2015). However, a decision was made to establish this breeding stock, and three captive populations were available or constituted in 2005: CEBC-CNRS (with 2–3 females only), La Haute Touche (MNHN, with 3–10 females), and CEPOP (see section Composition for the number of females in the breeding stock). The three centres produced a large number of eggs, with variable hatchability (see below, Table 1). We ended up mixing the second and third strategies: for example in 2005 and 2006, of the 206 total eggs, 108 were collected in the field (52.4%). The collection rate varied between sites and years (Table 1). Between 2005 and 2008, an experiment was conducted to preserve the clutches in the wild, in which eggs were removed from the nests when found, replaced by dummy eggs (which were incubated by the female), while the real eggs were incubated at the breeding centre. Fertile eggs were returned to the nests 24 h before hatching. During the trial, of 49 eggs (8 in 2005, 15 in 2006, 18 in 2007 and 8 in 2008), only three were abandoned by the female.

Egg Laying by Captive Females

The breeding period in captive females is rather extended, although not necessarily more than in the wild (see chapter "Breeding biology and demographic traits" and Bretagnolle et al. 2018). Indeed, based on available data, females in captivity lay their eggs at about the same time as wild ones since the extreme dates are almost identical to those obtained in the wild (Bretagnolle et al. 2018). There were also inter-annual differences in both the first laying date and span of the laying period (Table 2). Little bustards of Spanish origin in the French breeding centers tended to breed earlier than their French counterparts, by about 2 weeks, suggesting there may be a genetic component in breeding phenology.

The average number of eggs produced per captive female was also quite variable (Table 3). Care was taken to stop laying in females that had laid over ten eggs, in order to avoid exhaustion. The maximum number of eggs laid by a female was 19 in a single year and 55 in a lifespan.

Based on data collected from 2016 to 2019, there was weak evidence of age-related breeding investment in females, since from 2 to 5–6 years of age, females tended to increase their clutch investment, then it stabilized with a tendency to decrease, though non-significantly, after 10 years (Fig. 1), in a possible senescence pattern (Fig. 1a). There was also a weaker tendency for male age affecting female clutch size, but male and female ages in aviaries were correlated.

There was, however, an extreme heterogeneity between females in the schedule of laying, as shown below (Fig. 2).

The average volume of eggs from females in the captive stock is c.35.1 cm^3, i.e. about 3.0 cm^3 less than wild eggs. More importantly, the average volume of eggs produced by captive females has constantly been decreasing over the years (Fig. 3). Egg volume was determined by measuring the length and width of eggs and using the following equation: volume $=$ length $\times$ $(\text{width})^2$ $\times$ 0.507/1000 (Hoyt 1979). This negative trend highlights that the egg volume of captive females is smaller than that of wild females and that this difference increases with age.

Table 1 Summary statistics of origin and fate of 1387 eggs used in the French captive breeding programme in 2005–2019. Ratio Hatched to total (Hatched + Failed) presented as %, in bold.

	Eggs from the wild			Eggs from CEBC-CNRS			Eggs from MNHN			Eggs from CEPOP/Zoodyssée		
Year	Hatched	Failed	% success	Hatched	Failed	% success	Hatched	Failed	% success	Hatched	Failed	% success
2005	32	8	**80**	2	3	**40.0**						
2006	49	19	**72.1**	9	3	**75.0**						
2007	50	4	**92.6**	4	3	**57.1**	12	5	**57.1**	1	1	**50.0**
2008	31	4	**88.6**	17	3	**85.0**	20	17	**54.1**	8	4	**66.7**
2009	16	1	**94.1**	18	6	**75.0**	46	65	**41.4**	13	7	**65.0**
2010				9	10	**47.4**	11	25	**30.6**	25	22	**53.2**
2011	9	2	**81.8**	1	7	**12.5**	18	16	**52.9**	4	8	**33.3**
2012				4	6	**40.0**	1	7	**14.3**	20	30	**40.0**
2013	3	0	**100.0**				1	1	**50.0**	13	12	**52.0**
2014										40	54	**42.6**
2015	2	0	**100.0**							17	21	**44.7**
2016										34	42	**44.7**
2017	3	0	**100.0**							85	101	**45.7**
2018	2	0	**100.0**				0	13	**0**	34	67	**33.7**
2019	3	1	**75.0**				5	6	**45.5**	36	75	**32.1**
Total/average	199	39	**84.5**	64	41	**61.0**	114	155	**38.4**	330	444	**46.4**

Note that failed eggs also include eggs that were broken in the aviary or which were not located in the aviary immediately after laying, as well as infertile eggs

Table 2 Summary statistics of egg-laying period for females (only French origin) used in the French captive breeding programme in 2008–2019

	2008	2009	2010	2011	2012	2013	2014	2015	2016	2017	2018	2019	Average
First egg	22/05	18/05	09/05	15/05	29/05	25/05	21/05	30/05	01/06	17/05	24/05	01/06	23/05
Last egg	15/07	06/08	12/07	05/07	16/07	21/07	10/07	01/07	29/07	07/08	23/07	30/07	18/07
Laying period range (days)	54	80	64	51	48	57	50	32	58	82	60	59	57

Table 3 Average number of eggs produced per female that laid eggs in the French captive breeding programme (total number of females in brackets)

	2005	2006	2007	2008	2009	2010	2011	2012	2013	2014	2015	2016	2017	2018	2019
CEBC-CNRS	5(1)	6(2)	3.5(2)	10(2)	12(2)	6.3(3)	4(2)	5(2)							
CEPOP/Zoodyssée		6(1)	3(2)	5(3)	5.7(4)	5.9(9)	2.8(5)	5.6(10)	3.2(14)	6.7(17)	4.4(10)	5.2(14)	6.6(28)	4.8(21)	4.6(24)
MNHN			10.5(2)	6.8(6)	7.5(15)	7.2(5)	7.2(5)							1.9(7)	2.2(5)
Average	5.0(1)	6.0(3)	5.7(6)	6.9(11)	7.6(21)	6.4(17)	4.8(12)	5.5(12)	3.2(14)	6.7(17)	4.4(10)	5.2(14)	6.8(25)	4.4(24)	4.2(29)

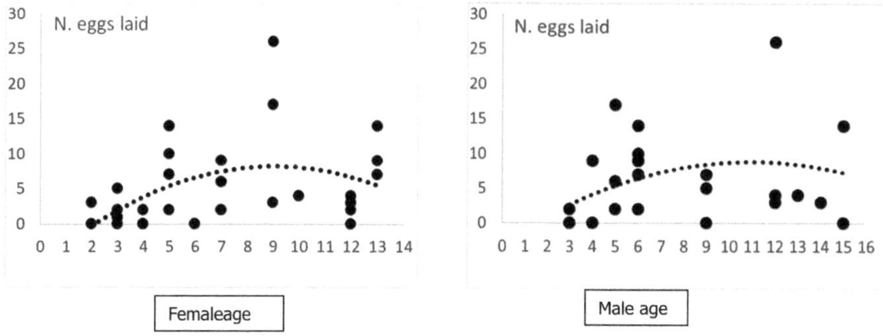

Fig. 1 Effects of female (left) and male (right) ages on the number of eggs laid per female (only cases of pairs, i.e. 1 male and 1 female per aviary, are shown here). Female/male age effects are not significant, but the trend is shown

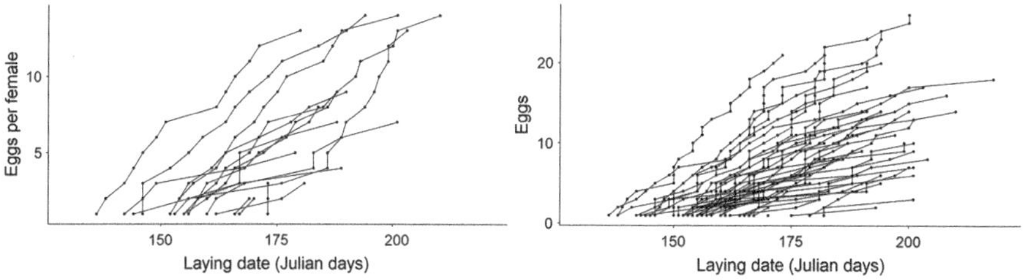

Fig. 2 Chronology of laying per individual female (2016–2018, left panel) and the total number of eggs laid per aviary (2016–2019, right panel)

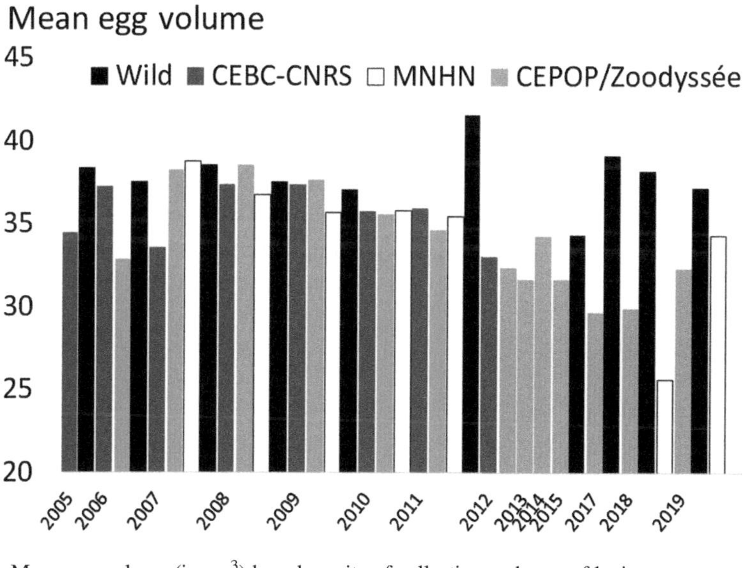

Fig. 3 Mean cgg volume (in cm^3) based on site of collection and year of laying

Incubation Techniques and Parameters

Eggs are laid in the aviaries, usually in concealed nests, so the precise date of laying is often unknown. Therefore, signs of females having laid (such as escaping from males) are used to detect the start of laying, and then, daily or every 2 days, inspections of aviaries are organized to collect eggs. After the first egg is laid, aviaries are inspected every 48 h or every 3 days (since little bustard females lay every 2 days); eggs are then removed from the aviaries and put into incubators. On day 10 of incubation, eggs are candled to see if they have been fertilized.

At La Haute Touche, eggs were incubated in automatic turning Grumbach incubators (chicks were later transferred to CEPOP). After disinfection and weighing, eggs were placed in a "dry" incubator for 3 days. The incubation temperature was 37.7 °C. In order to adjust the relative humidity of the incubators to an average weight loss of 15% over the duration of the incubation, one of the incubators was set for a relative humidity of 40% and the other to 50%. They were again weighed on the third day, and, in relation to percent weight loss, they were either kept in the same incubator or transferred to the wet incubator.

At CEPOP and then Zoodyssée, eggs were incubated in two Grumbach incubators with manual turning, and eggs were weighed every day to monitor weight loss and regulate temperature and humidity in the incubator. Large eggs were incubated at a higher temperature and a lower hygrometry than small eggs (T °C = 37.6 hygrometry = 40% vs. 37.5 and 50%) located in a ventilated and thermally stable room. The eggs are moved to a manual hatcher as soon as they are pipped, or the chicks' calls are audible through the shell (T °C = 37.5, hygrometry = 60%). Hatching usually occurs within 12 h of the eggs' pipping. After this time, the chicks are usually too weak to break the shell, and they die of exhaustion. In 2016 incubation parameters were set at 37.7 °C for all eggs, but in 2018 we changed the parameters to delay slightly the hatching date, which appeared to be too early (by c.1 day). Incubation temperature was reduced by only 0.1 and 0.2 °C, which delayed the hatching date to the expected correct day (21st day). Only 32% of eggs hatched at 21 days in 2016, 40% in 2017, 59% in 2018 and 67% in 2019, indicating that correcting for incubation temperature achieved the desired effect. Since 2018 the CO_2 level was also increased during the first 3 days of incubation because, in the wild, eggs are exposed to limited aeration with, therefore, a relatively high level of CO_2. The limitation of the oxygen level at the beginning of the embryonic development allows a better development of the chorioallantoic (vascular) membrane, the lungs and the heart.

Chicks and Chick Rearing

Hatching Rate and Hatching Failure

Hatching rates varied between centres and years; overall, hatching rates from eggs laid by captive females varied between 33.3 and 66.7% (see values in Table 1), which are low compared to eggs laid by wild females (see chapter "Breeding biology and demographic traits"). The eggs laid in the CNRS breeding centre had a higher hatching rate (65.4%), hence presumably higher fertilization rate, probably due to a more stable environment with little stress and the same person caring for the birds every day. Unfertilized eggs were removed from the incubators and systematically autopsied. Eight eggs from the wild were also autopsied, and they all were fertile. Between 2005 and 2019, 402 unhatched eggs were opened to check for embryo presence: 43.0% were fertile. Unfertilized eggs accounted for 41.9% in 2017 ($N = 186$), 41.4% in 2018 ($N = 87$) and 37.3% in 2019 ($N = 102$). Fertilization failure was possibly related to non-mating (no copulation), male sterility (either temporary or complete) or some form of genetic incompatibility due to inbreeding. Part of this might be accounted for by the senescence of both males and females from the breeding stock, as shown for houbara bustard (Preston et al. 2015). Senescence may also be at play in the little bustard. Indeed, the age of the birds from the breeding stock has markedly increased over the last 10 years, because only very few birds of wild origin joined the stock (see Table 4). Senescence is apparently quite strong in houbara bustard, affecting not only hatching rate but also chick growth curves and presumably survival; senescence is moreover detectable from the age of c.6 years old, or even earlier in males (Preston et al. 2015).

Based on 2016–2018 data ($N = 164$), we found a significant relationship between egg weight (at laying) and hatchling weight (Fig. 4), the former predicting the latter with over 75% confidence. Hatchling weight is an important feature since it predicts to some extent hatchling survival to fledging. In 2005–2015, the average weight at birth was 27.0 g ($N = 398$) for the chicks which survived at least 3 months and 26.2 g ($N = 127$) for those which did not. Using data from 2017 and 2018, an increase by 1 cm^3 in egg volume translates into 0.57 g increase in hatchling weight (0.41 g in 2018, 0.75 g in 2017). Using 2016–2019 data, we found that the hatchling weight of surviving chicks was 23.9 g ($N = 124$) while that of hatchlings that later died was 22.9 g ($N = 57$); in addition, while egg volume did not predict hatching rate (Estimate = −0.004, $Chi^2 = 0.03$, $p = 0.87$), hatchling weight did (estimate = 0.08, $Chi^2 = 4.43$, $p = 0.035$). Hatching at the right age (21 days) also increases the likelihood of survival. The hatchling survival rate is reduced by 5% if hatching occurs 1 day earlier and by 30% if 2 days earlier.

Table 4 Fate of chicks from the French captive breeding programme according to their origin. In total, 496 fledged. Of these, the percent that were released in the wild is given (bottom line of the table)

	Young birds from the wild			Young birds from CNRS eggs			Young birds from MNHN eggs			Young birds from CEPOP/ Zoodyssee eggs		
	Hatched	To captive stock	To release site	Hatched	To captive stock	To release site	Hatched	To captive stock	To release site	Hatched	To captive stock	To release site
2005	32	6	**22**	2		**2**						
2006	49	12	**30**	9	1	**4**						
2007	50	11	**35**	4	1	**2**	12	8		1		**1**
2008	31	1	**24**	17	2	**11**	20	6	**8**	8		**7**
2009	16	3	**12**	18	2	**9**	46	23	**13**	13		**12**
2010				9	5	**3**	11	6	**4**	25	8	**10**
2011	9	0	**3**	1		**1**	18	14	**1**	4		**1**
2012				4	1	**2**	1	1	**0**	20	4	**6**
2013	3						1	1	**0**	13	5	**2**
2014										40	15	**9**
2015	2	2								17	13	**0**
2016										34	20	**0**
2017	3	3	**0**							85	8	**42**
2018	2	2	**0**				0	0	**0**	34	5	**25**
2019	4	2	**0**				5	4	**0**	36	0	**25**
Total	201	42	**126 (63%)**	64	12	**34 (53%)**	114	63	**26 (23%)**	330	78	**140 (42%)**

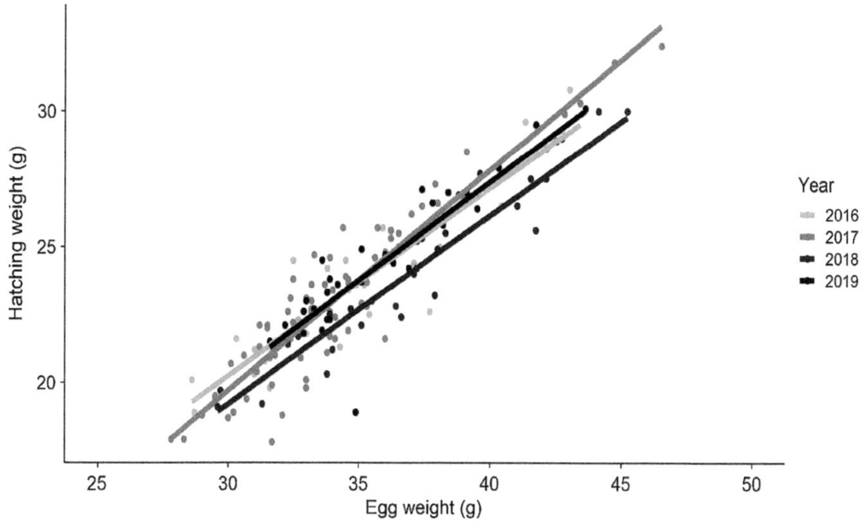

Fig. 4 Hatchling weight in relation to egg weight (all data from French captive breeding programme, 2016–2019)

Conditions for Caring Chicks According to Age

Once hatched and dry, the chicks are marked with coloured felt (since 2017, coloured elastic bands have been used), and placed in small equal-age groups of 2–5 (max 6), in individual boxes (l × w × h = 1 × 0.60 × 0.40 m). Boxes are lined with absorbent tissues. In the first 3 days, the chicks seek to snuggle together, and they are placed under a small cover at night before turning off the lights; since 2016, they have been placed next to an open washcloth. They are very sensitive to cooling during this period, and it is therefore important to avoid water presence within the box. Plastic carpets and grilled floors are also to be avoided, the first for reasons of hygiene and the second for safety. Boxes are equipped with a removable mesh cover, a heat lamp (35 °C) switched on 12 h/24 (24/24 for hatchlings), as well as, within the same box, a cooler place (c.24 °C) so they can self-regulate their temperature. The heat lamp is a UV lamp with a dose of about 150 UVA-B. The temperature is maintained between 35 and 37 °C during the first week after hatching and then regularly lowered during the second week until the lamp is removed when the chicks are 15 days old. Artificial lighting runs from 7 am to 9 pm. When they are 3 days old, chicks can stay outside when the air temperature allows it, but always with a heat lamp. By days 7–10 the chicks need exercise and should be moved outside in a larger enclosure, equipped with a shelter they can use during hot hours or in case of rain. At that age, groups should preferably contain chicks of various ages, the oldest showing the smallest how to feed themselves, for instance. Heating cannot be completely removed at night (given the latitude of Poitou-Charentes). When chicks are 20 days old, they are grouped (10 maximum) in an outdoor aviary of 10 × 6 m and height 2–2.5 m, with food and water at will.

Depending on the expected fate of the chick (i.e. either released in the wild or kept for the conservation centre), rearing differs quite dramatically. For birds scheduled for release, care is taken to avoid human contact completely from day 15, up to which time they have been held in wooden boxes. From 15 to 20 days old, they are transferred to outdoor aviaries, which are quite isolated and far from the building. Food is provided twice a day at this age. However, chicks are caught regularly (i.e. every 2–3 weeks) to monitor their biometrics. The rarity of handling events, coupled with the birds' isolation from humans, cause any familiarity with humans to disappear within a month, although only if chicks are reared in groups (chicks reared alone to learn more slowly to fear man). Around 60–80 days old, bustard chicks can be released into the natural environment and are usually placed in pre-release aviaries installed in fields (see below). For birds scheduled to be kept in the rearing centre, chicks are raised in boxes with transparent facades, they are handled before each meal, and a daily "game" is played when they are in outdoor aviaries. From 2005 to 2015, we also tried to have the females not only laying but also incubating their own eggs and rearing their chicks. Ten females successfully overcame the stress of captivity. The hatchability was 70% ($N = 40$ eggs), and the females were able to rear 19 youngsters.

Table 4 presents the fate of the chicks from hatching until they fledge, and they are either released in the wild or kept in the captive breeding stock. Of 709 hatched chicks, 496 (c.70%) survived until fledging, and 325 were released in the wild (i.e. 46% of hatchlings), while 195 were incorporated into the captive stock (this includes however many fledglings which could not be released because of injuries or malformation, and which did not survive their first winter).

Food for the Chicks

The diet has been progressively developed since 1996 at the CEBC, then at the CEPOP/Zoodyssée. This diet, mainly based on proteins, is supplemented by plants (alfalfa and clover) and calcium. For 16 years, we almost continuously experimented with combinations of food, with important constraints such as a diet being poor in salt and not too rich in carbohydrates (e.g. cereals) in order to avoid kidney and liver damage. Between 2006 and 2015 the birds were fed with insects, alfalfa and commercial cat (Flatazor), and parrot pellets (Nutribird) were added in 2014 which the chicks found very palatable. Until 2017 we provided the identical diet to chicks and adults, only the proportions of different foods differed between them.

During the 24–72 h following hatching, the chicks are fed exclusively on locusts, mealworms, and crickets, which are proportional in size to chick size. Insect heads are removed (to avoid ingestion of chitin) and are presented using a pair of tweezers. The chicks are stimulated frequently (i.e. one, up to five times per hour between 7 am and 9 pm, 8 am–8 pm since 2017), but later, to teach them to feed themselves, the feedings are spaced every hour at 5 days old and then every 2 h from 4–6 days old. Meals are progressively decreased from 10 to 4 meals with age. The birds are weighed every morning before the first feeding. At each feeding, they are stimulated by humans and food is distributed ad libitum in a dish. Under these conditions, chicks begin to feed alone from the seventh–ninth day, although they must regularly be encouraged for feeding four times a day until the age of 15, which allows checking it is feeding normally. The quantities of insects required to rear a chick are such (around 150–200 insects/day/chick) that it is necessary to find substitutes in the form of hydrated granules. By the third day, the diet was diversified with cat pellets, replaced since 2017 by a mixture of granules for pheasants and trout, plus mealworms and green vegetation (clover, alfalfa). Since 2018 a new diet has been specifically developed by WildNutrition for the little bustard, and these pellets are given to the chicks from their second day (in small proportions) and increased with age. Granules are made available to the chicks continuously, and insects are added. The ratio pellets/insects is around 0.66/0.33. Vitamin and calcium supplementation of Vitalos type is recommended. By the end of the first week, the insects are gradually replaced by pellets, but insects are maintained in the diet at low rate. On day 15, when transferred to outdoor aviaries, chicks feed alone. Alfalfa and grasses have been planted in aviaries and can be consumed at will as soon as the birds move into the outdoor aviary.

Problems Frequently Encountered During Chick Rearing

The main problems encountered during chick rearing include fractures (legs, wings), illness, and several anatomical and physiological weaknesses. Fractures may occur in a fairly large proportion of chicks in a given year (Table 5). From 2005–2019, fractures mainly occurred at 30–50 days old and mainly concerned wings. However, leg fractures have increased recently (since 2016), occurring at a younger age, on average 25 days old (Table 5). Fractures appear to be of pathological origin (vitamin D deficiency due to lack of UV exposure and/or poor phosphorus-calcium ratio caused by ingested insects). Fractures often lead to the death of the chick, not as a direct consequence but rather as a delayed effect (with euthanasia). A combination of UV lamp use, appropriate food and outdoor exercise is the best way to prevent fractures. In sunny years, even 1-day-old chicks should be put outside to sunbathe. If this is not possible, sunlight should be substituted by UV lamps and exercise to prevent bone weakness. The type of UV lamp is also important. Allowing chicks to go outside as soon and as often as possible and using proper UV light has reduced the number and proportion of fractures: 12 deaths in 2017 (22% of chicks born), one in 2018, and none in 2019 (despite 5 fractures). Recurrent disease (such as aspergillosis, which occurred in 2010 and 2016) can be lowered with the introduction of strict sanitary measures, such as complete sanitization of incubators and hatchers twice a week, and the wearing of gloves when manipulating eggs or feeding small chicks, etc. "Angel wings" may also occur in little bustard. This deformation of one or both

Table 5 Types and frequency of problems encountered by chicks reared under the French captive breeding programme 2004–2019

	No. chicks	Wing fracture	Leg fracture	% with problems	No. dead
2004	9	1	0	11.1	0
2005	34	1	2	8.8	2
2006	58	0	0	0	0
2007	67	0	0	0	0
2008	76	4	0	5.3	3
2009	93	3	1	4.3	2
2010	45	13	3	35.6	4
2011	32	6	1	21.9	3
2012	25	7	0	28.0	0
2013	14	0	1	7.1	0
2014	40	4	0	10.0	0
2015	17	0	0	0	0
2016	34	6	6	26.0	0
2017	85	25	26	40.0	12
2018	34	7	14	38.0	1
2019	36	1	4	12.0	0

wings during growth is attributed to a diet too high in carbohydrates and proteins, which is corrected by providing more green food. The weight of the growing flight feathers stresses the muscles, causing twisting. The problem was regularly encountered in fat chicks that experience temperature or food stress. Young birds can easily be treated in the early stages by binding the wing with a band-aid, but occasionally birds with this problem can break the wing.

Releasing Fledglings

Starting with two chicks in 1997, the release protocol for young bustards has been elaborated and improved from year to year (although further improvements can be made). Fledglings are released when they reach adult size, which is attained between days 50 and 60 (see also chapter "Breeding biology and demographic traits"), although releasing birds at later age is also possible. At this age, they are able to feed themselves and fly perfectly, but they, of course, need training before migration. They were systematically ringed and banded with coloured two-legged rings (since 2018, colour rings have been replaced by alphanumeric rings) and, for a proportion of them (depending on years), small VHF, GPS, or GPS-GSM transmitters. They are transported in fully closed cardboard (to prevent injury and death) boxes to the post-breeding flock site and are released at midday or in the morning. This hour was preferred since nightfall or very early morning releases may favour flying, which is to be avoided as they must remain in the group and be adopted by flocks of wild birds. Until 2018, pre-release aviaries were systematically used, identical to those used at the breeding centre. Release aviaries were tunnel greenhouses 20 × 6 m, and can accommodate at least 12–15 young bustards (up to 20). To avoid any risk of predation, each aviary was protected by an electric fence powered by a battery. Young bustards were housed in aviaries for up to 10 days. Pre-release aviaries were thought to improve familiarity with the release site, fixing them and allowing them to have a landmark after the release. Pre-release aviaries were abandoned in 2018 following a very high rate of predation in 2017 during three releases sessions, where a fox and several cats and dogs specialized in hunting released bustards around aviaries.

Released individuals have usually been monitored daily until the departure in migration, first of all, to check their location, survival, and fitness, but also their capacities to feed in the wild as well as to join with the groups of wild birds. Released birds were usually seen at the flocking site in flocks more than 40 days after release. Causes of casualties include predation (the main cause of death), shooting (few cases), collision with fences or powerlines, and the inability to stay in the wild (in 1998 and 1999, at least two birds returned to release aviary).

The French Captive Stock

Composition

The first little bustards kept in captivity for breeding were held at CEBC-CNRS, in 2004. There, up to six birds (three males and three females) were kept until 2012. The CEPOP captive stock started in 2005 (the first year of the LIFE project). In the first years, the captive stock was derived from wild eggs and hand-raised captive chicks. Later on, most of the new birds joining the captive stock came from captive-bred birds (i.e. eggs laid in the captive breeding centres). From 2014 on, we again tried to include wild birds (i.e. eggs laid in the wild) to diversify the gene pool. In 2019, the French captive breeding stock was composed of 42 birds held at Zoodyssée, 28 of which were born in the captive breeding centre and 14 from the wild (none of Spanish or hybrid origin), and 23 birds at La Haute Touche (of which 18 come from the Zoodyssée breeding centre and were moved in 2018). In total, therefore, the captive stock is 65 birds (33 males and 32 females, Table 6).

Food and Care of Adults in the Rearing Centre

Adults consume mainly pellets and vegetable matter, but this diet is supplemented by insects during the breeding season, which is supposedly important for female egg-laying. This seems to represent a balanced diet for the bustard and is mainly based on proteins (insects, cat granules) supplemented by parrot pellets, alfalfa, and

Table 6 Composition, by age and sex, of little bustard breeding stocks in the French captive breeding programme (only birds of French origin)

	2008						2009						2010						2011						2012					
	CNRS		CEPOP		MNHN		CNRS		CEPOP		MNHN		CNRS		CEPOP		MNHN		CNRS		CEPOP		MNHN		CNRS		CEPOP		MNHN	
Age	♂	♀	♂	♀	♂	♀	♂	♀	♂	♀	♂	♀	♂	♀	♂	♀	♂	♀	♂	♀	♂	♀	♂	♀	♂	♀	♂	♀	♂	♀
1			3	1	1	9		1		2						7					6	7	1	2				1		
2	1		2	2	3				3	1	1	7		1		1						7					4	8		
3					2	1	1		2	2	3				3	1	1	6		1		1						2		
4	2	2	1		2						2	1	1		2	2	3				2	2	1	6				1		
5							2	2	1		3						1	1	1		2	2	3				2	1	1	5
6													2	2	1		3					1	1		1		3	1	3	
7																			2	2	1		3						1	
8																									1	1	1		3	
9																														
10																														
11																														
12																														
>12																														
Total	3	2	6	3	8	10	3	3	6	5	9	8	3	3	6	11	8	7	3	3	11	20	9	8	2	1	10	14	8	5

(continued)

Table 6 (continued)

	2013				2014				2015				2016				2017				2018				2019			
	CEPOP		MNHN		CEPOP		MNHN		Zoodyssee		MNHN		Zoodyssee		MNHN		Zoodyssee		MNHN		Zoodyssee		MNHN		Zoodyssee		MNHN	
Age	♂	♀	♂	♀	♂	♀	♂	♀	♂	♀	♂	♀	♂	♀	♂	♀	♂	♀	♂	♀	♂	♀	♂	♀	♂	♀	♂	♀
1	1	4			2	2			4	4			3	4			12	8			3	4	3	0	2	5		
2		1				4			2	2			3	4			3	3			6	3	6	1	2	3	2	
3	4	6				1				2			2	2			3	4			1	3	2	4	6	2	6	3
4		1			4	5				1				3			2	2			1	3	1		1	3	2	1
5						1			4	3								3			1	1	1	1	1	3	1	1
6	2	1	1	4						1			4	3								3		1	1	1	1	1
7	2	1	2		2	1		3						1			2	3								3		
8					1	1	2		2	1		3						1			1	3						
9	2	1	2						1	1	2		2	1		3						1			1	2		
10					2	1	2							1	2		2	1		3						1		
11									2	1	2							1	2		2			2				
12													2	1	2							1	2		3			2
>12																	2	1	2		2	1	1		1	1	3	
Total	11	15	5	4	11	16	4	3	15	16	4	3	16	20	4	3	26	27	4	3	17	23	16	9	18	24	15	8

calcium. At La Haute Touche breeding centre, the diet of the bustards has been the subject of several modifications: in 2007, it was as developed at CEPOP, but in 2008 a new food composition was developed with a mixture of granules for cranes (Mazuri ©) and pellets developed for the houbara bustard by the company Guyomarch. The aviaries were also planted with grasses in the expectation the birds would eat them, and mealworms were distributed once a day during the breeding season. However, little bustards did not consume the houbara feed. Thus in 2009, the diet was modified by adopting a food formula based on a mixture of cat cakes "Flatazor ©" and granules for cranes "Mazuri ©." The granules were moistened before distribution, and cruciferous plants were added (e.g. broccoli), as well as mealworms and crickets at a rate of about 10 crickets/bird/day. Water and pellets were provided ad libitum and renewed every day. Sand and crushed oyster shells were also distributed in aviaries to provide grit and calcium intake, respectively, to avoid egg-laying deficiencies. At Zoodyssée, in 2018, new pellets confected by WildNutrition and St Laurent, consisting of alfalfa, oat, wheat, soya, peas, vitamins, and minerals, were tested in winter, while in summer, more animal proteins were added (worms, insects). In addition, fresh plants (broccoli, salad, young rapeseed, etc.) are also now provided.

Threats at the Captive Breeding Centre

Since its installation, the breeding centre has repeatedly faced the problem of predators, such as buzzards *B. buteo* that often forage near or even perch on the aviaries. Raptor presence causes stress to bustards: disturbed several times a day, they do not feel safe. Furthermore, they fly into the aviaries as soon as a buzzard appears, which may cause injury or even accidental death, in addition to severely disrupting reproduction, as happened in 2011. Other raptors may also frighten the birds, such as hen harrier *Circus cyaneus* or goshawks *A. gentilis*. In spring 2012, a fox managed to enter the centre and killed 14 bustards, including 8 females and 4 males, raised specifically for breeding. In summer 2015, another fox killed some young bustards about to be released.

Moreover, bustard food competitors may create serious problems: voles *Microtus arvalis* or house sparrows *Passer domesticus,* for instance may eat most food deposited for the bustards (pellets and insects), and the abundance of these small animals may even attract the attention of weasels. The only way to deal with all these issues is electric fences for larger carnivores, capture for smaller carnivores (and release elsewhere), vole trapping (and release elsewhere), and rodenticides (without poison). No strategy has been found for raptors.

Captive Breeding as a Conservation Strategy

Compatibility with IUCN Criteria

Captive breeding, reinforcement, or reintroduction projects all need to meet the criteria established by the Species Survival Commission of the International Union for Conservation of Nature (IUCN). Indeed, the feasibility study carried out in 2003–2004 strictly evaluated the IUCN criteria and how much the project of the captive breeding centre met them. A captive breeding programme should aim to strengthen populations that are in sharp decline and in danger of extinction in the near future. Releases should be done in areas still frequented by wild populations and under some degree of protection (e.g. NATURA 2000 sites or with high AES measures). Individuals in the breeding stock must not differ genetically from the wild population they are to join and must not be genetically impoverished. We, therefore, maintained a small number of birds of wild origin in the captive stock, if possible, every year. Finally, we determined that a reinforcement project would definitely reduce the significant risk of extinction that was facing the French populations of little bustards within the next 20 years. All these criteria were met during the course of the conservation project in place since 2004.

Impact of Egg Collection on Wild Populations

The impact of removing eggs (either temporarily or totally; see section "Legal aspects of breeding centres") was evaluated by CNRS during the LIFE project (2005–2010) and subsequently by two different approaches: first, by comparing the reproductive success of females that have not undergone egg removal compared to those that have; second, by estimating the probability of an egg, depending on the laying date and clutch size, producing a chick. This was analysed using data collected in 2005 and 2006 ($N = 61$ nests for which brood size was known). For 13 of them (21.3%), no collecting at all was done. We further removed from analyses nests completely destroyed by agricultural work. Thus, a final sample of 34 clutches was partially collected (9 clutches of 1 egg, 21 of 2, and 4 of 3). For comparison, only five broods with no collecting and with known fate for the whole family are available. Based on this limited sample size, we found no statistical effect of egg collecting. The hatching rate was 75% ($N = 24$ with collecting) or 80% ($N = 5$) without collecting. To investigate further, the larger CNRS database on bustard nests was used (Bretagnolle et al. 2018) over the period 1998–2004: hatching rate of this sample was 79% ($N = 28$), a value almost identical to the small sample size of 5 nests. It can therefore be reasonably concluded that the collecting of eggs (partial collection of eggs within a brood) had no detectable effect on the probability of hatching, nor on nest abandonment or predation rate (25% with egg collecting, 21% without sampling between 1998 and 2004; compared to 20% without collection in

2005 and 2006). At the egg level, out of 27 eggs (neither destroyed, predated, nor abandoned), 24 eggs (88%) hatched. Using data available from nests in 1998–2006 where no partial collecting was done, 67 of 73 eggs hatched (91%), a difference which is not significant ($Chi^2 = 0.20$, $p = 0.65$).

Therefore, removing part of the clutch for incubating eggs safely and hand-raise the chicks did not decrease, the hatching rate at brood or egg levels. On the other hand, it did not increase it either. Indeed, it might be expected that a female having to raise two chicks would have higher fledging success per capita than a female having to raise four chicks (with subsequent brood reduction). But this was not apparent from data collected in the field: of the 7 clutches monitored by CNRS, 4 families (1, 2, 2, 0; average 1.64) with partial collecting (2 eggs) were monitored throughout the breeding season. The family size of the non-collected broods (including some discovered at chick stage) was 1.5 chicks on average ($N = 6$), i.e. slightly lower. This difference was, however, not significant (Kruskal-Wallis test, $Chi^2 = 0.06$, $p = 0.8$), but sample sizes were very low. Before egg collecting was in place, between 2000 and 2004, the average family size was 1.84.

In conclusion, on the basis of these analyses, neither collecting full clutches for artificial incubation and bringing them back to the nest just before hatching nor partial collecting of a clutch, have little if any impact on the local bustard population. There was no impact on hatching rate and a slight positive (but not significant) effect on productivity. However, for the few nests where the dummy clutch was destroyed by agricultural works or predated, the full clutch was preserved in the rearing centre and the chicks hand-raised and eventually released in the field, and this was indeed a positive effect on population productivity. There was, therefore, no negative effect of nest searching and active conservation, at least at this single site in 2005 and 2006 and with the limited available data (23 eggs in total); and at this site, it is unlikely that in the absence of egg harvesting, the population growth rate would have been higher. Moreover, it does not seem strictly necessary to replace the eggs with dummy eggs since an experimental test showed that the females do not desert when some of their eggs are removed (irrespective of the stage of incubation, except during laying). However, fake eggs are needed when the whole brood is taken to be incubated safely and put back in the nest at hatching.

Survival of the Released Birds

The last but most important issue in a reinforcement programme is to judge its efficiency in regard to local population growth rate, which means ultimately to be able to evaluate, as accurately as possible, the survival probability and subsequent breeding of released birds. However, this simple task remains difficult to achieve, owing to the secretive habits of little bustards (in particular the females) but also to their highly dispersive behaviour. In addition, this requires constant search effort, both in space and time, to ensure a high probability of recapture. An alternative method is to use radio tags or even GPS, but this has potentially huge costs, and it may also impair the survival chances of the very birds whose survival we wish to measure, although Burnside et al. (2019) found no detectable effect of backpack satellite transmitters on any breeding parameters of houbara bustards. Below we present some lines of evidence that released birds survive quite well, at least under some conditions (in our case until 2016), breed, and thus contribute to local population dynamics. In addition, although the likelihood of resighting a released bird far from the release site (and the CNRS study site) is low, about 15% of the birds have been resighted outside their SPA release site, suggesting that releasing individuals also helps support neighbouring little bustard populations within the metapopulation, as was predicted (Bretagnolle and Inchausti 2005). The maximum distance travelled between the release site and display breeding site (for males) was 100 km.

The first line of evidence comes from simple visual recapture rates (i.e. sightings of colour-ringed released birds), which can be interpreted as the minimum number of bustards that have survived at least 1 year. This is a proxy of the survival rate. These data come from a file of over 5000 observations of which a third were provided by the CNRS team within the period 2005–2011. A total of 226 bustards were released in the wild during this period, of which 95 were resighted at least once and at least 1 year after release (see Table 7 for resighting rates in the first year after release). The observed survival rate is thus about 42% after 1 year. This rate, however, varies according to annual cohorts (groups of young bustards released each year; see Table 7). Average yearly recapture rates provide a slightly different value (38.3%), with high year-to-year variation: 18.8% in 2010 and 56% in 2008. From a subsample of 109 birds, 50% of the released bustards were resighted at one site, their release site, and 43% were contacted in two different SPAs.

The second line of evidence comes from radio-tracked birds. VHF radio was mostly used during the LIFE project to check whether birds migrated and where they spent winter. It showed that released birds migrated to Spain in areas where adults from French populations were also wintering (EL García de la Morena and MB Morales, unpubl. data). In addition, a combined dataset with released birds from the CNRS study site and re-sightings of these birds in this and other study sites (thus accounting to some extent for dispersal), allowed us to gather 10,717 resighting data with 442 individuals × sites for 379 different individual birds (323 released and 56 wild birds). Rémy Fay analysed data with Capture-Mark-Recapture software's RMark. He found that survival rates after 1 year, mixing both released and wild birds, differ between the sexes even when accounting for sex-biased recapture rate (males 0.605; females 0.495). There was little difference, however, between wild (0.513) and released birds (0.529). Together, these data indicate that the proxy of recovery rate is lower than survival rate estimated by the proper method, as expected. The survival rate of fledglings is about 0.5 and is not affected by captive rearing and release (actually released birds have slightly higher survival rates).

The third line of evidence comes from birds fitted with GPS or GPS/GSM loggers, which were only recently used. Since 2017, however, the predation rate at release sites has increased markedly, for reasons not yet understood. Such predation rate has been, in some years, over 50% (like in 2017), and survival rates of released birds from cohorts 2017–2019 is about 15%.

Table 7 Fate of fledglings released each year in the French little bustard captive breeding programme

	N. males released	N. females released	N. unsexed	Total	Fitted with GPS	Fitted with VHF	Dead in autumn	Resighted following year
1997			2 (16)	2	0	2	1	0
1998			11	11	0	11	3	1
1999			4	4	0	0	?	0
2000			0	0	0	0		
2001			0	0	0	0		
2002			7	7	1	4	?	1
2003	5	1	0	6	2	4	0	4
2004	7	2	0	9	0	0	?	0
2005	10	12	0	22	4	17	4	6
2006	19	14	0	33	0	25	3	11
2007	28	13	0	41	2	4	?	13
2008	29	21	0	50	0	0	?	23
2009	21	18	4	43	0	0	?	15
2010	9	3	3	15	0	0	?	1
2011	7	10	2	19	0	0	?	1
2012	4	1	3	0	0	0	0	0
2013	1	0	5	6	1	0	0	1
2014	6	4	1	18	0	0	?	1
2015	0	0	0	0	0	0		
2016	0	0	0	0	0	0		
2017	21	20	1	42	11	0	6	4
2018	11	11	3	25	14	0	8	3

Conclusions

Supportive breeding is particularly challenging when applied to migrating populations, given the risk of higher mortality rates as well as the loss of the migrating behaviour if entirely or partly learned from conspecifics, since in houbara bustards, at least, migrating behaviour has been shown to be partly altered by captive breeding (Burnside et al. 2017) and, despite care, there is an alteration of genetic traits (Chargé et al. 2014). Captive breeding in bustards is furthermore a difficult task, and this is the case for the little bustard. The main difficulties currently encountered concern egg fertility and fecundity, egg size (which predicts hatchling size, hence survival) and bird release. Up to now, no particular attention has been given to genetic issues (see Chargé et al. 2014), apart from excluding Spanish birds in the captive breeding stock and avoiding sib-sib and parent-offspring mating. One of the reasons for neglecting genetics is that the captive breeding stock is currently quite small, and short-term and may not currently be subject to these issues. Another reason is that other problems had to be solved. Releasing has been a particularly delicate task, since migratory behaviour is apparently partly innate but partly learned as well since solitary or late fledglings do not migrate (Villers et al. 2010), and the success of migration in captive-bred birds relies on whether they join a wild flock or not. Released birds which remain alone or in flocks of other released birds do not migrate, but if they do, they generally die in winter.

Predation has also been identified as a major threat to released little bustards, as it has been for houbara bustards, where the predation rate on released birds is about 38%, most being killed by mammalian predators, mainly red fox (van Heezik et al. 1999). Indeed, survival rates of released bustards from captive breeding stocks are usually low: in Asian houbara bustards, of 65 released birds, 58 died (27 in summer and 31 in winter), many prior to migration (Burnside et al. 2016). The proportion of surviving birds from release to next spring was only 0.108 (se 0.039), although the mean survival of released birds until migration was 58.5% (Burnside et al. 2016). Resident Asian houbara bustards released in Saudi Arabia had 47.5% survival probability in the first 3 months, despite predator-aversion training to increase survival probability (Saint-Jalme and van Heezik 1996). Survival of released captive-bred African houbara bustard at 3 months ranged between 19 and 93% (Hardouin et al. 2014). In addition, after 10 years in Abu Dhabi reserves, captive-bred houbara bustards were shown to survive (annual survival of 0.48 ± 0.12 in first year and 0.54 ± 0.08 thereafter) and breed (Azar et al. 2016, 2018). However, there is still uncertainty about the balance between the long-term efficiency of such costly (in both logistics and money) programmes, and indirect negative externalities such as a global fitness reduction of captive-bred birds released in the wild that may impact wild individuals. In the case of houbara bustards, several studies have warned about the use of annual supplementation to reinforce hunted populations over the long term (Burnside et al. 2017; Azar et al. 2018; Dolman et al. 2018).

Nevertheless, captive breeding and bird release in the Poitou-Charentes region has been successful, since, for instance, up to 12% of males displaying in the CEBC-CNRS study site came from the captive programme. These birds also dispersed to other places in the region, suggesting that the initial idea of releasing 100 fledglings per year (which has, however, never been attained) would be a significant asset in the conservation of little bustard in this and possibly other regions. However, so far there has been no significant impact of the reinforcement programme on the general population dynamics of little bustard in this region (although the aim of the reinforcement programme was simply to avoid extinction). To compare the alternative of finding nests and protecting them in order to improve breeding success, a cost-benefit analysis would be necessary since finding nests requires a team of four trained people working full time all over the season. Finding a nest with drones (equipped with infra-red cameras) is perhaps a promising alternative. Preliminary trials carried out in Poitou-Charentes by LPO with funds made available by the Action Plan (Ministry of Environment and DREAL Poitou-Charentes) is not yet fully conclusive.

Acknowledgments Many thanks to Christian Pacteau, Nadine and Noel Guillon, and several students who helped with the breeding centre at CNRS, which started in 1997. Special thanks to

Nadine Guillon, who cared for the bustards continuously until 2012. Michel Saint-Jalme, Rolland Simon and staff from La Haute Touche kindly provided data from the MNHN breeding centre. Over the years, many students, volunteers, fieldworkers, and contractors have helped with rearing bustards and their chicks without counting their hours. We also thank LPO (in particular Michel Métais, former Director), Conseil Départemental des Deux-Sèvres and DREAL Poitou-Charentes (now Nouvelle Aquitaine), in particular Catherine Ménard, for continuous funding support. Finally, we thank Philip Seddon and Robert Burnside for reviewing and improving this chapter.

References

Ardamatskaya TB (1992) Experience from work on the conservation of the Great Bustard abroad. Bustard Stud 5:107–121

Ashbrook K, Taylor A, Jane L, Carter I, Székely T (2016) Impacts of survival and reproductive success on the long-term population viability of reintroduced Great bustards *Otis tarda* in the UK. Oryx 50:583–592

Azar JF, Rautureau P, Lawrence M, Calabuig G, Hingrat Y (2016) Survival of reintroduced Asian houbara in United Arab Emirates' reserves. J Wildl Manag 80:1031–1039

Azar JF, Chalah T, Rautureau P, Lawrence M, Hingrat Y (2018) Breeding success and juvenile survival in a reintroduced captive-bred population of Asian houbara bustards in the United Arab Emirates. Endanger Species Res 35:59–70

Bacon L, Hingrat Y, Robert A (2017) Evidence of reproductive senescence of released individuals in a reinforced bird population. Biol Conserv 215:288–295

Bacon L, Robert A, Hingrat Y (2019) Long lasting breeding performance differences between wild-born and released females in a reinforced North African Houbara bustard (Chlamydotis undulataundulata) population: a matter of release strategy. Biodivers Conserv, Springer Verlag 28(3):553–570

Bretagnolle V, Inchausti P (2005) Modelling population reinforcement at a large spatial scale as a conservation strategy for the declining little bustard (*Tetrax tetrax*) in agricultural habitats. Anim Conserv. https://doi.org/10.1017/S1367943004001866

Bretagnolle V, Villers A, Denonfoux L, Cornulier T, Inchausti P, Badenhausser I (2011) Rapid recovery of a depleted population of Little Bustards *Tetrax tetrax* following provision of alfalfa through an agri-environment scheme. Ibis (Lond. 1859) 153:4–13

Bretagnolle V, Denonfoux L, Villers A (2018) Are farming and birds irreconcilable? A 21-year study of bustard nesting ecology in intensive agroecosystems. Biol Conserv 228:27–35

Burnside R, Carter I, Dawes A, Waters D, Lock L, Goriup P, Székely T (2012) The UK great bustard *Otis tarda* reintroduction trial: a 5-year progress report. Oryx 46:10

Burnside RJ, Collar NJ, Scotland KM, Dolman PM (2016) Survival rates of captive-bred Asian Houbara *Chlamydotis macqueenii* in a hunted migratory population. Ibis (Lond. 1859) 158: 353–361

Burnside RJ, Collar NJ, Dolman PM (2017) Comparative migration strategies of wild and captive-bred Asian Houbara *Chlamydotis macqueenii*. Ibis (Lond. 1859) 159:374–389

Burnside RJ, Guilherme JL, Collar NJ, Dolman PM (2019) Backpack-mounted satellite transmitters do not affect reproductive performance in a migratory bustard. Eur J Wildl Res 65:98

Chargé R, Sorci G, Saint Jalme M, Lesobre L, Hingrat Y, Lacroix F, Teplitsky C (2014) Does recognized genetic management in supportive breeding prevent genetic changes in life-history traits? Evol Appl 7:521–532

Dolman PM, Collar NJ, Scotland KM, Burnside RJ (2015) Ark or park: the need to predict relative effectiveness of ex situ and in situ conservation before attempting captive breeding. J Appl Ecol 52:841–850

Dolman PM, Collar NJ, Burnside RJ (2018) Captive breeding cannot sustain migratory Asian houbara *Chlamydotis macqueenii* without hunting controls. Biol Conserv 228:357–366

Gelinaud G, Combreau O, Seddon PJ (1997) First breeding by captive-bred houbara bustards introduced in Central Saudi Arabia. J Arid Environ 35:527–534

Gendre N, Eraud C, Bretagnolle V, Dalloyau S (2018) L'Outarde canepetière *Tetrax tetrax* en France en 2012 et 2016: effectifs et répartition. Ornithos 25:290–302

Hardouin LA, Robert A, Nevoux M, Gimenez O, Lacroix F, Hingrat Y (2014) Meteorological conditions influence short-term survival and dispersal in a reinforced bird population. J Appl Ecol 51:1494–1503

Hardouin LA, Hingrat Y, Nevoux M, Lacroix F, Robert A (2015) Survival and movement of translocated houbara bustards in a mixed conservation area. Anim Conserv 18:461–470

Hoyt DF (1979) Practical methods of estimating volume and fresh weight of bird eggs. Auk:73–77

Inchausti P, Bretagnolle V (2005) Predicting short-term extinction risk for the declining Little Bustard (*Tetrax tetrax*) in intensive agricultural habitats. Biol Conserv. https://doi.org/10.1016/j.biocon.2004.08.001

Jolivet C, Bretagnolle V (2002) L'Outarde canepetière *Tetrax tetrax* en France: Évolution récente des populations, bilan des mesures de sauvegarde et perspectives d'avenir. Alauda 70:93–96

Langgemach T (2008) Artificial incubation and rearing methods in the German Great Bustard (*Otis tarda*) conservation programme. Bustard Stud 7:5–17

Lesobre L, Lacroix F, Caizergues A, Hingrat Y, Chalah T, Saint Jalme M (2010) Conservation genetics of houbara bustard (*Chlamydotis undulata undulata*): population structure and its implications for the reinforcement of wild populations. Conserv Genet 11:1489–1497

Martin E, Alonso JA, Morales MB (1996) Evaluation of captive breeding as a method to conserve threatened Great Bustard (*Otis tarda*) populations. In: Sanz-Zuasti J, Fernandez-Gutierrez J (eds) Conservación de las Aves Esteparias y su Hábitat. Junta de Castilla y León, Valladolid, pp 131–138

Modlinger P, Chobot J, Peczely P (1998) Successful artificial insemination method for the captive breeding of the Great Bustard. Ostrich 69:239

Morales MB, Bretagnolle V, Arroyo B (2005) Viability of the endangered little bustard *Tetrax tetrax* population of western France. Biodivers Conserv. https://doi.org/10.1007/s10531-004-0382-z

Preston BT, Saint Jalme M, Hingrat Y, Lacroix F, Sorci G (2015) The sperm of aging male bustards retards their offspring's development. Nat Commun 6:1–9

Raab R, Kollar HP, Winkler H, Farago S, Spakovszky P, Chavko J, Maderic B, Skorpikova V, Patak E, Wurm H, Julius E, Raab S, S. C. (2010) Die Bestandsentwicklung westpannonischen Population der Grosstrappe, *Otis tarda* Linnaeus, 1758, von 1900 bis zum Winter 2008|2009. Egretta 51:74–99

Saint Jalme M, Combreau O, Seddon PJ, Paillat P, Gaucher P, Van Heezik Y (1996) Restoration of *Chlamydotis undulata* macqueenii (Houbara Bustard) populations in Saudi Arabia: a progress report. Restor Ecol 4:81–87

Saint-Jalme M, van Heezik Y (1996) Propagation of the houbara bustard. Kegan Paul International, London

van Heezik Y, Seddon PJ, Maloney RF (1999) Helping reintroduced houbara bustards avoid predation: effective anti-predator training and the predictive value of pre-release behaviour. Anim Conserv 2:155–163

van Heezik Y, Saint Jalme M, Hémon S, Seddon P (2002) Temperature and egg-laying experience influence breeding performance of captive female houbara bustards. J Avian Biol 33:63–70

Villers A, Millon A, Jiguet F, Lett JM, Attie C, Morales MB, Bretagnolle V (2010) Migration of wild and captive-bred Little Bustards *Tetrax tetrax*: releasing birds from Spain threatens attempts to conserve declining French populations. Ibis (Lond. 1859) 152:254–261

Yeatman L (1976) Atlas des oiseaux nicheurs de France. Société Ornithologique de France/Ministère de la Qualité de la Vie Environnement, Paris

Little Bustard Population Dynamics

Manuel B. Morales and Vincent Bretagnolle

Introduction

The study of population dynamics describes the way in which the size and structure of populations change over time (Fryxell et al. 2014). This process is largely determined by each species' characteristics, particularly life history traits, but also by the environmental conditions faced by each population, so that patterns of population growth and their underlying mechanisms may not only vary between species, but also between populations of a species. Therefore, to apprehend adequately a species' population dynamics, it is important to gain a good knowledge of its life history traits and the range of their variation, as well as of the heterogeneity of conditions under which the species may occur (De Roos et al. 2003). Likewise, our ability to predict what the future of a given population may be, in order to either manage or simply conserve it, requires such information to be as accurate as possible. Previous chapters have already tackled little bustard life history (chapter "Natural history of the little bustard: morphology, biometry, diet, sexual dimorphism, and social and breeding behaviour") and breeding biology and demography (chapter "Breeding biology and demographic traits"), as well as the range of environmental conditions in which the species is found (chapters "Habitat selection and space use" and "Migration, movements and

M. B. Morales (✉)
Terrestrial Ecology Group, Departamento de Ecología, Universidad Autónoma de Madrid, Madrid, Spain

Centro de Investigación en Biodiversidad y Cambio Global-CIBC, Universidad Autónoma de Madrid, Madrid, Spain
e-mail: manuel.morales@uam.es

V. Bretagnolle
Centre d'Etudes Biologiques de Chizé, CNRS & La Rochelle Université, Beauvoir sur Niort, France
e-mail: vincent.bretagnolle@cebc.cnrs.fr

V. Bretagnolle et al. (eds.), *Little Bustard: Ecology and Conservation*, Wildlife Research Monographs 5, https://doi.org/10.1007/978-3-030-84902-3_11

non-breeding ecology"). In the present chapter, we address little bustard population dynamics. We specifically examine population dynamic models in little bustard populations with different trends, that is, in expanding as well as in declining populations. We will review different types of density-dependent effects and evaluate the species' demographic strategy in the context of r and K selections. We describe the species' population phenology, particularly the pattern of territory occupation and the turnover of individuals within populations. Then we review the main results of different population viability analyses (PVA) performed on little bustard populations, identifying key parameters and thresholds, and synthesize results to extract lessons for conservation. We finally broaden the spatial scale of our review to examine the species' metapopulation dynamics.

Population Growth Models for the Little Bustard

Population Growth Rates in Expanding Little Bustard Populations

Breeding little bustards in southern France, and particularly in the SPA Costière Nimoise (Department of Gard), represent an extreme case of population increase in the species, because between 1998 and 2012 the number of males went from 130 to over 700 (Fig. 1). The little bustard in Gard had even been considered marginal (c. 10 males) until the mid-1990s (Devoucoux 2014), when the species' local population started a process of marked (exponential) increase, as in other areas of southeastern France (see chapter "The little bustard around the world: distribution, global conservation status, threats and population trends"). The transformation to cereal crops of parts of the grazed steppe of La Crau, known as *coussoul*, along with the abandonment of traditional vineyards and other cultivated areas in Gard, seems to have boosted the little bustard population growth in southern France (Wolff et al. 2001; Devoucoux 2014). Devoucoux (2014) examined this increase in Gard, evaluating how the relationship of population size with time fitted either the exponential or the logistic growth model, separately examining the breeding season (population size estimated as the number of displaying males) and winter (population size estimated as the total number of individuals). He found that in the period 1998–2014 the logistic model slightly better explained the growth of the little bustard breeding population in both Gard and the SPA, although the differences between models were not statistically significant. Interestingly, although it is quite reasonable to expect growth rate to slow down as the population approaches carrying capacity, the observed population size in 2012 slightly exceeded the carrying capacity estimated from the logistic model, before returning below it afterward, suggesting a density-dependent adjustment to habitat resource availability expected to generate yearly fluctuations around the predicted carrying capacity. After 2014, the observed population size, however, sharply decreased to levels below the

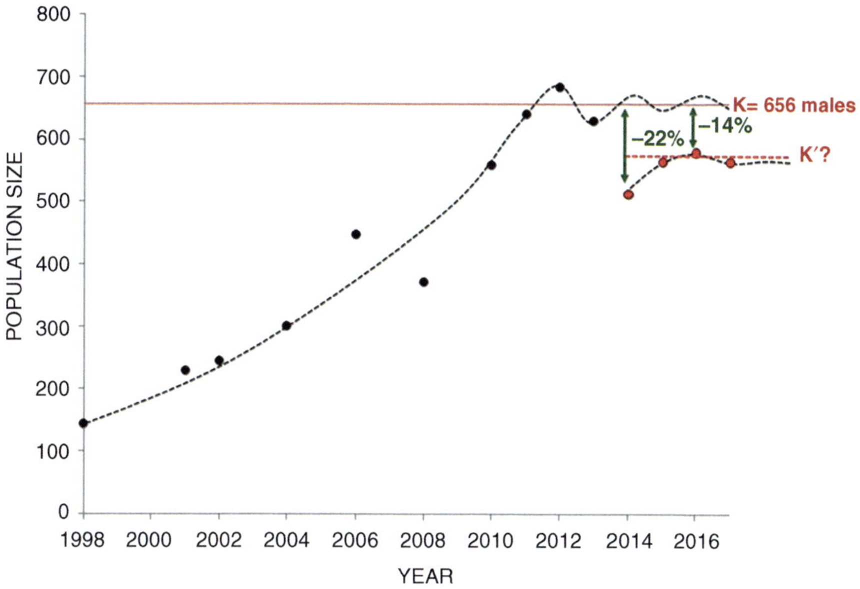

Fig. 1 Logistic growth of the little bustard population of Costière Nimoise SPA (Gard Department, southeastern France) between 1998 and 2016 following crop abandonment due to expropriation to build a high-speed railway. The adjustment of population carrying capacity (from K to K′) due to the start of works in 2012 and the resulting percentages of K reduction are illustrated. Black dots indicate population size up to 2012 and show an oscillation around K (656 males) from 2010 to 2012. Red dots represent population counts from 2013 to 2016 and show a stabilization of population size at ca. 580 males (K′). Based on Devoucoux (2014) and updated by P. Devoucoux (unpubl. data)

estimated carrying capacity, but this was indeed related to habitat degradation associated with the start of a large infrastructural installation (a high-speed railway). Approximately 20% of the male population (150–170 males) was lost and numbers never recovered to the levels of the former carrying capacity (Fig. 1). This population response, therefore, suggests an adjustment to a lower habitat carrying capacity, which would also support a logistic-growth response of the population.

Devoucoux (2014) further contrasted exponential and logistic models to fit growth curves back to 1980 and up to 2020. Over this longer period (40 years), the logistic model showed a better fit to the data, predicting a rapid stabilization of population growth. When he evaluated the growth of the wintering population, he also found a better fit with the logistic model, particularly in the SPA, although this was also true at the larger spatial scale of Gard department. In fact, winter population sizes exceeded estimated carrying capacity in three different winters, suggesting that favourable years can boost population growth beyond local carrying capacity, which should be expectedly followed by density-dependent adjustments.

The population dynamics revealed by Devoucoux (2014) indicate that little bustard populations can grow quite rapidly when conditions are favourable. They can reach extremely high densities from very low numbers in less than two decades, with a population growth rate of $\lambda = 1.08$ and even 1.2 from year to year, so that growth curves tend to show a marked exponential phase. Density-dependent effects may occur later, with a slowing down of growth rate and the stabilization of the population at carrying capacity, as expected from logistic models. The rapid recovery of some populations in central-western France following the application of adequate agri-environmental measures in the early 2000s (Bretagnolle et al. 2011) points in the same direction, although in that case, populations were simultaneously being reinforced with captive-bred individuals.

Population Growth Rates in Declining Little Bustard Populations

Between the late 1970s and 2000, the little bustard population breeding in the intensive cereal croplands of central-western France experienced one of the steepest decreases ever documented in a bird species (Jolivet and Bretagnolle 2002). Using national census data, we have modelled such decline over the period 1978–2012 (see data presented in chapter "Natural history of the little bustard: morphology, biometry, diet, sexual dimorphism, and social and breeding behaviour"), based on the exponential growth model (Fig. 2). According to this model, the population declined at a yearly discrete rate (lambda) of $\lambda = 0.926$, which corresponds to an intrinsic growth rate of $r = -0.077$ individuals/individual and year and an annual decrease percent rate of −7.4%. As shown in Fig. 2a, the values recorded over this period for the whole of central-western France, obtained in national censuses in 1979, 1982, 1986, 1996, 2000, 2004, and 2008–2017, fitted the population size trajectory predicted by this negative exponential growth model, showing a highly significant correlation with predicted population sizes (simple regression GLM, $R^2 = 0.98$,

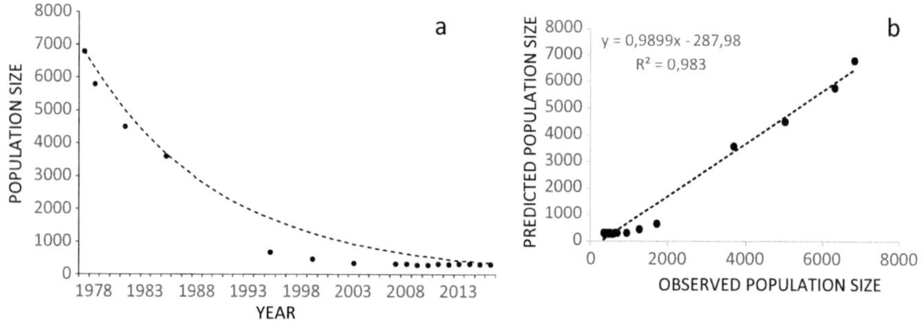

Fig. 2 Decline of the little bustard population of central-western France, modelled with the exponential growth model ($\lambda = 0.926$): (**a**) Observed population size estimated from national censuses (black dots) in relation to the predicted population trajectory (dashed line); (**b**) Linear regression fit between population size observed in national censuses from 1978, 1985, 2000, 2004, 2008 and 2012–2017, and values predicted by the exponential growth model for those years. Regression curve (dashed line) and fitted GLM equation are shown

$p < 0.001$, Fig. 2b). Moreover, the weighted mean yearly lambda calculated from the size of seven local populations annually monitored from 1997 to 2000 (V. Bretagnolle unpublished data) also yielded a value of 0.918 ($r = -0.086$ individuals/ individual and year) and a weighted mean annual decrease percent rate of −8.2% (although each population's individual trajectory varied in sign and slope; two of them remained stable and a third one slightly increased). These are similar values to those yielded by the exponential model for the entire region over the whole 40-year period (1978–2017), mentioned above.

The strong decline of the species' populations across Spain between 2005 and 2016 corroborates this pattern, with several regional populations declining at yearly rates over 10% and a global overall decrease of 59% in that 11-year period (García de la Morena et al. 2018). These results indicate that little bustard population decreases tend to be markedly exponential rather than steady or slowly accelerated when environmental conditions become adverse to their reproductive success and survival, which poses a problem to conservation strategies since declines may be faster than expected by managers, particularly in their initial stages. And more generally, increasing and decreasing population growth rate models indicate that this species can show extremely rapid and even abrupt population changes according to habitat quality, which is probably related to their rather large clutch size for a bird of this size (four eggs) but at the same time a rather high sensitivity to adult mortality (see below).

Density Dependent Effects on Little Bustard Population Parameters and Behaviour

Many aspects of little bustard biology and behaviour seem to be density dependent, that is, sensitive to changes in population density (e.g. Newton 1998) For example, the above-mentioned levelling-off of the exponential population growth of the little bustard population in the Costière Nimoise as it approached habitat saturation is a typical density-dependent response (Devoucoux 2014). However, the effect of change in population size on little bustard behaviour and population dynamics was best explored by Villers (2010). He analyzed the change over time of different behavioural traits and population parameters over an 11-year period (1998–2009) in a French little bustard population going through two contrasting population growth stages: a declining phase followed by a period of recovery. Various life history traits were inspected for signs of density dependence. He found that the sex ratio in that population became more male biased as the population increased (recovery phase). He also found that this recovery was mostly due to local recruitment, which suggested that male bias was largely due to female dispersal outside the study area, or possibly to differential mortality. The number of long-staying territorial males (i.e. those which remained in the study site for a month or more) also increased as the population recovered, implying less turnover in the leks. However, with the number of long-staying males, short stayers also increased, which is consistent with the key role of conspecific attraction expected in lek formation (Widemo and Owens 1995), a typical density-dependent process (e.g. Wagner and Danchin 2003) that increases male attractiveness and lek attendance (see also Jiguet and Bretagnolle 2006). The fact that no significant change in the frequency of different quality-related male phenotypic traits could be detected within any of the two phases suggests that the observed increase in male attractiveness (measured as a number of accompanying females) was due to female density rather than improved male attractiveness. Unsurprisingly, Villers (2010) found that snort-call rate, a display mainly involved in male–male territorial interactions (Jiguet and Bretagnolle 2001), was lowest when the population reached its minimum. In addition, the degree of male aggregation decreased during the declining stage and became stronger as the population recovered. Altogether, these findings show the density dependence of male reproductive success.

The Allee effect (e.g. Courchamp et al. 1999) is a particular case of positive density dependence in declining populations. It has not been described so far in the Little Bustard, although it was considered when modelling little bustard populations (see Inchausti and Bretagnolle 2005) and was actually shown to increase extinction rate. There is indirect evidence that an Allee effect may occur in this species. For instance, solitary males (i.e. bustard lek with a single male) display much less than those with neighbour males (Jiguet and Bretagnolle 2006; see also chapter "Behavioural ecology of the little bustard: sexual selection and mating systems"), and therefore are presumably less attractive to females. Indeed Jiguet and Bretagnolle (2006, 2014) have experimentally shown using decoys that leks composed of single males were far less attractive to females than leks of four males (average size of leks in Deux-Sèvres, France).

The density dependence of little bustard breeding behaviour was also shown in the experimental study of Morales et al. (2014). They placed male decoys in the territory of individually identified males to evaluate their response in relation to immediate conspecific (males and females) abundance and territory habitat features. The time taken by males to return to their territories after experimental decoy intrusion was shorter when the local abundance of males and females was higher. Consistently, the intensity of their aggressive response to decoys increased with the abundance of conspecifics of both sexes. The snort-call rate of experimental males decreased with their degree of clustering with neighbours. Moreover, more aggressive males called at lower rates, which suggests that snort-call rate could also function as a cue of territory safety for females (as in other lekking species, female harassment is frequent in the little bustard). Nevertheless, the study by Morales et al. (2014) also showed that habitat features played a role in male territorial response, whose intensity was positively correlated with habitat quality indicators such as herbaceous vegetation cover.

As a concluding remark on population growth rates in the little bustard, many aspects of the species' ecology are largely governed by density dependence. Accordingly, the little bustard is close to typical K-strategists as defined in the theoretical framework developed by MacArthur and Wilson (1967) and Pianka (1970).

However, the fast recovery capacity shown by the species when favourable conditions are met would place it closer to r-strategists according to the same framework. Of course, these kinds of classifications must always be used with caution and in relative terms, but they are useful to evaluate how a species is placed along the r–K population strategy gradient, in relation to other ecologically similar species. Ambivalence in the classification of the little bustard within the r–K gradient may be paralleled with the ambivalence within the specialist–generalist classification in regard to habitat: little bustard is a specialist of grassland-dominated landscapes (and thus limited by the availability of these landscapes) at a large scale, but a generalist species at a smaller scale, being able to exploit a variety of habitats providing suitable conditions (see Villers 2010; Bretagnolle et al. 2018).

Territory Turn Over, Natal and Adult Dispersal

Delgado et al. (2010) report a maximum length of male territory occupation of, respectively, 9 and 11 weeks in two consecutive breeding seasons in a population in central Spain, so that territories were occupied from the last week of March to the last week of May in the first year, and from the last week of March to the second week of June in the second (see also chapter "Breeding biology and demographic traits"). The maximum proportion of territories occupied during those periods was reached in the seventh and eighth weeks (second and third week of May, respectively). After those peaks, territories began to be more or less gradually deserted by males. Similar values were obtained in Deux-Sèvres (France), although about a third of the males had polynuclear territories (i.e. constituted by several disjoint areas, see Jiguet et al. 2000). Males usually display within a very limited space within fields, and most surprisingly male identification (through colour rings or comparing digital photographs) has revealed that often two or even three males may use exactly the same location consecutively during a breeding season. Territory switch may be the rule rather than the exception, as further revealed by GPS tracking, which suggests that males, even though faithful to their main territory during a complete breeding season, regularly visit other males' territories, leading to fights between males and escort flights.

Although the between-year territory fidelity of males is relatively high, some inter-annual turnover of breeding males also takes place. In a breeding population of west-central France, Jiguet and Ollivier (2002) found male territory fidelity of 57 and 53% in 2 consecutive years, while recruitment rate was only 17–19%. GPS tracking performed in Deux-Sèvres (France) confirmed that males can sometimes shift to a distant secondary territory, often 5 km but up to 20 km or more. Thus, males shifting between populations are not rare. Such movements are also documented in females, based on colour ringing, with up to 80 km distance within and between years (V. Bretagnolle, unpubl. data).

Natal dispersal has only been documented based on colour ringing programs involving captive-reared released birds, within the framework of the reinforcement program in Poitou-Charentes (see also chapter "Captive breeding, handling and care, and the impact of releases on wild populations"). Despite very high heterogeneity in recapture rate and observation effort, there are many cases of long-range natal dispersal (20–80 km), both in males and in females (V. Bretagnolle and A. Villers unpublished data). Given the higher probability of recovering or re-sighting of males compared to females, and despite the fact that more males were re-sighted at longer distances than females, females might likely disperse more than males in this species.

Lessons from Population Viability Analyses

Key Parameters and Thresholds

As a threatened species, the little bustard has been the subject of different Population Viability Analyses (PVAs) with the aim to identify the population parameters that are central in explaining population trends and dynamics, as well as the threshold values of those parameters that have to be avoided or reached in order to ensure population persistence. These PVAs were also used to predict population trajectories under various scenarios of landscape and climate change, public policies or conservation strategies.

Two studies on the breeding populations of central-western France (Morales et al. 2005a; Inchausti and Bretagnolle 2005) identified local fecundity, estimated as the number of fledglings (chicks surviving at least 30 days) per adult female, adult survival and initial population size as strongly influencing the extinction risk of populations. Using these three parameters, Delgado et al. (2009) simulated different scenarios for two populations in central Spain, additively increasing the value of one parameter at a time. The parameter that most positively impacted population survival when its corresponding increase was added in the PVA model was fecundity, suggesting its critical importance in population persistence. Morales et al. (2005a) and Inchausti and Bretagnolle (2005) also found that, in the case of west-central France, fecundity presented high capacity to reverse population trends although the capacity of adult survival was even higher according to sensitivity analyses. In a conservation context, however, these authors acknowledged the practical difficulties of managing populations to guarantee high adult survival rates, and concluded that conservation strategies should better focus on fecundity parameters.

According to the sensitivity analyses performed by Morales et al. (2005a), adult survival values between 60 and 80% increased the probability of population persistence from less than 0.3 to 0.98, above which improvements are obviously marginal. Therefore, survival rates over 60% would be required for avoiding rapid population extinction, if other parameters are held at viable values. In this respect, a PVA of Spanish populations (both at national and at regional levels) based on the results of the second national census (García de la Morena et al. 2018) showed that improving adult female survival by 25% would halt current declines in all populations over the

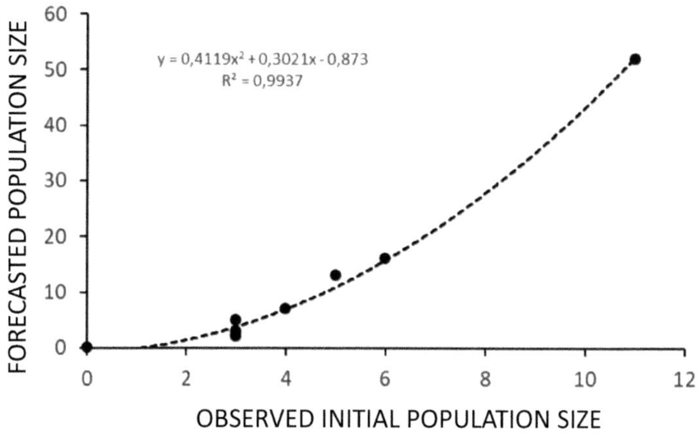

Fig. 3 Relationship between population sizes predicted by PVA after 7 years for nine populations of central Spain and the corresponding observed initial sizes, all other parameters kept at viability values. The quadratic fit line and equation are also shown. Based on Morales et al. (2008a)

Table 1 Minimum threshold values of key population parameters identified by means of PVA to guarantee little bustard population persistence (see text for details)

Population parameter	Threshold value
Adult survival	60%
Fecundity	0.9 Fledglings/female
Population size	20 Individuals
Sex ratio	0.3–0.5 Males/total individuals

Based on Morales et al. (2005a)

minimum viable population size (see below; Traba et al. 2020, unpublished report). A fecundity rate of one fledgling/female would assure a population survival probability very close to 100%, while rates below 0.9 fledglings/female would lead to a marked fall, so that for rates of 0.4 fledglings/female population survival is already under 10%. The PVA based on the second Spanish census consistently showed that increasing fecundity by 50%, combined with a 25% female survival increase, would reverse the current declining trends to positive growth, except for populations below the minimum viability size (Traba et al. 2020, unpublished report). Indeed, population persistence increases rapidly with population size, so that a population of only 20 individuals reaches a nearly 100% survival probability, if other parameters remain at viability values.

The importance of initial population size is illustrated in Fig. 3, which shows how the final population size forecasted by PVA for nine nuclei in central Spain (Morales et al. 2008a) increases exponentially with observed initial size, if all other parameters remain at viability values. These thresholds, summarized in Table 1, seem consistent with the classification of the little bustard as a K-strategist whose population growth rate is based on high adult survival and low reproductive rates, as well as with the importance of density dependence in the species' ecology and population dynamics mentioned above. In this latter respect, Inchausti and Bretagnolle (2005) showed that the lek mating system of the little bustard implies a significant increase of population extinction risk in relation to a scenario in which the social structure was not taken into account, which points to an Allee effect on breeding rate mediated by conspecific attraction due to the dispersal of individuals when total abundance is too low. Indeed, Inchausti and Bretagnolle (2005) reported the abandonment of leks in populations where lek size was below six individuals (males and females cumulated). This kind of response would thus be typical of lekking species (Alonso et al. 2004; Morales et al. 2001; Dale 2001), resulting in the extinction of small populations and the accumulation of individuals in the largest nuclei, as described in the great bustard (Alonso et al. 2004; Pinto et al. 2005). In the sensitivity analysis of Morales et al. (2005a), the degree of polygyny, i.e. the proportion of males fertilizing females, had little influence on population viability, which is also consistent with the strong mating bias usually found in lek mating systems (Höglund and Alatalo 1995).

Other parameters identified as relevant in PVA sensitivity analyses were juvenile survival and sex ratio (Morales et al. 2005a). Population viability increases steadily (almost monotonically) with juvenile survival, and thus it is not possible to identify thresholds. On the other hand, sex ratio, calculated as the number of males relative to total population size, had a non-linear influence on population survival rates, which showed significant decreases for values in the proportion of males below 0.3 and above 0.5 (Table 1), while it remained very close to 100% between those values. Again, consistent with lek dynamics, little bustard populations seem to tolerate relatively small male numbers, but they rapidly approach extinction when even a small shortage of females occurs.

Metapopulation Dynamics

Populations tend to be connected to greater or lesser degrees through the dispersal of individuals of both sexes and different ages, which is the core process in metapopulation dynamics (Hanski 1999). This happens to be the case for the little bustard as suggested by PVAs, which depict quite dramatically different trajectories when involving metapopulation dynamics or not. As predicted by the metapopulation theory, dispersal is the key driver of such dynamics, and opposed trajectories occur when the dispersal rate increases and exceeds a threshold value. Morales et al. (2005a) addressed this issue and analyzed the effect of population connectivity by simulating the impact in population survival of different migration rates between two hypothetical populations, associated with different survival rates during migration. They found that connection favoured metapopulation persistence if survival was over 90%. However, metapopulation persistence dropped significantly when migration survival was below that value, particularly for high migration rates, a pattern consistent with the existence of ecological traps (Fig. 4). In a more elaborated simulation accounting for the level of environmental correlation between the two hypothetical populations, as well as for the potential differences in

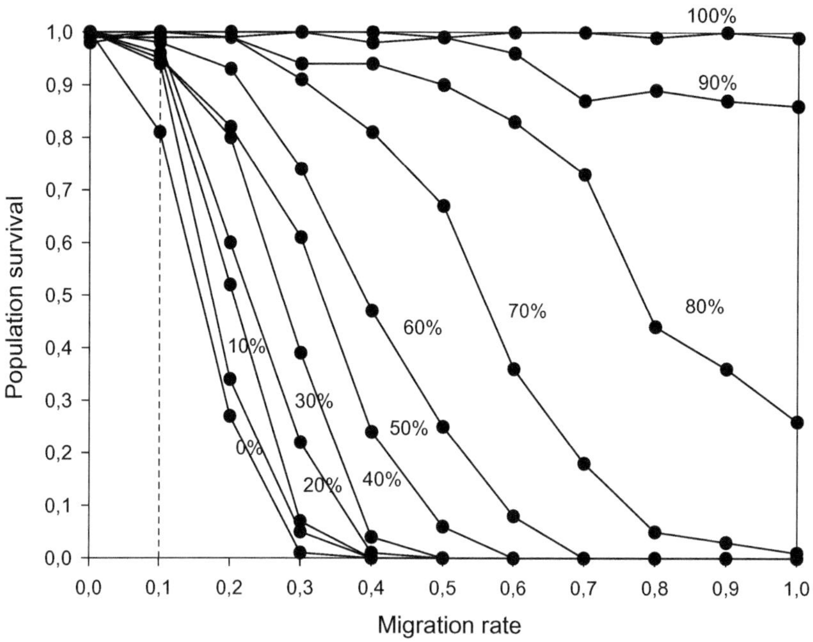

Fig. 4 Relationship of population survival probability with the rate of migration between two hypothetical little bustard populations. Each curve corresponds to a different rate of survival associated with migration, indicated by the corresponding percentage value. Based on Morales et al. (2005a)

population dynamics (a source and a sink population vs. two populations with similar trends), Inchausti and Bretagnolle (2005) found that the extinction risk of the metapopulation decreased with dispersal of both adults and young, regardless of sexual differences in dispersal rate. However, when dispersal was incorporated into the PVA for the real little bustard populations of west-central France, metapopulation extinction risk was higher than for the model without dispersal. They explained such an effect by the fact that, in the absence of dispersal, the largest subpopulations do not function as a source of juveniles for other nuclei that would otherwise act as sinks. At the local population level, they found that for small populations with low fecundity dispersal decreased extinction risk, while for large populations with high fecundity (and thus functioning as sources), dispersal increased extinction risk.

Altogether, these results illustrate the relevance of population connectivity in the dynamics of both little bustard local populations and metapopulations. Dispersal can be detrimental for a source population if it is not compensated by local recruitment or immigration, but contributes to the rescue of sink populations that otherwise would decline to extinction. Therefore, these classic metapopulation dynamics (see Hanski 1999) seem to adequately describe what has been observed in the species. They also highlight the importance of guaranteeing a secure landscape matrix for dispersing birds, given the impact of mortality during the dispersal process (see results in Marcelino et al. 2017). However, a source-sink or a metapopulation model may become an ecological trap model if a given habitat shows very high attraction while being unable to provide sufficient numbers of fledglings. Such a pattern has been described for the little bustard south of Deux-Sèvres (Bretagnolle et al. 2018) and is further discussed in section "Predicted impact of land use and climate change on observed and modelled population dynamics".

Predicted Impact of Land Use and Climate Change on Observed and Modelled Population Dynamics

We have already referred to the capacity of little bustard populations to recover when habitat improves and thus its carrying capacity for the species increases. The spectacularly high densities reached in Costière Nimoise after agricultural abandonment, or the rapid recovery of populations in Poitou-Charentes following the spread of alfalfa (Bretagnolle et al. 2011), are good examples and bring hope for currently declining little bustard populations. However, land use changes do not only lead to changes in the trends of particular populations, but may also have larger-scale (i.e. regional) effects concerning metapopulation dynamics. For example, the comparison of the species' distribution in Spain at 10 × 10 km square resolution between the first and second national censuses, carried out in 2005 and 2016, respectively, indicates not only a range contraction due to a global population decline but also the concentration of high-abundance squares in regions where suitable habitat has persisted, as well as the appearance of presence squares in areas where agricultural abandonment has favoured the spread of uncultivated grassy habitat (SEO/BirdLife, unpublished data). These large-scale changes can have effects on metapopulation dynamics, for example if dispersing individuals cannot reach peripheral or isolated populations, leading to the latter's extinction if local recruitment is insufficient, or halting their functioning as population sources.

Such processes can be further intensified by density-dependent mechanisms. Morales et al. (2008a) reported on the decline of 8 little bustard populations in the region of Madrid after comparing local censuses carried out, respectively, in 2000 and 2007 (mean decline of 61.5%, range 30–100%, which included a local extinction event), while another population, the largest one surveyed in 2000, increased by 29%. This suggested that the regional metapopulation could be undergoing a density-dependent concentration in the largest nuclei, which was supported by the location in the latter population of some radio-tracked individuals captured in other populations in previous years. Nevertheless, it is important to bear in mind that little bustard abundance can vary annually within and across neighbouring nuclei following changes in the availability of suitable habitat, as shown by Morales et al. (2005b), also in central Spain, where permanent grassland habitat is scarce and birds rely on rotational fallow, so that part of the differences observed when comparing censuses from different years could be attributed to these rather cyclic changes.

But inter-annual variation in little bustard abundance does not only respond to land use change. Another environmental factor known to exert such effects is weather and, in the long run, climate. Based on a 7-year series of census data, Delgado et al. (2009) showed the positive influence of total rainfall in each year's October–May period on little bustard abundance in two populations of central Spain, as well as a negative effect of mean April temperature in one of them. These results highlight the importance of primary production for the species as the main bottom-up ecosystem process governing the availability of food resources, either plant or animal (i.e. invertebrates). This is particularly true for Mediterranean populations, where rainfall is limited to certain periods of the year, and it is further supported by the role of April temperature in one of the study sites, whose calcareous soil favours water drainage and thus greater substrate drought. Because of the importance of vegetation structure for little bustard microhabitat selection (Morales et al. 2008b; Silva et al. 2014), weather, and more precisely precipitation, also influences the species' habitat choice and space use. For example, in the Mediterranean grasslands of southern Portugal, the response of little bustard abundance to stocking rates depends on each year's rainfall, so that females and chicks are more abundant in lightly to moderately grazed pastures in dry years, but shift to more heavily grazed fields in wetter ones (Faria and Morales 2017). Therefore, we should keep in mind that management of habitat for little bustard breeding success and thus population growth needs to be sensitive to weather conditions, particularly in Mediterranean regions where between-year weather variability is high.

Temperature has also been shown to influence little bustard breeding performance. Silva et al. (2015) showed that the display activity of GPS-tagged breeding males markedly dropped above 25 °C. Climate models for the coming decades forecast a marked decrease in mean precipitation and an increase in mean temperature in Iberia and other parts of the species distribution range, and this is expected to have a negative impact on its breeding success and population trends. Consequently, stocking rates should be lowered in drier years in order to favour juvenile survival. The provision of permanent grassland habitat such as rain-fed alfalfa fields in sites with deeper and wetter soils is an interesting alternative in Mediterranean areas with suitable climate conditions for this culture.

Conclusions: Conservation Strategies and Demographic Ecological Traps

Ecological traps have received more theoretical attention (Battin 2004; Robertson and Hutto 2006) than empirical support (review in Suvorov and Svobodová 2012), despite their strong conservation consequences (Delibes et al. 2001). Ecological traps are expected to occur in rapidly changing habitats under strong anthropogenic influence (Suvorov and Svobodová 2012). Farmland is one of these habitats and ecological traps are strongly suspected for farmland species like skylarks *Alauda arvensis* breeding in cereals (Poulsen et al. 1998; Donald et al. 2002), meadow birds such as corncrake *Crex crex* or whinchat *Saxicola rubetra* (Grüebler et al. 2012), or medium-sized mammals such as hare *Lepus* spp. (Reid et al. 2010).

In response to the little bustard decline in Deux-Sèvres (France), which has mainly been attributed to a food deficit for chicks (Jiguet 2002) and destruction of clutches, broods or even incubating females during mowing (Bretagnolle et al. 2011, 2018), a land-sharing conservation strategy was set up mostly through the application of agri-environmental schemes (AES) (see chapters "Threats affecting little bustards: human impacts" and "Little bustard and humans: conservation and management", Bretagnolle et al. 2018). Little bustards selected grasslands (i.e. alfalfa fields and prairies) for breeding, which represented less than 15% of available habitat, and avoided all other habitats irrespective of traits such as vegetation height, distance to field border or plant species composition (Bretagnolle et al. 2018). The quantity and quality of this rare selected habitat (grasslands in general) has shrunk (Silva et al. 2018; Traba and Morales 2019), suggesting the species is now strongly habitat-limited.

Female bustards, therefore, are much attracted to this rare but sub-optimal (because grasslands, such as alfalfa, are regularly mown) habitat. They behave as "grassland specialists" in regard to available habitat, but as "grassland generalists" in regard to the characteristics of the grasslands preferred, which leads to temporary grasslands such as alfalfa acting as ecological traps, since they are mown every 4 weeks on average (Faria et al. 2016; Kershner and Bollinger 1996). In addition, frequently mown grasslands are usually poor in prey items (especially large insects; Badenhausser et al. 2009). Such grasslands can be considered to be attractive sinks because there is a mismatch between habitat selection and habitat quality. Mowing every 4 weeks is too high for allowing females to breed successfully (Bretagnolle et al. 2018) and therefore females breeding in temporary grasslands do not produce enough chicks in the long term, since half of the nests are destroyed, and brood reduction further decreases family size to 0.3 female fledglings per adult female, which is approximately one-third of the replacement rate (Bretagnolle and Inchausti 2005). The temporary grasslands habitat is thus not a sink but a trap, since females prefer temporary grasslands to other habitats (they could, for instance, lay eggs in the safer cereal crop habitat, then rear their chicks in temporary grasslands). Although the AES strategy had positive results, since nests in AES fields were rarely destroyed by mowing (Bretagnolle et al. 2011), there are no significant effects of the AESs on clutch size, egg volume and habitat choice, suggesting habitat quality was not improved by the AESs (Bretagnolle et al. 2018), and females did not prefer fields under AES for breeding. Unsurprisingly, female little bustards still bred in remaining non-AES grassland fields, which reduced breeding success as they were not protected. Therefore, even AES grasslands function as ecological traps, because their proportion in the landscape is very low and thus they do not allow sustainable populations. A similar ecological trap could be functioning in Iberia in fallows that are selected for nesting and later ploughed to prepare the next crop cycle (Morales et al. 2013).

Acknowledgements We are thankful to Alex Villers for reviewing and commenting on initial versions of the chapter. The comments and suggestions by Juan Traba further improved the manuscript.

References

Alonso JC, Martin CA, Alonso JA, Palacin C, Magaña M, Lane SJ (2004) Distribution dynamics of a great bustard metapopulation throughout a decade: influence of conspecific attraction and recruitment. Biodivers Conserv 13:1659–1674

Badenhausser I, Amouroux P, Lerin J, Bretagnolle V (2009) Acridid (Orthoptera: Acrididae) abundance in Western European grasslands: sampling methodology and temporal fluctuations. J of Appl Entomol 133:720–732

Battin J (2004) When good animals love bad habitats: ecological traps and the conservation of animal populations. Conserv Biol 18:1482–1491

Bretagnolle V, Inchausti P (2005) Modelling population reinforcement at a large spatial scale as a conservation strategy for the declining little bustard (*Tetrax tetrax*) in agricultural habitats. Anim Conserv 8:59–68

Bretagnolle V, Villers A, Denonfoux L, Cornulier T, Inchausti P, Badenhausser I (2011) Rapid recovery of a depleted population of little bustards *Tetrax Tetrax* following provision of alfalfa through an agri-environment scheme. Ibis 153:4–13

Bretagnolle V, Denonfoux L, Villers A (2018) Are farming and birds irreconcilable? A 21-year study of bustard nesting ecology in intensive agroecosystems. Biol Conserv 228:27–35

Courchamp F, Clutton-Brock T, Grenfell B (1999) Inverse density dependence and the Allee effect. TREE 14:405–410

Dale S (2001) Female-biased, low-female recruitment, unpaired males, and the extinction of small and isolated populations. Oikos 92:344–356

De Roos AM, Persson L, McCauley E (2003) The influence of size-dependent life-history traits on the structure and dynamics of populations and communities. Ecol Lett 6:473–487

Delgado MP, Morales MB, Traba J, García de la Morena EL (2009) Determining the effects of habitat management and climate on the population trends of a declining steppe bird. Ibis 151: 440–451

Delgado MP, Traba J, García de la Morena EL, Morales MB (2010) Habitat selection and density-dependent relationships in spatial occupancy by male Little Bustards *Tetrax tetrax*. Ardea 98: 185–194

Delibes M, Ferreras P, Gaona P (2001) Attractive sinks, or how individual behavioural decisions determine source–sink dynamics. Ecol Lett 4:401–403

Devoucoux P (2014) Conséquences et impacts prévisibles d'une perte d'habitat majeure sur une espèce menacée aux exigences écologiques complexes : effets de la mise en place du contournement ferroviaire à grande vitesse Nîmes-Montpellier sur la dynamique de la population d'outarde canepetière des Costières de Nîmes. Dissertation, University of Poitiers

Donald PF, Evans AD, Muirhead LB, Buckingham DL (2002) Survival rates, causes of failure and productivity of Skylark *Alauda arvensis* nests on lowland farmland. Ibis 144:652–664

Faria N, Morales MB (2017) Population productivity and late breeding habitat selection by the threatened little bustard: the importance of grassland management. Bird Conserv Int. https://doi.org/10.1017/S0959270917000387

Faria N, Morales MB, Rabaca JE (2016) Exploring nest destruction and bird mortality in mown Mediterranean dry grasslands: an increasing threat to grassland bird conservation. Eur J Wildl Res 62:663–670

Fryxell JM, Sinclair ARE, Caughley G (2014) Wildlife ecology, conservation, and management, 3rd edn. Wiley, Oxford

García de la Morena EL, Bota G, Mañosa S, Morales MB (2018) El sisón común en España. II Censo Nacional (2016). SEO/Birdlife, Madrid

Grüebler MU, Schuler H, Horch P, Spaar R (2012) The effectiveness of conservation measures to enhance nest survival in a meadow bird suffering from anthropogenic nest loss. Biol Conserv 14:197–203

Hanski I (1999) Metapopulation dynamics. Oxford University Press, Oxford

Höglund J, Alatalo RV (1995) Leks. Princeton University Press, Princeton, NJ

Inchausti P, Bretagnolle V (2005) Predicting short-term extinction risk for the declining Little Bustard (*Tetrax tetrax*) in intensive agricultural habitats. Biol Conserv 122:375–384

Jiguet F (2002) Arthropods in diet of Little Bustards *Tetrax tetrax* during the breeding season in western France. Bird Study 49:105–109

Jiguet F, Bretagnolle V (2001) Courtship behaviour in a lekking species: individual variations and settlement tactics in male little bustard. Behav Process 55:107–118

Jiguet F, Bretagnolle V (2006) Manipulating lek size and composition using decoys: an experimental investigation of lek evolution models. Am Nat 168:758–768

Jiguet F, Bretagnolle V (2014) Sexy males and choosy females on exploded leks: correlates of male attractiveness in the Little Bustard. Behav Process. https://doi.org/10.1016/j.beproc.2014.01.008

Jiguet F, Ollivier D (2002) Male phenotypic repeatability in the threatened Little Bustard *Tetrax tetrax*: a tool to estimate turnover and dispersal. Ardea 90:43–50

Jiguet F, Arroyo B, Bretagnolle V (2000) Lek mating systems: a case study in the Little Bustard *Tetrax tetrax*. Behav Process 51:63–82

Jolivet C, Bretagnolle V (2002) L'outarde canepetière en France : évolution récente des populations, bilan des mesures de sauvegarde et perspectives d'avenir. Alauda 70(2002):93–96

Kershner EL, Bollinger EK (1996) Reproductive success of grassland birds at East-Central Illinois airports. Am Midl Nat 136:358–366

MacArthur RH, Wilson EO (1967) The theory of island biogeography. Princeton University Press, Princeton, NJ

Marcelino J, Moreira F, Mañosa S, Cuscó F, Morales MB, García de la Morena EL, Bota G, Palmeirim JM, Silva J (2017) Tracking data of the little bustard (*Tetrax tetrax*) in Iberia shows high anthropogenic mortality. Bird Conserv Int. https://doi.org/10.1017/S095927091700051X

Morales MB, Jiguet F, Arroyo B (2001) Exploded leks: what bustards can teach us. Ardeola 48:85–98

Morales MB, Bretagnolle V, Arroyo B (2005a) Viability of the endangered little bustard *Tetrax tetrax* population of western France. Biodivers Conserv 14:3135–3150

Morales MB, García JT, Arroyo B (2005b) Can landscape composition changes predict spatial and annual variation of little bustard male abundance? Anim Conserv 8:167–174

Morales MB, García de la Morena EL, Delgado MP, Traba J (2008a) Tendencia reciente y viabilidad futura de las poblaciones de sisón común *Tetrax tetrax* en la Comunidad de Madrid. Anu Ornitol Madr 11:42–55

Morales MB, Traba J, Carriles E, Delgado MP, García de la Morena EL (2008b) Sexual differences in microhabitat selection of breeding Little Bustards *Tetrax tetrax*: ecological segregation based on vegetation structure. Acta Oecol 34:345–353

Morales MB, Traba J, Delgado MP, García de la Morena EL (2013) The use of fallows by nesting little bustard *Tetrax tetrax* females: implications for conservation in mosaic cereal farmland. Ardeola 60:85–97

Morales MB, Casas F, García de la Morena EL, Ponjoan A, Calabuig G, Martínez-Padilla J, García JT, Mañosa S, Viñuela J, Bota G (2014) Density dependence and habitat quality modulate the intensity of display territory defence in an exploded lekking species. Behav Ecol Sociobiol 68: 1493–1504

Newton I (1998) Population limitation in birds. Academic Press, San Diego

Pianka ER (1970) On r- and k-selection. Am Nat 104:592–597

Pinto M, Rocha P, Moreira F (2005) Long-term trends in great bustard (*Otis tarda*) populations in Portugal suggest concentration in single high quality area. Biol Conserv 124:415–423

Poulsen JG, Sotherton NW, Aebischer NJ (1998) Comparative nesting and feeding ecology of skylarks *Alauda arvensis* on arable farmland in southern England with special reference to set-aside. J Appl Ecol 35:131–147

Reid N, McDonald RA, Montgomery WI (2010) Homogeneous habitat can meet the discrete and varied resource requirements of hares but may set an ecological trap. Biol Conserv 143:1701–1706

Robertson B, Hutto R (2006) A framework for understanding ecological traps and an evaluation of existing evidence. Ecology 87:1075–1085

Silva JP, Estanque B, Moreira F, Palmeirim JM (2014) Population density and use of grasslands by female Little Bustards during lek attendance, nesting and brood-rearing. J Ornithol 155:53–63

Silva JP, Catry I, Palmeirim JM, Moreira F (2015) Freezing heat: thermally imposed constraints on the daily activity patterns of a free-ranging grassland bird. Ecosphere 6:article number 119. https://doi.org/10.1890/ES14-00454.1

Silva JP, Correia R, Alonso H, Martins RC, D'Amico M, Delgado A, Sampaio H, Godinho C, Moreira F (2018) EU protected area network did not prevent a country wide population decline in a threatened grassland bird. PeerJ 6:e4284. https://doi.org/10.7717/peerj.4284

Suvorov P, Svobodová J (2012) The occurrence of ecological traps in bird populations: is our knowledge sufficient? A review. J Landsc Ecol 5:36–56

Traba J, Morales MB (2019) The decline of farmland birds in Spain is strongly associated to the loss of fallowland. Sci Rep 9:9473. https://doi.org/10.1038/s41598-019-45854-0

Traba J, Bota G, Mañosa S, García de la Morena EL, Morales MB (2020) Análisis de viabilidad de la población de sisón común (*Tetrax tetrax*) en España. Anexo a la solicitud de recatalogación del sisón común en el Catálogo Español de Especies Amenazadas. Unpublished report

Villers A (2010) Ecologie spatiale, processus comportementaux et dynamiques des populations d'une espèce menacée, l'Outarde Canepetière. PhD Thesis. University of Paris 6, Paris

Wagner RH, Danchin E (2003) Conspecific copying: a general mechanism of social aggregation. Anim Behav 65:405–408

Widemo F, Owens PF (1995) Lek size, male mating skew and the evolution of lekking. Nature 373: 148–151

Wolff A, Paul J, Martin JL, Bretagnolle V (2001) The benefits of extensive agriculture to birds: the case of the little bustard. J Appl Ecol 38:963–975

Threats Affecting Little Bustards: Human Impacts

João Paulo Silva, Beatriz Arroyo, Ana Teresa Marques, Manuel B. Morales, Pierrick Devoucoux, and François Mougeot

João Paulo Silva and Beatriz Arroyo contributed equally to this chapter.

J. P. Silva (✉)
CIBIO/InBIO, Centro de Investigação em Biodiversidade e Recursos Genéticos, Laboratório Associado, Universidade do Porto, Campus Agrário de Vairão, Vairão, Portugal

CIBIO/InBIO, Centro de Investigação em Biodiversidade e Recursos Genéticos, Laboratório Associado, Instituto Superior de Agronomia, Universidade de Lisboa, Lisboa, Portugal
e-mail: jpsilva@cibio.up.pt

B. Arroyo (✉) · F. Mougeot
Instituto de Investigación en Recursos Cinegéticos, IREC (CSIC-UCLM-JCCM), Ciudad Real, Spain
e-mail: beatriz.arroyo@uclm.es; francois.mougeot@uclm.es

A. T. Marques
CIBIO/InBIO, Centro de Investigação em Biodiversidade e Recursos Genéticos, Laboratório Associado, Universidade do Porto, Campus Agrário de Vairão, Vairão, Portugal

CIBIO/InBIO, Centro de Investigação em Biodiversidade e Recursos Genéticos, Laboratório Associado, Instituto Superior de Agronomia, Universidade de Lisboa, Lisboa, Portugal

cE3c, Centro de Ecologia, Evolução e Alterações Ambientais, Faculdade de Ciências da Universidade de Lisboa, Lisbon, Portugal

M. B. Morales
Terrestrial Ecology Group, Departamento de Ecología, Universidad Autónoma de Madrid (TEG-UAM), Madrid, Spain

Centro de Investigación en Biodiversidad y Cambio Global (CIBC-UAM), Universidad Autónoma de Madrid, Madrid, Spain
e-mail: manuel.morales@uam.es

P. Devoucoux
Centre d'Etudes Biologiques de Chizé, CNRS & La Rochelle Université, Beauvoir sur Niort, France

V. Bretagnolle et al. (eds.), *Little Bustard: Ecology and Conservation*, Wildlife Research Monographs 5, https://doi.org/10.1007/978-3-030-84902-3_12

Introduction

Most studies identifying the major threats to the little bustard have been carried out in western Europe, mainly in France, Spain and Portugal. Much attention has been given to the species in these countries, owing to a severe decline observed in France at the end of the last century (Jolivet 2001; Bretagnolle and Inchausti 2005; Inchausti and Bretagnolle 2005). This decline prompted the need to promote its conservation also in Iberia, which was considered at the time as the species world's stronghold (De Juana and Martínez 1996, 2001). The population decline in France (92% in only 18 years) was halted at the beginning of the 2000s (Bretagnolle and Inchausti 2005; Inchausti and Bretagnolle 2005). Since then and until 2012 the French population slowly recovered, mainly in the southern subpopulations, after which a slow decline started again (minus 6% between 2012 and 2016; Gendre et al. 2019). Simultaneously, little bustards have steeply declined in Portugal and Spain, with a 50% loss over a 10-year period (García de la Morena et al. 2018; Silva et al. 2018). Eastern populations, mainly those located in Russia and Kazakhstan, are thought to have increased due to greater availability of breeding habitats after the dismantling of the USSR (Kamp et al. 2011), but they are nevertheless thought to be particularly vulnerable on their wintering grounds due to hunting pressure (Kamp et al. 2011; Collar et al. 2018; Yousefi et al. 2018). Overall, the little bustard is a highly vulnerable species and improving our understanding of its main threats is essential to plan sound conservation strategies. In this chapter, we outline the main threats known for the species, grouped in four categories: habitat loss and degradation, impacts of infrastructures, climate change and hunting. We highlight that most of these threats are related to human activities.

Habitat Loss and Degradation

Habitat loss and degradation are major causes of little bustard decline throughout its entire range (Iñigo and Barov 2010; BirdLife International 2021), although the drivers are still not fully understood (see also chapter "Migration, movements and non-breeding ecology" for an overview of the little bustard winter conservation).

In western Europe the little bustard is well adapted to extensive farming systems, traditionally presenting a diversity of land uses including cereal fields, legume crops and grazed fallows, which would be rotated over time (e.g. Martínez 1994; Salamolard and Moreau 1999; Morales et al. 2005b; Chapter "Habitat selection and space use"). These systems provide suitable breeding and non-breeding habitat throughout the year (Silva et al. 2004, 2007; Faria and Silva 2010). However, they are economically uncompetitive and marginal (Suárez et al. 1997), and therefore highly vulnerable to either agricultural intensification or abandonment. This habitat is prone to land use changes, driven principally by agricultural markets, resulting sometimes in complete habitat loss from one year to the next. In western Europe, the

Common Agricultural Policy (CAP), which was created to increase self-sufficiency in food, has been the main driver of habitat loss and degradation for farmland birds, by providing incentives for farmers to adopt new technologies and methods, expand average farm size, encourage specialization and increase yields of all supported crops, mostly cereals, but also fodder for livestock for the production of meat and milk (Pain and Dixon 1997; van Vliet et al. 2015; Reif and Vermouzek 2019).

The eastern populations of Russia and Kazakhstan seem to have benefited from agricultural extensification at the end of the last millennium as a consequence of the dismantling of the USSR, and of significant cultivation of legume crops (Collar et al. 2018) and large-scale farmland abandonment, thus favouring a reversion to breeding habitat (Lesiv et al. 2018). However, current and future trends towards the intensification of agriculture in those areas may also lead to the decline of grassland species in the near future (Kamp et al. 2011), and farmland abandonment may ultimately lead to habitat encroachment, thus evolving towards inappropriate vegetation structure for little bustard breeding (Fonderflick et al. 2010; Devoucoux 2014).

Agricultural Intensification

Agricultural intensification has enabled remarkable advances in food production globally, but it has also led to biodiversity loss (Matson et al. 1997; Emmerson et al. 2016). The western range of the little bustard has suffered major agricultural intensification, mainly driven as previously mentioned by the CAP (Santos and Suárez 2005). Agricultural intensification leads to landscape simplification, monoculture, increased use of pesticides and fertilizers, increased irrigation schemes, intensification of the agricultural rotation, and suppression of fallow land (Matson et al. 1997), all of which may or do affect little bustards. Additionally, it may allow for the appearance or expansion of certain crops more profitable under current markets, such as the significant expansion of permanent crops in Iberia, which we frame as habitat conversion, as it radically changes the habitat.

Intensification of Traditional Dry Cereal Agriculture

The traditional cereal rotation scheme ensured the presence of fallow land in the landscape, fundamental for little bustard breeding (e.g. Morales et al. 2005b; Silva et al. 2010a; chapter "Habitat selection and space use"). Traditionally, fallow land has been extensively grazed and specific plots were often maintained as fallow for 2 consecutive years in many areas. One common effect of agricultural intensification is the implementation of faster cereal/fallow rotations (through the increased use of fertilizers), which reduce or even suppress the use of fallow in the rotation system, depending on the productivity of the soils. This in turn leads to landscape homogenization and the loss of the main breeding habitat for little bustards. In France, intensification driven by CAP has been linked to a 92% decline of the national population over an 18-year period (Inchausti and Bretagnolle 2005). The species' decline in Spain is associated at the national scale with a loss of 1.1 million ha of fallow across the country in 15 years (Traba and Morales 2019). The maintenance of fallows and the promotion of their 'extensive' (non-intensive) management have been shown to benefit farmland birds (Ribeiro et al. 2014; Sanz-Pérez et al. 2019) and should be promoted in future CAPs to halt farmland bird declines and assist little bustard population recoveries (Tarjuelo et al. 2020).

Within the little bustard's range in Spain (principally in Castilla–La Mancha and Extremadura) it is still relatively common to find dry cereal farming using a short rotational scheme of cereal/fallow/legume. However, annual fallows there are ploughed between April or May (to prevent weed growth), i.e. during the little bustard's nesting period. Such intensification of fallow management is now widespread and is likely to have a major impact on breeding populations, as it turns fallow into an ecological trap: birds are attracted early in the season to this land use for breeding, where subsequently habitat and nests are destroyed through ploughing (Morales et al. 2013). Replacement clutches are more likely to be unsuccessful (Cuscó Martinez 2019), in part as a mismatch with trophic availability for adults and chicks, as vegetation dries out during late spring and summer and cereal harvest takes place. Conservation measures should include delaying the ploughing of fallows at least until the end of June, when most chicks have left the nest.

The use of more economically profitable crop varieties such as fodder can also lead to earlier harvesting/mowing, conflicting with the incubation and chick-rearing stages. Furthermore, modern machinery can operate at high speed, so its use can also be a source of chick and female mortality. In France, at least 25% of the egg loss in some regions and overall 36% of nest failure occurs in cultivated habitats (Inchausti and Bretagnolle 2005), mainly due to alfalfa mowing operations (70%) or abandonment by females after vegetation loss around the nest during harvest (Bretagnolle et al. 2011, 2018). Moreover, earlier ripening of cereals (and thus earlier cereal harvest dates) has been observed in recent decades (Berger-Geiger et al. 2019). After harvesting, stubbles that are a preferred winter habitat (Silva et al. 2004) now represent a shorter sward, which compromises their quality as post-breeding and winter habitat because the low vegetation increases exposure to predators.

Additionally, public subsidies promoted by the CAP have encouraged the amalgamation of landholdings within landscapes previously dominated by small fields, thereby reducing landscape diversity and field edge density, which provide important nesting habitat and trophic resources for chicks (Wolff 2001, 2004; Morales et al. 2005b).

Increased Use of Agrochemicals

Pesticide use is a major cause of farmland bird declines, either through indirect effects on food supply or through direct toxic effects on the survival, health and/or reproduction of birds (Geiger et al. 2010; Mineau and Whiteside 2013; Hallmann et al. 2014).

Arthropods are crucial for the development of little bustard chicks during the first 2/3 weeks of life (Jiguet 2002). Reduced availability of arthropods (particularly beetles and grasshoppers) can lead to the starvation of chicks (Bretagnolle et al. 2011) and is likely to be a limiting factor for productivity (Traba et al. 2008). Moreover, food shortage may limit the ability of females to meet the energetic demands made by egg laying and chick rearing. A 15-year study in France confirmed that the provision of targeted agri-environmental schemes, increasing plant species diversity and grasshopper abundance, reversed the decline of a little bustard population there, thus confirming that the intensification of agricultural practices and use of agrochemicals represent the major drivers of the decline (Bretagnolle et al. 2011).

Modern techniques of minimal soil mobilization and direct sowing, which attempt to reduce soil erosion, are concomitant with the widespread application of herbicides for winter and spring crops. These practices can critically reduce trophic resources for adults, which feed on green plants. Nevertheless, Faria and Morales (2019) reported a greater abundance of little bustard males on unploughed fields of Évora region (Portugal), but only in dry years, when such habitat provides cover and food availability.

Nothing is known about the direct lethal or sublethal effects of agrochemicals on little bustards. However, studies conducted on other farmland birds, particularly red-legged partridges *Alectoris rufa*, have documented a wide range of sublethal effects (Lopez-Antia et al. 2013, 2015a, b), including important reductions in productivity (Lopez-Antia et al. 2018). Little bustards are potentially exposed to agrochemicals sprayed over crops (fungicides, herbicides) or used as seed coating (which are subsequently present in seedlings). Future studies should assess pesticide exposure levels and their potential adverse effects on wild little bustards.

Conversion to Unsuitable Land Uses

The landscape transformation of extensive agricultural farming is a consequence of both agricultural intensification (in most areas) and abandonment (in the less productive ones), driven by agricultural policies such as the CAP and market demands (Fig. 1). The level and type of transformation are highly dependent on the quality of the soil for agriculture and on whether it can be irrigated.

Irrigation of Annual Crops

Irrigation is an important driver of habitat intensification within Mediterranean countries. It not only increases the density and yield of traditional crops such as winter cereal, but also enables the production of spring/summer industrial horticultural crops such as tomato but also corn (the latter usually used as fodder). Most irrigated crops, being denser, form an unsuitable breeding habitat for the little bustard (see chapter "Habitat selection and space use"). The level of pesticides and

Fig. 1 Example of land use conversion over an 11-year period of an area near Évora, Portugal. The landscape was dominated by extensive cereal farming and pastures in 2006 (left) but was rapidly converted to a landscape dominated by irrigation of annual and permanent crops by 2017 (right). Based on Google Earth imagery

fertilizers used is often greater than for dry cereal farming, and irrigated land usually involves high levels of human disturbance, as more interventions occur throughout the agronomic cycle (Suárez et al. 1997). However, the irrigation of legume crops such as alfalfa can improve the quality of this habitat for little bustards by providing cover, food and water during the critical summer season, as well as in winter (Sehhatisabet et al. 2012; García de la Morena et al. 2015; Yousefi et al. 2018). These irrigated legume crops usually occur within a matrix of land uses that include cereal and fallows which may also provide cover but which have low food availability in the dry summers (Fig. 2). Irrigated alfalfa fields can also be used during the breeding season for nesting by re-laying females, as found in Catalonia (north-eastern Spain). But in such cases, the probabilities of success are very small owing to the frequent mowing of this crop, so these alfalfa fields can become ecological traps (Cuscó Martinez 2019).

The relationship between irrigation, farming landscapes and little bustards is relatively well documented in the Iberian Peninsula, but less known in France. However, there is at least one population where impacts have been described: the irrigation of hayfields by flooding in La Crau, southern France, prevented its use by nesting females but promoted suitable habitat during the non-breeding season (Wolff 2001, 2004).

Shift to Permanent Crops

Conversion of cereal fields or other extensive annual crops to permanent woody crops is a trend that comes with market demands and fluctuations, facilitated by agricultural intensification, as it is a form of agricultural specialization in which irrigation allows for denser yields. Although woody crops in the Mediterranean region have been traditionally rainfed, the trend now is towards the use of irrigation to maximize production (Ribeiro et al. 2014). The commonest permanent crops replacing cereal fields are vineyards and olive groves (FAO 2017), and more recently also almond and pistachio trees (Díaz et al. 2012; Rabadán et al. 2017). This

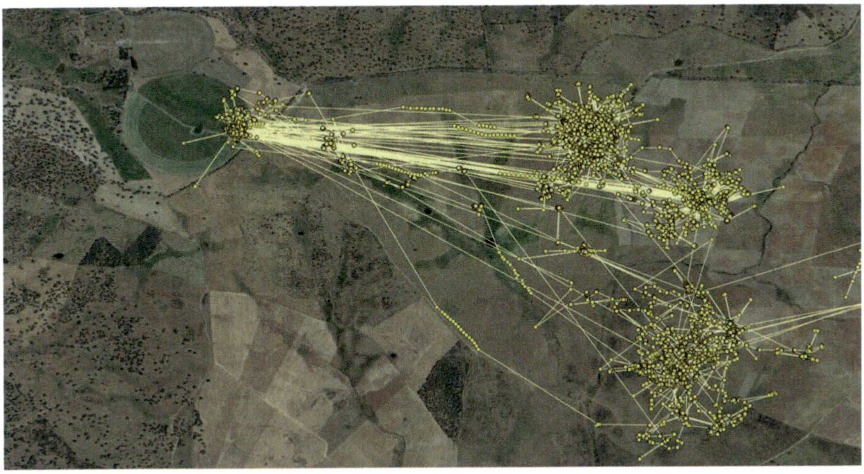

Fig. 2 Example of how irrigation can benefit the little bustard: movements of an adult male in Extremadura (Spain) during the post-breeding summer period (Silva, in prep.), where there is a marked complementary strategy of habitat use between stubbles and pastures (on the right of the picture), which offer cover, and irrigated alfalfa (on the left), offering higher trophic availability. Based on Google Earth imagery

conversion tends to occur on more productive soils, which in Mediterranean countries usually coincide with land that can be irrigated. Permanent crops radically change the habitat, by shifting towards a tree-dominated landscape, representing, very rapidly, a total habitat loss for steppe birds.

Permanent Pastures

Large areas of dry cereal farming covering most of the Alentejo, Portugal, and parts of Spain, principally Extremadura, have been converted to permanent pastures over the last 30 years (Fig. 3), mostly for beef production (Faria 2015), concomitant with an increase in livestock density, particularly over the last 20 years (Silva et al. 2018). Permanent pastures are grazed all year round, being managed in rotational schemes that result in variable grazing intensities (Faria et al. 2012). Additionally, to ensure supplementary food during winter or in drought periods, livestock farms increasingly produce hay or silage. Pastures tend to be overgrazed, thus providing limited breeding and post-breeding habitat for the little bustards because of the low height of the vegetation. Females may use hay or fodder fields (mostly seeded barley fields) for nesting, but their mowing in late April–early May can threaten nests and nesting females. In rainy years, grasslands can also be fertilized and mowed during the critical nesting period. In the Alentejo, chick and adult mortality by farm machinery in fodder crops has been found to be a serious threat (Faria et al. 2016). In France, mowing machinery caused 40% of clutch failures prior to implementing management measures (Iñigo and Barov 2010; Bretagnolle et al. 2011, 2018). Another

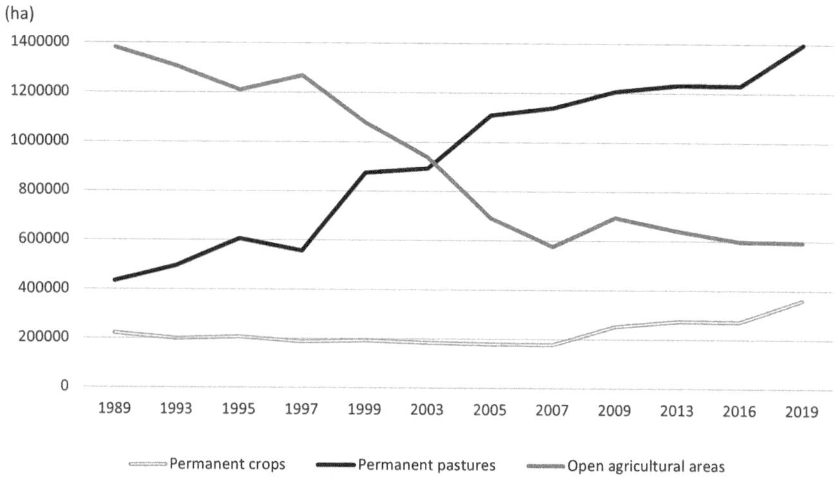

Fig. 3 Land use dynamics in the open farmland area of Alentejo, Portugal, between 1989 and 2019 (INE 2021). Most open, nonirrigated agricultural areas in Alentejo were replaced by permanent pastures over the course of the last 30 years. Permanent pastures more than tripled since 1989, while open agricultural areas decreased 2.3 times compared to the area it occupied in 1989. Permanent crops (mostly olive groves, orchards and vineyards) doubled its area between 2007 and 2019

consequence of conversion to pastures is a very significant increase in fencing that might cause mortality by collision. The little bustard's negative trends in Portugal between 2003–2006 and 2016 are significantly related to the shift of livestock species, from sheep to cattle, and also to an intensification of grazing (Marques et al. 2020b).

However, medium densities of livestock can favour little bustard breeding habitat (Wolff et al. 2002; Faria et al. 2012). Extremely high densities of both breeding males and nesting females are found in permanent pastures that are not grazed during the breeding season in consecutive years (Silva et al. 2010a, 2014).

Afforestation and Farmland Abandonment

During the late 1980s and 1990s, the CAP promoted the afforestation of agricultural areas in EU countries through economic incentives. This occurred mainly in areas where the soils were less productive, but often of high conservation value, such as at Castro Verde in southern Portugal. Additionally, there has been a trend towards agricultural abandonment in less productive areas, which may change the landscape physiognomy. As seen in the south of France, at a first stage agricultural abandonment can lead to high density of breeding numbers for the species (Devoucoux 2014; Devoucoux et al. 2019), but over time, as the habitat evolves into shrubland, it leads to habitat loss (e.g. Fonderflick et al. 2010).

Linear Infrastructures

The little bustard's European action plan identified industrial and urban development as a threat to the species, as new human facilities grow rapidly in the countryside. Such developments may cause a reduction in overall population density and lead to local extinctions. To begin with, at the beginning of the millennium, linear infrastructures were considered as having a low impact on the little bustard (De Juana and Martínez 2001), but a decade later they were classified as having a medium impact (Iñigo and Barov 2010).

Linear infrastructures like roads and powerlines are usually long, occur in all types of landscapes and are omnipresent, crossing many natural and semi-natural habitats, including key areas for nature conservation, such as Special Protection Areas designated for steppe bird conservation. Moreover, these infrastructures are often spatially clustered, as successive developments tend to occupy existing corridors or adjacent areas, and new facilities usually require new linear infrastructures like powerlines and roads.

There is evidence that linear infrastructures magnify the negative effects of land use changes on little bustard habitats. Other than direct effects, linear infrastructures cause discontinuities in open landscapes, contributing to their deterioration and negatively affecting the species (García de la Morena et al. 2007; Santos et al. 2016).

Power Lines

Overhead cables, mainly transmission and distribution powerlines, are responsible for a high rate of bird mortality due to collision (Loss et al. 2014). Such fatalities occur when birds fail to detect these infrastructures and fly into wires.

Bustards are among the bird group most vulnerable to collision with overhead wires. Their morphological features, particularly high wing loading (ratio of bird weight to wing area) and low-medium wing aspect ratio (ratio of wingspan squared to wing area), are identified as one of the main factors influencing collision risk (Bevanger 1998; Janss 2000). Bird sensorial perception is also a key factor. Eye characteristics and position in the head determine the visual field of an animal. Bustards, with eyes located laterally, have broad visual coverage of their surroundings but have a blind frontal area above the binocular field, meaning that these birds have a restricted ability to detect obstacles that lie ahead of them (Martin and Shaw 2010; Martin 2011). Due to these family-specific features, the little bustard can be considered a collision-prone species.

Little bustard deaths by collision with powerlines are documented throughout the species' range (Table 1, Fig. 4). Although the majority of studies focusing on this source of mortality are from the Iberian Peninsula, several authors report little bustard fatalities by collision in France, Italy and the eastern range, in Kazakhstan and Azerbaijan. A recent study pooling data from tracked birds over a 12-year period

Table 1 Studies reporting little bustard fatality by collision with overhead wires, with location, line type (*T* transmission, *D* distribution), survey effort, study duration, visit interval, number of collisions and data source

Country	Location	Line type	Type of register	Survey effort (km)	Study duration (months)	Visit interval (days)	No. of collisions	Source
Spain	Cáceres	D	Systematic survey	3.9	24	30–60	25	Janss and Ferrer (1998)
Spain	Cáceres	T	Systematic survey	45	24	30–60	1	Janss and Ferrer (1998)
Spain	Rosalejo	T	Systematic survey	10	12	15	12	Alonso and Alonso (1999)
Spain	Almaraz	T	Systematic survey	10	12	15	2	Alonso and Alonso (1999)
Spain	Puerto Lápice	T	Systematic survey	10	12	15	2	Alonso and Alonso (1999)
Portugal	Alentejo region	T	Systematic survey	48	12	c.30	19	Neves et al. (2005)
Portugal	Castro Verde	T	Systematic survey	11	16	15	26	Marques et al. (2007)
Portugal	Castro Verde	T	Systematic survey	11.2	29	15	30	Marques et al. (2007)
Portugal	Ervidel	T	Systematic survey	5.7	29	30	1	Marques et al. (2007)
Portugal	Castro Verde	D	Systematic survey	50	12	15	15	Marques et al. (2007)
Portugal	Castro Verde	D	Systematic survey	29.7	8–31	15	28	LPN (2012)
Portugal	Piçarras	T	Systematic survey	2.7	12	30	2	Marques pers.data

Portugal	Ferreira do Alentejo	T	Systematic survey	41	12	30	11	Ecossistema (2007)
Portugal	Alqueva	T	Systematic survey	25	67	15/30	6	Procesl (2010)
Portugal	Sines	T	Systematic survey	19.5	24	30/90	5	Ecosativa (2009)
Portugal	Ferreira do Alentejo/Évora	T	Systematic survey	61.3	24	30	2	Infante (2011)
France	Vienne	T	Unsystematic registers	n.a.	n.a.	n.a.	3	Poyrel, pers. comm.
France	Maine et Loire	T	Unsystematic registers	n.a.	n.a.	n.a.	7	Guillou, pers. comm.
France	Bouches-du-Rhône	T	Unsystematic registers	n.a.	n.a.	n.a.	1	Wolff, pers. comm.
France	Gard	T	Unsystematic registers	n.a.	n.a.	n.a.	3	Devoucoux pers. data
France	Indre et Loire	T	Unsystematic registers	n.a.	n.a.	n.a.	2	Favier, pers. comm.
Kazakhstan	Karaganda and Kostanai	D	Short assessment	680	Spring and Summer	n.a.	5	Vorona et al. (2012)
Italy	Italy	T&D	Compilation of studies	27.7	30–730	n.a.	1	Rubolini et al. (2005)
Azerbaijan	Kura-Arax and Greater Caucasus	T&D	Unsystematic registers	n.a.	n.a.	n.a.	3	Gauger (2007)
Azerbaijan	n.a.	Telegraph wires	Unsystematic registers	n.a.	n.a.	n.a.	10 per km	Iwanow and Prilonskij (1965) in Gauger (2007)

These values correspond to raw data and do not account for methodological bias, such as search efficiency, carcass removal or intervals between field surveys

Fig. 4 Little bustard deaths by collision with powerlines. Patch of feathers and a carcass next to powerlines—presumed mortalities caused by the utility infrastructure (photos: A.T. Marques)

in Iberia found that collision with powerlines is the main anthropogenic threat for the adult population, with an estimated adult yearly mortality rate of 3.4–3.8% (Marcelino et al. 2017). This is one of the highest mortality rates due to collision with powerlines ever recorded for a species.

Collision risk with powerlines changes significantly both spatially and temporally with seasonal changes in the species' behaviour and movements (Silva et al. 2014). A study pooling mortality data recorded from 2003 to 2015 along 280 km of transmission powerlines in Portugal found that little bustard collisions show two major annual peaks: one in the post-breeding season (mainly during the driest months), when the birds move longer distances and perform more frequent flights at powerline collision height; and another at the beginning of the breeding season when females perform movements between lekking areas (Marques et al. 2020a).

To date, burying a powerline is the only solution that completely prevents birds collisions. This option is usually discarded owing to technical constraints and high financial costs (APLIC 2012; Bernardino et al. 2018). The use of technical configurations with smaller pylons, fewer cables displaced vertically and smaller distances between top and bottom wires help reduce collision risk (Marques et al. 2020a). However, wire marking, aiming to increase wire visibility, is the mitigation strategy most commonly used. Marking is achieved with bird flight diverters like spirals, plates, flappers, swivels or spheres (Barrientos et al. 2011; APLIC 2012). Several studies have shown that flight diverters can decrease bird collisions, but their effectiveness still seems to be species-specific and apparently low for collision-prone species like bustards (Alonso et al. 1994; Janss and Ferrer 1998; Jenkins et al. 2010; Barrientos et al. 2012). Therefore, avoiding the construction of aerial powerlines in important areas for bustards, including breeding, post-breeding and wintering grounds, should be a priority (Marques et al. 2020a).

The disturbance caused by these structures and their avoidance also have key effects on little bustards. Transmission powerlines affect little bustard habitat selection, with birds avoiding the areas closest to these structures even when they are otherwise highly suitable for the species (Silva et al. 2010b; Santos et al. 2016). Such behaviour may occur due to increased perceived predation risk, as tall structures are usually used by predators to perch, or neophobia, as birds may respond negatively to a novelty in their environment (Walters et al. 2014).

Roads

Roads are recognized as being responsible for habitat fragmentation, mortality, sensory disturbance and chemical pollution, among other impacts (Fahrig and Rytwinski 2009). Wildlife abundance may decrease near roads and/or in landscapes with high road density either because of high mortality rates through collisions with vehicles or because animals avoid these locations due to traffic disturbances like noise, lights, pollution or motion (Fahrig and Rytwinski 2009). Little bustards may be more affected by the latter than the former, as deaths by collision with vehicles have not been identified as a serious source of mortality for the species (Marcelino et al. 2017). Nevertheless, roads and paths with low traffic intensity may represent a risk of roadkill, as they are sometimes used by displaying males due to their increased visibility, or by females and fledglings as corridors between feeding sites (Fig. 5).

Over recent decades, the development of motorways and highways has negatively affected the suitability of important areas for little bustards in Spain, Portugal and France (Hedo et al. 1999; Garcia 2004; Devoucoux 2014). Construction work, although temporary, brings additional disturbance and may have an important effect during the breeding season, if no mitigation is undertaken.

Most studies dealing with habitat selection have found the little bustard avoiding the vicinity of roads during the breeding season (Suárez-Seoane et al. 2002; García et al. 2007; Osborne and Suárez-Seoane 2007; Santangeli and Dolman 2011; Santos et al. 2016) and at stopover sites during post-breeding movements (Alonso et al. 2020) and wintering sites. However, other studies have not found this relationship during the breeding season (Martínez 1994) or in winter (Silva et al. 2004; Suárez-Seoane et al. 2008). Road size and traffic intensity may play an important role in

Fig. 5 Little bustard male displaying on a track in southern France and a roadkilled young little bustard in western France (photos: P. Devoucoux)

determining the occurrence or intensity of road avoidance (Rytwinski and Fahrig 2015; Devoucoux 2017), and may explain the differences between these studies. However, to date no studies with little bustards have focused on such variables.

The mechanisms underlying road avoidance by birds are still not understood. Traffic noise is frequently reported as the main driver of such behaviour, mainly in songbirds, but most studies on this topic failed to discard confounding effects such as visual disturbance, collisions or chemical pollution (McClure et al. 2013). A recent study conducted by Martínez-Marivela et al. (2018) did not find an effect of traffic noise on male little bustard habitat selection and spatial distribution during the breeding season. This may be because the male territorial call frequency range is only masked to a small degree by traffic noise. Moreover, some recent results suggest that little bustards may adjust their sexual display to anthropogenic road noise. Using directional recordings and noise mapping, Barrero et al. (in press) determined that little bustard males showed higher call rates at sites exposed to higher traffic noise levels. Moreover, call duration decreased under higher levels of traffic noise before calling. These results suggest that little bustards are able to make fine-scale adjustment of signals, despite some behavioural constraints that may compromise communication with low-pitched calls in noisy environments.

Railways

The effects caused by railways are quite similar to those of roads, although they are assumed to be lower due to the lower traffic flow and narrower corridors of railways (Borda-de-Água et al. 2017). However, unlike roads, railways are usually coupled with aerial wires for energy supply, the catenary, which is an additional collision risk factor.

Railways generally have high technical layout constraints as they require gentle slopes and smooth turns. These features make new high-speed trains developments likely to impact important areas for wildlife and steppe-birds in particular.

Transport infrastructures are in steady development in European countries, and the construction of new high-speed trains in important areas for the little bustard in Spain, France or Portugal has been authorized despite their great potential impact (Hedo et al. 1999; Devoucoux 2014; le Coent et al. 2017; Malo et al. 2017).

Malo et al. (2017) assessed occupation and bird behaviour in farmlands crossed by different high-speed railways in central Spain. An exclusion effect is shown in bustard species, including the little bustard. The density of this species decreased after the construction of these infrastructures, and birds significantly increased their distance (by 60–150 m) from the railway once the trains started running. Also, the little bustard exhibits frequent flights (ca. 40% of the flights recorded for the species) through the collision risk area (i.e. under or between the wires of the catenary), revealing that the species is highly vulnerable to collision with the trains or with the aerial wires of the catenary. In southern France, a new high-speed train crossing Natura 2000 key sites for little bustard conservation has also induced an avoidance effect on displaying males (Devoucoux 2017). The latter also showed higher impacts of the high-speed railway compared to traditional railway, being exclusion distances about twice larger for females than for males (Devoucoux 2014).

Other Infrastructures

Habitat loss and disturbance caused by proximity to urban areas are important issues for the conservation of little bustards in the western population. For example, in southern France, where there have been very few changes in agricultural practices, urban encroachment is considered a major threat to the species (Devoucoux 2014). In Portugal, Silva et al. (2004) found that little bustards are sensitive to inhabited houses, with occurrence probability increasing with distances from them. However, such effects may be context dependent, as other studies have not found any response of little bustards to buildings (Martínez 1994; Faria and Rabaça 2004).

Airports

Airports impose costly environmental disturbances including habitat loss, noise and reduced air quality. For birds, such infrastructures pose an additional risk of mortality through collision with airplanes during take-off or landing (Blackwell et al. 2009). Besides occupying large areas, airports often represent major land use changes beyond their boundaries, as complementary infrastructures such as highways, railways and new urban areas are usually constructed in the surroundings. Such changes render occupied land basically unsuitable for grassland and steppe birds like the little bustard. Moreover, they can favour changes in space and habitat use patterns associated to, for example, land abandonment near the airport, as described in the great bustard *Otis tarda* (López-Jamar et al. 2011), which may give rise to ecological traps.

On the reverse, airport and the way vegetation is managed may attract little bustards and cause problems with flight activities. The vegetation of airport fields is typically managed to assure unattractive habitats to wildlife, aiming to reduce long-term hazard to aviation (Blackwell et al. 2009). However, due to the characteristics (height and structure) of the vegetation at these places, some airports in North America have been identified as key areas for the conservation of arthropods and grassland bird species, when certain management actions are undertaken (Blackwell et al. 2013). Even so, grassland areas in most airports are typically patchy and non-contiguous, hence too small to support viable populations of grassland bird species (Blackwell et al. 2013) like the little bustard. Still these habitats may favour the occurrence of wintering flocks, representing a major risk for aviation safety. A good example is provided by the military airport of Getafe (Madrid, Spain), where the presence of little bustard wintering flocks of up to 400 individuals has been reported for more than 15 years, along with a few breeding birds both males and females (Cano 2009). In more recent years these numbers have declined to zero (J. Cano, pers. com.). The presence of large winter flocks has historically been observed in Madrid's Barajas Airport and was the original cause for the use of falconry as a means of bird collision control there (Cano 2009).

In France, breeding and wintering little bustards can be observed at airports, next to the runways (Devoucoux and Wolff, in prep.). Vegetation height, arthropod availability and lack of predators (like dogs and foxes, which are managed by airport staff) seem to be key factors attracting the birds, despite the disturbance caused by plane movements and the noisy environment. For these reasons, conservation measures targeting little bustards were promoted inside airports since the start of the millennium. In 2012, in the Provence–Alpes–Côte d'Azur region (Mediterranean France), approximately 20% of the regional breeding population (number of displaying males) was making use of civil or military airports. Attraction of airports to little bustards has resulted in at least 94 collisions (and death) between 2010 and 2019 at Marignane, Marseille Airport. The Marseille airport has been a particularly concerning issue: after two major bird strikes at this and another airport in 2013, the bustards were then disturbed and flushed, using falconry and dogs, in order to prevent their further use of airport fields (Devoucoux and Wolff, in prep.). A culling program has even been implemented, with up to 71 little bustards being killed between 2013 and 2019 by legal authorities (with a ministerial authorization).

Wind Farms

Wind energy developments have increased immensely in recent decades, mainly due to their contribution in preventing anthropogenic climate change (IPCC 2011). Despite its overall environmental benefits, four adverse effects of wind energy on birds have been identified: death by collision with wind turbines, displacement due to disturbance, barrier effects and habitat loss (Drewitt and Langston 2006).

Little information is available on the effect of wind farms on the little bustard, possibly because few monitoring programme results are published or accessible (Marques et al. 2014) or because few places used by this industry overlap with the species' distribution range so far. Nevertheless, in a study that determined a global vulnerability index to collision mortality at wind farms of over 9000 bird species based on ecological, morphological and life-history traits, the little bustard was found among the most vulnerable species, ranking 150th (among 9568 species) to such a mortality source (Thaxter et al. 2017). Even so, to date we have only found a single record of mortality of this species due to collision with wind turbines, in Albacete, south-east Spain (Atienza et al. 2012) and one possible case in Poitou-Charente Region, France, in spring 2019.

Monitoring programmes of four wind farms within the little bustard's western range (one in the Important Bird Area 'São Pedro de Sólis', southern Portugal; one in La Crau plain, France; and two in the Languedoc–Roussillon region, Aude department, France) did not find mortality or avoidance effects (CEN-PACA 2009; Bioinsight 2016; Mathieu Bourgeois, pers. comm.). This is, however, a small dataset, and results should be analyzed with caution, as direct and indirect impacts are still greatly unknown. Still, suitable habitats that favour the occurrence of little bustards under these plants should be avoided, as they may act as ecological traps by increasing collision risk with wind turbines. In any case, an expected impact of wind farms is related to the transmission or distribution powerlines that connect these structures to the general electric grid, which increase mortality risk due to collision (see section "Power Lines"). This collision risk could be higher than expected if wind farms are installed in remote sites, far from urban areas, so that long stretches of transmission lines are needed to connect with the power grid (Piwko et al. 2005).

Solar Facilities/Solar Power Plants

Habitat loss and fragmentation are probably the largest impacts associated with the construction, operation and maintenance of solar energy facilities. These infrastructures are enclosed by a fence and the vegetation below the solar panels is typically removed or altered to accommodate the considerable amount of infrastructure required at these facilities (Turney and Fthenakis 2011). Altogether solar farms render large areas of the landscape unsuitable for bustards.

The flat open farmland areas of the Iberian Peninsula where the little bustard typically occurs seem highly suitable for this industry, due to their smooth topography and open land and high insolation levels. Although single developments may cause relatively little habitat loss for the species, cumulative effects may be important. Portugal's Alentejo is one of the regions in Europe with the highest photovoltaic solar electricity potential (Botelho et al. 2017), and it supplied ca. 41% of the solar production of mainland Portugal in 2016 (DGEG 2016). Moreover, 103 of the nation's 168 photovoltaic projects licenced or in licencing process are located there (DGEG 2017). This is also the main area of occurrence of steppe birds in the country, including the little bustard (Equipa Atlas 2008). Important areas for the post-breeding and wintering seasons are the most vulnerable to such developments, as they typically lack adequate conservation status. Some examples are the solar power plants at Amareleja (250 ha, 46 MW peak capacity), Ferreira do Alentejo (94 ha) and Brinches (60 ha, 11 MW peak capacity) (ACCIONA Energy 2017; Botelho et al. 2017; Catavento 2017).

The solar power sector, principally solar photovoltaic and thermo-solar plants, has also witnessed rapid development in open areas in Spain. In southern France, the solar panel industry is also increasing, and solar farms are interfering with little bustard areas. The current advice of environmental authorities is for solar parks mainly to avoid SPAs, but as demands increase pressure to build inside SPA limits will also increase. Indeed, some projects have already been approved and constructed inside SPA limits in Spain (Andrés-Ruiz et al. 2015).

Other potential impacts on little bustards from this industry are still unknown. There are records of other species of bird being attracted by the reflective surfaces of the heliostats or the photovoltaic panels and ending up colliding with these structures, and of birds being burned when they fly within the beam of concentrated sunlight generated in solar plants (McCrary et al. 1986). It is however unknown whether such things may affect little bustards.

As in the case of wind farms or airports, the effects of solar developments on the little bustard may extend far beyond those posed by the solar power plant itself, as these projects require additional infrastructures, principally powerlines (often several kilometres in length) and new roads.

Climate Change

Climate change is raising average temperatures within the little bustard's range and increasing the frequency and intensity of extreme events such as droughts and heat waves. Even though the species' range is expected to shift gradually over time, the little bustard is not predicted to disappear from its current distribution within the next 100 years (Estrada et al. 2016). Still, climate change is expected to alter the species' behavioural patterns. Notably, little bustard activity has been found to be significantly constrained by elevated temperatures during the day, possibly related to a thermoregulatory response essential to reduce metabolic rates (Silva et al. 2015). Inactivity levels are thus expected to increase by 37% and 59% during the breeding and post-breeding season, respectively, during the hottest hours of the day in the Iberian Peninsula, under a scenario of 5 °C temperature rise (Silva et al. 2015). This will likely affect breeding activity as well as foraging patterns, and thus potentially affect breeding success and population dynamics (Silva et al. 2015; Gudka et al. 2019). Impacts are expected to be exacerbated by heatwaves, reducing the activity time window even more, which is likely to compromise birds' fitness seriously. When heatwaves occur during drought episodes, which also significantly limit food availability, the bustards' survival may be even more jeopardized by the overall cumulative impacts. Extreme drought events, such as those that occurred in 2005 or 2017 in different parts of Iberia, led to reduced time windows for reproduction that could have compromised the breeding season altogether, particularly in areas where soils are less productive (Juhlin and Silva, n.d.).

Droughts also pose a very serious threat for little bustard habitat, particularly in pastures, as vegetation productivity will be reduced, and vegetation structure may be further simplified if grazed under a stocking density geared to normal weather years. Consequently, during the post-breeding and winter seasons the little bustard will be vulnerable to higher levels of predation owing to degraded habitat and limited trophic availability.

Finally, climate change may also produce shifts in breeding phenology. In western France, the average lay date now occurs earlier than 20 years ago (Bretagnolle et al. 2018). On the other hand, in Mediterranean countries climate change is expected to lead to an earlier end to the breeding season.

Hunting

Hunting is a widespread activity in many landscapes occupied by little bustards, including the agricultural areas of western Europe, where it can provide an added value to rural economies. There is growing evidence that hunting is a significant source of non-natural mortality and disturbance for declining little bustards (Casas et al. 2009; Tarjuelo et al. 2015; Marcelino et al. 2017; Yousefi et al. 2018; see also chapters "Interspecific relationships: predation, competition or coexistence, parasites" and "Migration, movements and non-breeding ecology"). Direct mortality by hunting (shooting) is an important threat in the species' western stronghold, in Iberia, and the eastern wintering quarters, principally Azerbaijan and Iran, where most of the population is thought to congregate (Iñigo and Barov 2010). Given that adult and juvenile survival are key demographic parameters (Morales et al. 2005a), any increase in anthropogenic mortality can have a major negative impact on the viability of little bustard populations, particularly those with low breeding productivity (Bretagnolle et al. 2011; Lapiedra et al. 2011). In Iberia, where productivity is currently low (see chapters "Breeding biology and demographic traits" and "Little bustard population dynamic"), mortality from shooting has recently been found to be an important cause of mortality, even though it is not a game species. In fact, mortality from shooting represented 2.4–3% of annual mortality in Iberia, so it could still be a relevant driver of the decline of little bustards at least locally (Marcelino et al. 2017). In the eastern range, hunting is an important source of mortality and disturbance during the winter, so this factor could threaten the populations (Gauger 2007; Sehhatisabet et al. 2012). In Iran, wintering populations of the little bustard are thought to have responded to heavy persecution by concentrating next to the international border where strict exclusion zones imposed by the military provide safe areas for the species (Yousefi et al. 2018).

In addition to direct effects, hunting may negatively influence little bustards through disturbance during hunting activities (Mougeot and Arroyo 2017). Animals often perceive humans as potential predators and alter their behaviour accordingly (Beale and Monaghan 2004). Increases in human population density and leisure activities have amplified the potential consequences of human disturbance on wildlife (Blanc et al. 2006; Arlettaz et al. 2007), and hunting as a leisure activity has been shown to affect non-target protected species (Arroyo and Razin 2006; Thiel et al. 2007; Sastre et al. 2009). Two studies have quantified this in little bustards (Casas et al. 2009; Tarjuelo et al. 2015). In western France, hunting activities have been shown to increase flight frequency and time spent vigilance to the detriment of resting or feeding, and on hunting days little bustards concentrated in hunt-free areas (Casas et al. 2009). In central Spain, disturbance due to hunting was also shown to affect the behaviour and physiological stress of wintering little bustards (Tarjuelo et al. 2015). During weekends, when hunting activities took place, little bustards flew more frequently (which can also lead to higher collision risks, if linear infrastructures are present), spent more time vigilant, and changed their habitat use, with an increased use of stubble fields. Greater physiological stress levels, measured through glucocorticoid metabolite concentrations in faeces, were also associated with increased human disturbance during hunting weekends (Tarjuelo et al. 2015). Glucocorticoid metabolite levels were affected by hunting shot rates more than by other human activities. Faecal corticosterone metabolites in little bustards remained high on Mondays after hunting weekends, suggesting that the negative effects may persist for some time (Tarjuelo et al. 2015). Hunting disturbance may thus imply energetic costs, which coupled with the shorter time available for foraging may result in reduced condition and increased mortality risk through natural causes (e.g. predation, diseases). Indeed, after hunting weekends, a change in little bustard foraging patterns was observed which pointed to a strategy for recovering energy stores after disturbance (Tarjuelo et al. 2015).

Suitable winter habitat and adequate management (in terms of its size, location, network structure and vegetation cover) of hunt-free reserves could reduce the impact of hunting disturbance on little bustards. When hunt-free reserves occur in the landscape, as in western France, little bustards almost exclusively use them during the hunting season, even when they are very small in size, confirming their importance (Casas et al. 2009). Birds confined to hunt-free reserves might however have a reduced choice of feeding habitat. For little bustards, crops such as rape-seed or alfalfa are particularly important for foraging during the non-breeding season (Wolff et al. 2002), probably because they provide relatively high energy as compared with other available crops. Therefore, habitat availability inside and outside hunt-free reserves should be an important factor to consider in their design.

On the other hand, hunting management can also benefit little bustards through the maintenance or restoration of suitable habitats, or the provision of key resources (Arroyo et al. 2013). It has been shown, for example that little bustard abundance in hunting estates in central Spain is higher with increasing levels of fox control on the estate, which suggests that predator suppression carried out to benefit partridges may indirectly benefit little bustards (Estrada et al. 2016).

Concluding Remarks

In this review, we describe and update the main threats known for the little bustard, highlight new threats and direct attention to the little-known eastern populations. Looking back to the first version of the European action plan written nearly 20 years ago (De Juana and Martinez 1997), there is now a much deeper understanding of the scale of the threats and how they can impact populations.

Notably, the CAP continues to be a major driver of changes in farmland leading to habitat loss and degradation for little bustards. There is an urgent need for the implementation of an agricultural policy that takes into account biodiversity conservation needs, such as fallow maintenance and effective agri-environment schemes that promote appropriate habitat conservation and management.

Energy development in steppic habitats will likely increase in the future and solar plants will clearly have a more direct impact on the little bustard, as these infrastructures occupy large areas in open landscapes, particularly those located on poorer, less productive soils where breeding populations persist. These energy developments require a significant expansion of the electricity grid, posing a greater risk of collision events.

We would like to stress that threats should not be analyzed in isolation, as many of them occur synchronously and may have cumulative (even multiplicative) effects on the species. For example, more intensified farmland requires more infrastructure such as roads and powerlines, so the impacts go beyond land use change, bringing more frequent disturbances and greater risk of collision with powerlines.

A key question relates to whether many of the threats to the western European populations also apply to the eastern populations, which now hold the world's highest numbers. There have been increases there, possibly due to the greater availability of breeding habitat, but there is now a risk of fresh declines again with intensifying farming methods. Hunting and poaching, moreover, appear also to be threatening the eastern population, principally in the wintering grounds of Azerbaijan and Iran. Hunting pressure is such that disturbance and persecution seem to be major drivers of the species' winter distribution in the eastern range, where birds tend to concentrate in areas with less human pressure. Moreover, direct and indirect effects of hunting in the western range seem important enough to deserve further attention.

Acknowledgements We are thankful to Rocío Tarjuelo for her review and comments on a previous version of this chapter and to Nigel J. Collar for kindly reviewing the English.

References

ACCIONA Energy (2017) Amareleja photovoltaic plant. Available at: http://www.acciona-energia.com/areas-of-activity/photovoltaic/major-projects/amareleja-photovoltaic-plant/. Accessed 25 Aug 2017

Alonso JA, Alonso JC (1999) Collision of birds with overhead transmission lines in Spain. In: Ferrer M, Janns GF (eds) Birds and power lines: collision, electrocution, and breeding. Quercus, Madrid, pp 57–82

Alonso JC, Alonso JA, Muñoz-Pulido R (1994) Mitigation of bird collisions with transmission lines through groundwire marking. Biol Conserv 67:129–134

Alonso H, Correia RA, Marques AT, Palmeirim JM, Moreira F, Silva JP (2020) Male post-breeding movements and stopover habitat selection of an endangered short-distance migrant, the Little Bustard *Tetrax tetrax*. Ibis 162:279–292

Andrés-Ruiz C, Iranzo-García E, Espejo-Marín C (2015) Solar thermoelectric power landscapes in Spain. In: Frolova M, Prados MJ, Nadaï A (eds) Renewable energies and European landscapes. Springer, Netherlands, pp 237–254

APLIC (2012) Reducing avian collisions with power lines: the state of the art in 2012. Washington, DC

Arlettaz R, Patthey P, Baltic M, Leu T, Schaub M, Palme R, Jenni-Eiermann S (2007) Spreading free-riding snow sports represent a novel serious threat for wildlife. Proc R Soc B Biol Sci 274(1614):1219–1224. https://doi.org/10.1098/rspb.2006.0434

Arroyo B, Razin M (2006) Effect of human activities on bearded vulture behaviour and breeding success in the French Pyrenees. Biol Conserv 128(2):276–284. https://doi.org/10.1016/j.biocon.2005.09.035

Arroyo B, Delibes-Mateos M, Caro J, Estrada A, Mougeot F, Díaz-Fernández S, Casas F, Viñuela J (2013) Efecto de la gestión para las especies de caza menor sobre la fauna no cinegética. Ecosistemas 22:27–32

Atienza JC, Fierro IM, Infante O, Valls J, Dominguez J (2012) In: SEO/BirdLife (ed) Directrices para la evaluaci{ó}n del impacto de los parques e{ó}licos en aves y murci{é}lagos (versi{ó}n 3.0). SEO/BirdLife, Madrid. Available at: http://www.seo.org/media/docs/Manual%7B_%7Dmolinos.pdf

Barrientos R, Alonso JC, Ponce C, Palacín C (2011) Meta-analysis of the effectiveness of marked wire in reducing avian collisions with power lines. Conserv Biol 25(5):893–903. Available at: http://www.ncbi.nlm.nih.gov/pubmed/21676031. Accessed 19 Aug 2013

Barrientos R, Ponce C, Palacín C, Martín CA, Martín B, Alonso JC (2012) Wire marking results in a small but significant reduction in avian mortality at power lines: a BACI designed study. PLoS One 7(3):e32569. https://doi.org/10.1371/journal.pone.0032569

Beale CM, Monaghan P (2004) Human disturbance: people as predation-free predators? J Appl Ecol 41(2):335–343. https://doi.org/10.1111/j.0021-8901.2004.00900.x

Berger-Geiger B, Galizia CG, Arroyo B (2019) Montagu's Harrier breeding parameters in relation to weather, colony size and nest protection schemes: a long-term study in Extremadura, Spain. J Ornithol 160(2):429–441. https://doi.org/10.1007/s10336-018-1618-0

Bernardino J, Bevanger K, Barrientos R, Dwyer JF, Marques AT, Martins RC, Shaw JM, Silva JP, Moreira F (2018) Bird collisions with power lines: state of the art and priority areas for research. Biol Conserv 222. https://doi.org/10.1016/j.biocon.2018.02.029

Bevanger K (1998) Biological and conservation aspects of bird mortality caused by electricity power lines: a review Biol Conserv, 86(1):67–76. Available at: http://linkinghub.elsevier.com/retrieve/pii/S0006320797001766. Accessed 23 Aug 2013

Bioinsight (2016) Sub-Parque Eólico de Mértola: Monitorização da comunidade de Aves. Relatório IV (Fase de exploração – Ano 2015/2016)

BirdLife International (2021) Species factsheet: Tetrax tetrax. Downloaded from http://www.birdlife.org. Accessed 01 Aug 2020

Blackwell BF, DeVault TL, Fernández-Juricic E, Dolbeer RA (2009) Wildlife collisions with aircraft: a missing component of land-use planning for airports. Landsc Urban Plan 93(1):1–9. https://doi.org/10.1016/j.landurbplan.2009.07.005

Blackwell BF, Seamans TW, Schmidt PM, De Vault TL, Belant JL, Whittingham MJ, Martin JA, Fernández-Juricic E (2013) A framework for managing airport grasslands and birds amidst conflicting priorities. Ibis 155(1):199–203. https://doi.org/10.1111/ibi.12011

Blanc R, Guillemain M, Mouronval J, Desmonts D, Fritz H (2006) Effects of non-consumptive leisure disturbances to wildlife. Rev Ecol Terre Vie 61:117–133

Borda-de-Água L, Barrientos R, Beja P, Pereira HM (2017) Railway ecology. Springer, Cham, pp 3–9

Botelho A, Lourenço-Gomes L, Pinto L, Sousa S, Valente M (2017) Accounting for local impacts of photovoltaic farms: the application of two stated preferences approaches to a case-study in Portugal. Energy Policy 109:191–198. https://doi.org/10.1016/j.enpol.2017.06.065

Bretagnolle V, Inchausti P (2005) Modelling population reinforcement at a large spatial scale as a conservation strategy for the declining little bustard (*Tetrax tetrax*) in agricultural habitats. Anim Conserv 8(1):59–68. https://doi.org/10.1017/S1367943004001866

Bretagnolle V, Villers A, Denonfoux L, Cornulier T, Inchausti P, Badenhausser I (2011) Rapid recovery of a depleted population of Little Bustards *Tetrax tetrax* following provision of alfalfa through an agri-environment scheme. Ibis 153(1):4–13. https://doi.org/10.1111/j.1474-919X.2010.01092.x

Bretagnolle V, Denonfoux L, Villers A (2018) Are farming and birds irreconcilable? A 21-year study of bustard nesting ecology in intensive agroecosystems. Biol Conserv 228:27–35

Cano J (2009) La población de sisón común (*Tetrax tetrax*) en la base aérea de Getafe (Madrid). Anu Ornitol Madr 12:92–115

Casas F, Mougeot F, Viñuela J, Bretagnolle V (2009) Effects of hunting on the behaviour and spatial distribution of farmland birds: importance of hunting-free refuges in agricultural areas. Anim Conserv 12(4):346–354. https://doi.org/10.1111/j.1469-1795.2009.00259.x

Catavento (2017) Hércules Solar Park/Solar Serpa. Available at: http://www.catavento.pt/uk/projectos_06.asp.html. Accessed 25 Aug 2017

CEN-PACA (2009) Inventaire des populations nicheuses d'outarde canepetière (*Tetrax tetrax*) et d'Oedicnème criard (*Burhinus oedicnemus*) sur le site éolien du mas de Leuze. Année 2008

Collar NJ, Baral HS, Batbayer N, Bhardwaj GS, Brahma N, Burnside RJ, Choudhury AU, Combreau O, Dolman PM, Donald PF, Dutta S, Gadhavi D, Gore K, Goroshko OA, Hong C, Jather GA, Jha RRS, Jhala YV, Koshkin MA, Lahkar BP, Liu G, Mahood SP, Morales MB, Narwade SS, Natsagdorj T, Nefedov AA, Silva JP, Thakuri JJ, Wang M, Zhang Y, Kessler AE (2018) Averting the extinction of bustards in Asia. Forktail 33:1–26

Cuscó Martinez F (2019) De l'ecologia del comportament a la demografia: lliçons per a la conservació del sisó comú (*Tetrax tetrax*). PhD Thesis. Universitat de Barcelona

De Juana E, Martínez C (1996) Distribution and conservation status of Little Bustard *Tetrax tetrax* in the Iberian Peninsula. Ardeola 43:157–167

De Juana E, Martinez C (1997) European Union species action plan for Little Bustard (*Tetrax tetrax*)

De Juana E, Martínez C (2001) Little bustard (*Tetrax tetrax*). In: Schaffer N, Gallo-Orsi U (eds) European Union Action Plans for eight priority bird species. European Commission, Luxembourg

Devoucoux P (2014) Conséquences et impacts prévisibles d'une perte d'habitat majeure sur une espèce menacée aux exigences écologiques complexes: Effets de la mise en place du contournement ferroviaire à grande vitesse Nîmes-Montpellier sur la dynamique de la population d'O. PhD Thesis, Université de Poitiers, 185pp

Devoucoux P (2017) Suivi de l'impact du Contournement ferroviaire Nîmes-Montpellier sur la dynamique de la population d'Outarde canepetière des Costières de Nîmes, préconisations de gestion 2018–2037

Devoucoux P, Besnard A, Bretagnolle V (2019) Sex-dependent habitat selection in a high-density Little Bustard *Tetrax tetrax* population in southern France, and the implications for conservation. Ibis 161(2):310–324. https://doi.org/10.1111/ibi.12606

DGEG (2016) Renováveis – estatísticas rápidas, n° 146. Lisboa. Available at: http://www.apren.pt/fotos/newsletter/conteudos/estatisticas_rapidas-renovaveis_dezembro_2016_dgeg_1488900704.pdf

DGEG (2017) Geoportal da DGEG. Available at: http://geoportal.dgeg.pt/geoportal/catalog/search/search.page. Accessed 28 Aug 2017

Díaz AR, Hernández CM, Serrato FB (2012) Cambios de usos del suelo en la región de Murcia. el almendro como cultivo de referencia y su relación con los procesos de erosión. Nimbus 29–30: 607–626

Drewitt AL, Langston RHW (2006) Assessing the impacts of wind farms on birds. Ibis 148:29–42. https://doi.org/10.1111/j.1474-919X.2006.00516.x

Ecosativa (2009) Monitorização de Vertebrados Voadores – Linha Sines-Portimão 3, a 400 kV. Relatório Final. São Teotónio

Ecossistema (2007) Linhas de alta tensão Alqueva – Ferreira do Alentejo e Ferreira do Alentejo – Sines 2, a 400kV. Monitorização de impactes na avifauna. Lisboa

Emmerson M, Morales MB, Oñate JJ, Batáry P, Berendse F, Liira J, Aavik T, Guerrero I, Bommarco R, Eggers S, Pärt T, Tscharntke T, Weisser W, Clement L, Bengtsson J (2016) How agricultural intensification affects biodiversity and ecosystem services. Adv Ecol Res 55: 43–97. https://doi.org/10.1016/bs.aecr.2016.08.005

Equipa Atlas (2008) Atlas das aves nidificantes em Portugal (1999–2005). Instituto da Conservação da Natureza e da Biodiversidade, Sociedade Portuguesa para o Estudo das Aves, Parque Natural da Madeira e Secretaria Regional do Ambiente e do Mar
Estrada A, Delgado MP, Arroyo B, Traba J, Morales MB (2016) Forecasting large-scale habitat suitability of European bustards under climate change: the role of environmental and geographic variables. PLoS One 11(3):e0149810. https://doi.org/10.1371/journal.pone.0149810
Fahrig L, Rytwinski T (2009) Effects of roads on animal abundance: an empirical review and synthesis. Ecol Soci 14(1):21
FAO (2017) Land use. Available at: http://www.fao.org/faostat/en/#data/EL/visualize
Faria N (2015) Implications of dry grassland management in the ecology and conservation of grassland birds in South Portugal. Universidad Autónoma de Madrid
Faria N, Morales MB (2019) Effects of soil tillage regime and frequency of cultivation on grassland bird assemblages in Mediterranean drylands. J Environ Manag 233:211–217. https://doi.org/10.1016/j.jenvman.2018.12.038
Faria N, Rabaça JE (2004) Breeding habitat modelling of the little bustard *Tetrax tetrax* in the site of community importance of Cabrela (Portugal). Ardeola 51(2):331–343
Faria N, Silva JP (2010) Habitat selection of the little bustard during the beginning of an agricultural year. Ardeola 57:363–373
Faria N, Rabaça JE, Morales MB (2012) The importance of grazing regime in the provision of breeding habitat for grassland birds: the case of the endangered little bustard (*Tetrax tetrax*). J Nat Conserv 20(4):211–218. https://doi.org/10.1016/j.jnc.2012.03.003
Faria N, Morales MB, Rabaça JE (2016) Exploring nest destruction and bird mortality in mown Mediterranean dry grasslands: an increasing threat to grassland bird conservation. Eur J Wildl Res 62(6):663–671. https://doi.org/10.1007/s10344-016-1039-4
Fonderflick J, Caplat P, Lovaty F, Thévenot M, Prodon R (2010) Avifauna trends following changes in a Mediterranean upland pastoral system. Agric Ecosyst Environ 137(3–4): 337–347. https://doi.org/10.1016/j.agee.2010.03.004
Garcia GMD (2004) Impacte da Auto-Estrada A2 na Zona de Protecção Especial de Castro Verde: Efeitos na Distribuição e Densidade de Sisão (*Tetrax tetrax*). Universidade Técnica de Lisboa. Instituto Superior de Agronomia
García de la Morena EL, Morales MB, de Juana E, Suárez F (2007) Surveys of wintering Little Bustards *Tetrax tetrax* in Central Spain: distribution and population estimates at a regional scale. Bird Conserv Int 17(01):23. https://doi.org/10.1017/S0959270906000608
García de la Morena EL, Morales MB, Bota G, Silva JP, Ponjoan A, Suárez F, Mañosa S, De Juana E (2015) Migration patterns of iberian Little Bustards *Tetrax tetrax*. Ardeola 62(1):95–112. https://doi.org/10.13157/arla.62.1.2015.95
García de la Morena EL, Bota G, Mañosa S, Morales MB (2018) El sisón común en España. II Censo Nacional (2016). SEO/BirdLife. Madrid
García J, Suárez-Seoane S, Miguélez D, Osborne PE, Zumalacárregui C (2007) Spatial analysis of habitat quality in a fragmented population of little bustard (*Tetrax tetrax*): implications for conservation. Biol Conserv 137(1):45–56. https://doi.org/10.1016/j.biocon.2007.01.013
Gauger K (2007) Occurrence, ecology and conservation of wintering Little Bustards *Tetrax tetrax* in Azerbaijan. Arch Nat Landsc 46(2):5–27
Geiger F, Bengtsson J, Berendse F, Weisser WW, Emmerson M, Morales MB, Ceryngier P, Liira J, Tscharntke T, Winqvist C, Eggers S, Bommarco R, Pärt T, Bretagnolle V, Plantegenest M, Clement LW, Dennis C, Palmer C, Oñate JJ, Guerrero I, Hawro V, Aavik T, Thies C, Flohre A, Hänke S, Fischer C, Goedhart PW, Inchausti P (2010) Persistent negative effects of pesticides on biodiversity and biological control potential on European farmland. Basic Appl Ecol 11(2): 97–105. https://doi.org/10.1016/j.baae.2009.12.001
Gendre N, Eraud C, Bretagnolle V, Dalloyau S (2019) L'Outarde canepetière *Tetrax tetrax* en France en 2012 et 2016: effectifs et repartition. Ornithos 25:290–302

Gudka M, Santos CD, Dolman PM, Abad-Gómez JM, Silva JP (2019) Feeling the heat: elevated temperature affects male display activity of a lekking grassland bird. PLoS One 14(9): e0221999. https://doi.org/10.1371/journal.pone.0221999
Hallmann CA, Foppen RPB, van Turnhout CAM, de Kroon H, Jongejans E (2014) Declines in insectivorous birds are associated with high neonicotinoid concentrations. Nature 511(7509): 341–343. https://doi.org/10.1038/nature13531
Hedo D, Atienza JC, Ruiz A, Naveso MA, Martí R (1999) Improving performance of roads and railways for wildlife conservation: showing results with two case studies in Central Spain. In: Third International Conference on Wildlife Ecology and Transportation. Available at: http://www.icoet.net/downloads/99paper18.pdf
Inchausti P, Bretagnolle V (2005) Predicting short-term extinction risk for the declining Little Bustard (*Tetrax tetrax*) in intensive agricultural habitats. Biol Conserv 122(3):375–384. https://doi.org/10.1016/j.biocon.2004.08.001
Infante S (Coord) (2011) Estudo de avaliação da eficácia de dispositivos anti colisão – Relatório Final. Castelo Branco, Portuga
Iñigo A, Barov B (2010) Action plan for the Little Bustard *Tetrax tetrax* in the European Union. SEO/BirdLife and BirdLife International for the European Commission
Instituto Nacional de Estatística (INE) (2021) Estatísticas agrícolas de base. Downloaded from http://www.ine.pt. Accessed 20 Sep 2021
IPCC (2011) IPCC special report on renewable energy sources and climate change mitigation: summary for policymakers. Cambridge. Available at: http://www.uncclearn.org/sites/default/files/inventory/ipcc15.pdf
Janss GFE (2000) Avian mortality from power lines: a morphologic approach of a species-specific mortality. Biol Conserv 95(3):353–359. Available at: http://linkinghub.elsevier.com/retrieve/pii/S0006320700000215. Accessed 23 Aug 2013
Janss GF, Ferrer M (1998) Rate of bird collision with power lines: effects of conductor-marking and static wire-marking. J Field Ornithol 69(1):8–17
Jenkins AR, Smallie JJ, Diamond M (2010) Avian collisions with power lines: a global review of causes and mitigation with a South African perspective. Bird Conserv Int 20(03):263–278. https://doi.org/10.1017/S0959270910000122
Jiguet F (2002) Arthropods in diet of Little Bustards *Tetrax tetrax* during the breeding season in western France. Bird Study 49(2):105–109. https://doi.org/10.1080/00063650209461253
Jolivet C (2001) L'Outarde canepetière *Tetrax tetrax* en France. Statut de l'espèce à la fin du xxe siècle. Ornithos 8:89–95
Juhlin C, Silva JP (n.d.) Climate change impact on grassland birds in the Mediterranean basin: a case study with the Little Bustard
Kamp J, Urazaliev R, Donald PF, Hölzel N (2011) Post-Soviet agricultural change predicts future declines after recent recovery in Eurasian steppe bird populations. Biol Conserv 144(11): 2607–2614. https://doi.org/10.1016/j.biocon.2011.07.010
Lapiedra O, Ponjoan A, Gamero A, Bota G, Mañosa S (2011) Brood ranging behaviour and breeding success of the threatened little bustard in an intensified cereal farmland area. Biol Conserv 144(12):2882–2890. https://doi.org/10.1016/j.biocon.2011.08.005
le Coent P, Préget R, Thoyer S (2017) Compensating environmental losses versus creating environmental gains: implications for biodiversity offsets. Ecol Econ 142:120–129. Available at: https://hal.archives-ouvertes.fr/hal-01487105/document
Lesiv M, Schepaschenko D, Moltchanova E, Bun R, Dürauer M, Prishchepov AV, Schierhorn F, Estel S, Kuemmerle T, Alcántara C, Kussul N, Shchepashchenko M, Kutovaya O, Martynenko O, Karminov V, Shvidenko A, Havlik P, Kraxner F, See L, Fritz S (2018) Spatial distribution of arable and abandoned land across former Soviet Union countries. Sci Data 5(1): 180056. https://doi.org/10.1038/sdata.2018.56
Lopez-Antia A, Ortiz-Santaliestra ME, Mougeot F, Mateo R (2013) Experimental exposure of red-legged partridges (*Alectoris rufa*) to seeds coated with imidacloprid, thiram and difenoconazole. Ecotoxicology 22(1):125–138. https://doi.org/10.1007/s10646-012-1009-x

Lopez-Antia A, Ortiz-Santaliestra ME, Camarero PR, Mougeot F, Mateo R (2015a) Assessing the risk of fipronil-treated seed ingestion and associated adverse effects in the red-legged partridge. Environ Sci Technol 49(22):13649–13657. https://doi.org/10.1021/acs.est.5b03822

Lopez-Antia A, Ortiz-Santaliestra ME, Mougeot F, Mateo R (2015b) Imidacloprid-treated seed ingestion has lethal effect on adult partridges and reduces both breeding investment and offspring immunity. Environ Res 136:97–107. https://doi.org/10.1016/j.envres.2014.10.023

Lopez-Antia A, Ortiz-Santaliestra ME, Mougeot F, Camarero PR, Mateo R (2018) Brood size is reduced by half in birds feeding on flutriafol-treated seeds below the recommended application rate. Environ Pollut 243:418–426. https://doi.org/10.1016/j.envpol.2018.08.078

López-Jamar J, Casas F, Díaz M, Morales MB (2011) Local differences in habitat selection by Great Bustards *Otis tarda* in changing agricultural landscapes: implications for farmland bird conservation. Bird Conserv Int 21:328–341

Loss SR, Will T, Marra PP (2014) Refining estimates of bird collision and electrocution mortality at power lines in the United States. PLoS One 9(7):e101565. https://doi.org/10.1371/journal.pone.0101565

LPN (2012) Project LIFE Estepárias—conservation of the Great Bustard, Little Bustard and Lesser Kestrel in the Cereal Fields of Baixo Alentejo

Malo JE, García de la Morena EL, Hervás I, Mata C, Herranz J (2017) Cross-scale changes in bird behavior around a high speed railway: from landscape occupation to infrastructure use and collision risk. In: Railway ecology. Springer, Cham, pp 117–134

Marcelino J, Moreira F, Mañosa S, Cuscó F, Morales MB, García de la Morena EL, Bota G, Palmeirim JM, Silva JP (2017) Tracking data of the Little Bustard *Tetrax tetrax* in Iberia shows high anthropogenic mortality. Bird Conserv Int:1–12. https://doi.org/10.1017/S095927091700051X

Marques AT, Rocha P, Silva JP (2007) Monitorização dos Efeitos da Linha de Muito Alta Tensão Ferreira do Alentejo/Ourique sobre Espécies Prioritárias, Mortalidade e Taxas de Voo. Instituto para a Conservação da Natureza, Lisboa

Marques AT, Batalha H, Rodrigues S, Costa H, Pereira MJR, Fonseca C, Mascarenhas M, Bernardino J (2014) 'Understanding bird collisions at wind farms: an updated review on the causes and possible mitigation strategies. Biol Conserv 179:40–52. https://doi.org/10.1016/j.biocon.2014.08.017

Marques AT, Martins RC, Silva JP, Palmeirim JM, Moreira F (2020a) Power line routing and configuration as major drivers of collision risk in two bustard species. Oryx:1–10. https://doi.org/10.1017/S0030605319000292

Marques AT, Moreira F, Alcazar R, Delgado A, Godinho C, Sampaio H, Rocha P, Sequeira N, Palmeirim JM, Silva JP (2020b) Changes in drassland management and infrastructures contribute to the decline of an endangered bird population. Sci Rep 10:15150. https://doi.org/10.1038/s41598-020-72154-9

Martin GR (2011) Understanding bird collisions with man-made objects: a sensory ecology approach. Ibis 153(2):239–254. Available at: http://doi.wiley.com/10.1111/j.1474-919X.2011.01117.x. Accessed 26 Aug 2013

Martin GR, Shaw JM (2010) Bird collisions with power lines: failing to see the way ahead? Biol Conserv 143(11):2695–2702. Available at: http://linkinghub.elsevier.com/retrieve/pii/S0006320710003204. Accessed 10 Aug 2013

Martínez C (1994) Habitat selection by the Little Bustard *Tetrax tetrax* in cultivated areas of Central Spain. Biol Conserv 67(2):125–128

Martínez-Marivela I, Morales MB, Iglesias-Merchán C, Delgado MP, Tarjuelo R, Traba J (2018) Traffic noise pollution does not influence habitat selection in the endangered little bustard. Ardeola 65(2):261–270. https://doi.org/10.13157/arla.65.2.2018.ra5

Matson PA, Parton WJ, Power AG, Swift MJ (1997) Agricultural intensification and ecosystem properties. Science 277(5325):504–509. https://doi.org/10.1126/science.277.5325.504

McClure CJW, Ware HE, Carlisle J, Kaltenecker G, Barber JR (2013) An experimental investigation into the effects of traffic noise on distributions of birds: avoiding the phantom road. Proc R Soc B Biol Sci 280(1773):20132290. https://doi.org/10.1098/rspb.2013.2290

McCrary MD, McKernan RL, Schreiber RW, Wagner WD, Sciarrotta TC (1986) Avian mortality at a solar energy power plant. J Field Ornithol 57(2):135–141. Available at: http://www.jstor.org/stable/4513113

Mineau P, Whiteside M (2013) Pesticide acute toxicity is a better correlate of U.S. grassland bird declines than agricultural intensification. PLoS One 8(2):e57457. https://doi.org/10.1371/journal.pone.0057457

Morales MB, Bretagnolle V, Arroyo B (2005a) Viability of the endangered Little Bustard *Tetrax tetrax* population of Western France. Biodivers Conserv 14(13):3135–3150. https://doi.org/10.1007/s10531-004-0382-z

Morales MB, García JT, Arroyo B (2005b) Can landscape composition changes predict spatial and annual variation of little bustard male abundance? Anim Conserv 8(2):167–174. https://doi.org/10.1017/S1367943005001988

Morales MB, Traba J, Delgado MP, García de la Morena EL (2013) The use of fallows by nesting Little Bustard *Tetrax tetrax* females: implications for conservation in mosaic cereal farmland. Ardeola 60(1):85–97. https://doi.org/10.13157/arla.60.1.2012.85

Mougeot F, Arroyo B (2017) Behavioural responses to human activities and implications for conservation. Ecosistemas 26(3):5–12

Neves J, Infante S, Ministro J, Brandão R (2005) Estudo sobre o Impacto das Linhas Eléctricas de Muito Alta Tensão na Avifauna em Portugal (in Portuguese). Castelo Branco, Portugal

Osborne PE, Suárez-Seoane S (2007) Identifying core areas in a species' range using temporal suitability analysis: an example using Little Bustards *Tetrax Tetrax L.* in Spain. Biodivers Conserv 16(12):3505–3518. https://doi.org/10.1007/s10531-006-9058-1

Pain DJ, Dixon J (1997) Why farming and birds in Europe? In: Pain D, Pienkowski MW (eds) Farming and birds in Europe: the common agricultural policy and its implications for bird conservation. Academic Press, London

Piwko R, Miller N, Sanchez-Gasca J, Yuan X, Dai R, Lyons J (2005) Integrating large wind farms into weak power grids with long transmission lines. In: IEEE/PES (ed) IEEE/PES transmission and distribution conference and exhibition: Asia and Pacific, pp 1–7. Available at: http://www.science.smith.edu/~jcardell/Readings/Wind/Piwko-integr wkTx.pdf

Procesl (2010) Linha de transporte de energia Alqueva – Fronteira Espanhola, a 400kV. Relatório final de monitorização, Alfragide

Rabadán A, Álvarez-Ortí M, Pardo JE, Gómez R, Pardo-Giménez A, Olmeda M (2017) A comprehensive approach to pistachio oil production. Br Food J 119(4):921–933. https://doi.org/10.1108/BFJ-08-2016-0373

Reif J, Vermouzek Z (2019) Collapse of farmland bird populations in an Eastern European country following its EU accession. Conserv Lett 12(1):e12585. https://doi.org/10.1111/conl.12585

Ribeiro PF, Santos JL, Bugalho MN, Santana J, Reino L, Beja P, Moreira F (2014) Modelling farming system dynamics in High Nature Value Farmland under policy change. Agric Ecosyst Environ 183:138–144. https://doi.org/10.1016/j.agee.2013.11.002

Rubolini D, Gustin M, Bogliani G, Garavaglia R (2005) Birds and powerlines in Italy: an assessment. Bird Conserv Int 15(02). https://doi.org/10.1017/S0959270905000109

Rytwinski T, Fahrig L (2015) The impacts of roads and traffic on terrestrial animal populations. In: van der Ree R, Smith DJ, Grilo C (eds) Handbook of road ecology. Wiley, Chichester, pp 237–246

Salamolard M, Moreau C (1999) Habitat selection by Little Bustard *Tetrax tetrax* in a cultivated area of France. Bird Study 46(1):25–33. https://doi.org/10.1080/00063659909461112

Santangeli A, Dolman PM (2011) Density and habitat preferences of male little bustard across contrasting agro-pastoral landscapes in Sardinia (Italy). Eur J Wildl Res 57(4):805–815. https://doi.org/10.1007/s10344-011-0492-3

Santos T, Suárez F (2005) Biogeography and population trends of steppe land birds. In: Bota G, Morales M, Mañosa S, Camprodon J (eds) Ecology and conservation of steppe-land birds. Barcelona, Lynx Edicions, pp 69–102

Santos M, Bessa R, Cabral JA, Pacheco FAL, Leitão D, Moreira F, Pinto M, Lecoq M, Silva JP (2016) Impacts of land use and infrastructural changes on threatened Little Bustard *Tetrax tetrax* breeding populations: quantitative assessments using a recently developed spatially explicit dynamic modelling framework. Bird Conserv Int 26(04):418–435. https://doi.org/10.1017/S0959270915000258

Sanz-Pérez A, Giralt D, Robleño I, Bota G, Milleret C, Mañosa S, Sardà-Palomera F (2019) 'Fallow management increases habitat suitability for endangered steppe bird species through changes in vegetation structure. J Appl Ecol 56(9):2166–2175. https://doi.org/10.1111/1365-2664.13450

Sastre P, Ponce C, Palacín C, Martín CA, Alonso JC (2009) Disturbances to great bustards (*Otis tarda*) in Central Spain: human activities, bird responses and management implications. Eur J Wildl Res 55(4):425–432. https://doi.org/10.1007/s10344-009-0254-7

Sehhatisabet ME, Abdi F, Ashoori A, Khaleghizadeh A, Khani A, Rabiei K, Shakiba M (2012) Preliminary assessment of distribution and population size of wintering Little Bustards *Tetrax tetrax* in Iran. Bird Conserv Int 22(3):279–287. https://doi.org/10.1017/S0959270911000281

Silva JP, Pinto M, Palmeirim JM (2004) Managing landscapes for the Little Bustard *Tetrax tetrax*: lessons from the study of winter habitat selection. Biol Conserv 117(5):521–528. https://doi.org/10.1016/j.biocon.2003.09.001

Silva JP, Faria N, Catry T (2007) Summer habitat selection of the threatened little bustard Tetrax tetrax in Iberian agricultural landscapes. Biol Conserv 139:186–194. https://doi.org/10.1016/j.biocon.2007.06.013

Silva JP, Palmeirim JM, Moreira F (2010a) Higher breeding densities of the threatened Little Bustard *Tetrax tetrax* occur in larger grassland fields: implications for conservation. Biol Conserv 143(11):2553–2558. https://doi.org/10.1016/j.biocon.2010.06.023

Silva JP, Santos M, Queirós L, Leitão D, Moreira F, Pinto M, Leqoc M, Cabral JA (2010b) Estimating the influence of overhead transmission power lines and landscape context on the density of Little Bustard *Tetrax tetrax* breeding populations. Ecol Model 221(16):1954–1963. https://doi.org/10.1016/j.ecolmodel.2010.03.027

Silva JP, Palmeirim JM, Alcazar R, Correia R, Delgado A, Moreira F (2014) A spatially explicit approach to assess the collision risk between birds and overhead power lines: a case study with the little bustard. Biol Conserv 170:256–263. https://doi.org/10.1016/j.biocon.2013.12.026

Silva JP, Catry I, Palmeirim JM, Moreira F (2015) Freezing heat: thermally imposed constraints on the daily activity patterns of a free-ranging grassland bird. Ecosphere 6(7):art119. https://doi.org/10.1890/ES14-00454.1

Silva JP, Correia R, Alonso H, Martins RC, D'Amico M, Delgado A, Sampaio H, Godinho C, Moreira F (2018) EU protected area network did not prevent a country wide population decline in a threatened grassland bird. PeerJ 6:e4284. https://doi.org/10.7717/peerj.4284

Suárez F, Naveso MA, De Juana E (1997) Farming in the drylands of Spain: birds of the pseudosteppes. In: Pain D, Pienkowsky M (eds) Farming and birds in Europe. The common agricultural policy and its implications for bird conservation. Academic Press, London, pp 297–330

Suárez-Seoane S, Osborne PE, Alonso JC (2002) Large-scale habitat selection by agricultural steppe birds in Spain: identifying species-habitat responses using generalized additive models. J Appl Ecol 39(5):755–771. https://doi.org/10.1046/j.1365-2664.2002.00751.x

Suárez-Seoane S, García de la Morena EL, Morales Prieto MB, Osborne PE, de Juana E (2008) Maximum entropy niche-based modelling of seasonal changes in little bustard (*Tetrax tetrax*) distribution. Ecol Model 219(1–2):17–29. https://doi.org/10.1016/j.ecolmodel.2008.07.035

Tarjuelo R, Barja I, Morales MB, Traba J, Benítez-López A, Casas F, Arroyo B, Delgado MP, Mougeot F (2015) Effects of human activity on physiological and behavioral responses of an endangered steppe bird. Behav Ecol 26(3):828–838. https://doi.org/10.1093/beheco/arv016

Tarjuelo R, Margalida A, Mougeot F (2020) Changing the fallow paradigm: a win–win strategy for the post-2020 common agricultural policy to halt farmland bird declines. J Appl Ecol 57(3): 642–649. https://doi.org/10.1111/1365-2664.13570
Thaxter CB, Buchanan GM, Carr J, Butchart SHM, Newbold T, Green RE, Tobias JA, Foden WB, O'Brien S, Pearce-Higgins JW (2017) 'Bird and bat species' global vulnerability to collision mortality at wind farms revealed through a trait-based assessment. Proc R Soc B Biol Sci 284(1862):20170829. https://doi.org/10.1098/rspb.2017.0829
Thiel D, Ménoni E, Brenot J-F, Jenni L (2007) Effects of recreation and hunting on Flushing distance of capercaillie. J Wildl Manag 71(6):1784–1792. https://doi.org/10.2193/2006-268
Traba J, Morales MB (2019) The decline of farmland birds in Spain is strongly associated to the loss of fallowland. Sci Rep 9(1):9473. https://doi.org/10.1038/s41598-019-45854-0
Traba J, Morales MB, García de la Morena EL, Delgado M-P, Krištín A (2008) Selection of breeding territory by little bustard (*Tetrax tetrax*) males in Central Spain: the role of arthropod availability. Ecol Res 23(3):615–622. https://doi.org/10.1007/s11284-007-0418-4
Turney D, Fthenakis V (2011) Environmental impacts from the installation and operation of large-scale solar power plants. Renew Sust Energ Rev 15(6):3261–3270. https://doi.org/10.1016/j.rser.2011.04.023
van Vliet J, de Groot HLF, Rietveld P, Verburg PH (2015) Manifestations and underlying drivers of agricultural land use change in Europe. Landsc Urban Plan 133:24–36. https://doi.org/10.1016/j.landurbplan.2014.09.001
Vorona VV, Pulikova GI, Kim KK, Andreeva EV, Bekker VR, Aitbaev T (2012) The impact of power lines on bird mortality in Central Kazakhstan. Rap Conserv 24:52–60
Walters K, Kosciuch K, Jones J (2014) Can the effect of tall structures on birds be isolated from other aspects of development? Wildl Soc Bull 38(2):250–256. https://doi.org/10.1002/wsb.394
Wolff A (2001) Changements agricoles et conservation de la grande avifaune de plaine : étude des relations espèce-habitats à différentes échelles chez l'Outarde canepetière. Université Montpellier II
Wolff A (2004) Influence of the habitats mosaic on the ecology and distribution of the little bustard in the Crau. Ecol Mediterr 30:111–132
Wolff A, Paul J-P, Martin J-L, Bretagnolle V (2002) The benefits of extensive agriculture to birds: the case of the little bustard. J Appl Ecol 38(5):963–975. https://doi.org/10.1046/j.1365-2664.2001.00651.x
Yousefi M, Kafash A, Malakoutikhah S, Ashoori A, Khani A, Mehdizade Y, Ataei F, Ilanloo S, Rezaei HR, Silva JP (2018) Distance to international border shapes the distribution pattern of the growing Little Bustard *Tetrax tetrax* winter population in Northern Iran. Bird Conserv Int 28(4): 499–508. https://doi.org/10.1017/S0959270917000181

Little Bustard and Humans: Conservation and Management

Juan Traba, Manuel B. Morales, Nuno Faria, Vincent Bretagnolle, Francisco Moreira, and Gerard Bota

Introduction

A thorough review of the world conservation status of little bustard is provided in chapter "The little bustard around the world: distribution, global conservation status, threats and population trends": in summary, the species shows declining trends over

J. Traba (✉) · M. B. Morales
Terrestrial Ecology Group, Departamento de Ecología, Universidad Autónoma de Madrid (TEG-UAM), Madrid, Spain

Centro de Investigación en Biodiversidad y Cambio Global (CIBC-UAM), Universidad Autónoma de Madrid, Madrid, Spain
e-mail: juan.traba@uam.es; manuel.morales@uam.es

N. Faria
Department of Pest/Diseases Monitoring and Diagnosis, InnovPlantProtect (InPP), Elvas, Portugal

LabOr – Laboratório de Ornitologia, MED – Instituto Mediterrâneo para a Agricultura, Ambiente e Desenvolvimento, Universidade de Évora, Évora, Portugal
e-mail: farnuno@uevora.pt

V. Bretagnolle
Centre d'Etudes Biologiques de Chizé, CNRS & La Rochelle Université, Beauvoir sur Niort, France
e-mail: vincent.bretagnolle@cebc.cnrs.fr

F. Moreira
CIBIO/InBIO – Centro de Investigação em Biodiversidade e Recursos Genéticos, Universidade do Porto, Campus Agrário de Vairão, Vairão, Portugal
e-mail: fmoreira@cibio.up.pt

G. Bota
Landscape Dynamics and Biodiversity Programme, Forest Sciences Center of Catalonia (CTFC), Solsona, Catalonia, Spain
e-mail: gerard.bota@ctfc.es

V. Bretagnolle et al. (eds.), *Little Bustard: Ecology and Conservation*, Wildlife Research Monographs 5, https://doi.org/10.1007/978-3-030-84902-3_13

all its western distribution range, while trends in the eastern part are less well known but the species has increased over the last two decades, although it is suspected to have declined again in some areas very recently (see chapter "The little bustard around the world: distribution, global conservation status, threats and population trends"). The causes of decline in the western range are now well documented (see chapters "The little bustard around the world: distribution, global conservation status, threats and population trends", "Habitat selection and space use" and "Threats affecting little bustards: human impacts"), and are mainly attributable to major changes in European farmland landscapes, which are currently intensively managed at both regional and field scales. Landscape homogenization, shifts from traditional to more productive crops, the disappearance of fallows and meadows, irrigation, mechanization and the use of agrochemicals have all drastically reduced food and nesting site availability (Collar et al. 2016; Traba and Morales 2019), and increased additive mortality (Inchausti and Bretagnolle 2005; Faria et al. 2016). In the near future, the conservation status of the species needs to be reviewed and updated (see Morales and Bretagnolle 2021), urgent actions in regard to protection and management, and periodic censuses assessing the situation of the species in the various parts of its range need to be undertaken to reverse the drivers of decline.

The dramatic decrease of the little bustard in Europe (see chapter "The little bustard around the world: distribution, global conservation status, threats and population trends") has prompted the inclusion of the species in several conservation and action plans, both at European and at national levels, and the funding of projects aiming to improve the conservation status of the species. However, the population trends of the species have not reversed over the past 20 years, or only locally, despite these efforts. So far, the strategy for little bustard conservation in Europe has been almost exclusively restricted to protected areas (e.g. Natura 2000). This may work in countries like France, where 80% of little bustards in western France are now located in such networks (see below), but is less effective for instance in Iberia where the species is still widespread in major tracts of unprotected land. There, little bustard conservation may require applying measures directly targeting the species' potential habitat (Wolff et al. 2002; Ausden 2007), whether it is included in the protected area network or not. This latter strategy would better connect with Common Agricultural Policy (CAP) implementation, both within and outside protected areas. The new CAP (after 2020) should offer alternatives for many of the problems already identified, although conflicts with ever more intensive agriculture are to be expected. Several projects aimed at the conservation of the little bustard, among other pseudo-steppe birds, have been carried out in recent years in Spain, France and Portugal. But most projects, benefitting from EU funds, concerned Natura 2000 sites, leaving out unprotected areas important for both breeding and wintering sites, which are critical to the maintenance of current populations (chapter "Migration, movements and non-breeding ecology"). Aside from habitat management for little bustards, reintroduction and reinforcement projects have also been carried out in the last 20 years, especially in France but also in Italy (see chapter "Captive breeding, handling and care, and the impact of releases on wild populations").

While chapter "Threats affecting little bustards: human impacts" makes a relevant synthesis about threats to the little bustard over the whole distribution area, in this chapter we provide a general review of the global, European and national conservation activities dedicated to this species, evaluate its level of protection in different countries, examine some of the most relevant management measures carried out to improve its conservation status, and propose several new ones to reverse its precipitate decline.

Legal Protection and Conservation Status

The little bustard is included in CITES Appendix II (species that are not necessarily threatened with extinction but that may become so, unless trade is closely controlled). At present (2022), the species is included on Appendices I and II of the Convention on the Conservation of Migratory Species of Wild Animals. Listing on Appendix II promotes international cooperation to protect this migratory bird. The species is included in Annex II of the Bern Convention, as well as in Annex I of the European Union Birds Directive (2009/147/CE Birds Directive), which includes species that are threatened with extinction, are rare or require special attention due to the vulnerability or specificity of their habitats, and should be subjected to special conservation measures regarding their habitat in order to ensure their survival and reproduction in their distribution area.

The little bustard is included in 12 national red lists or national catalogues of threatened species under IUCN criteria (Table 1) and listed as *Vulnerable* and *Near Threatened* under the European and Global IUCN Red Lists, respectively. An upgrading of the European status into the *Endangered* category, due to rapid overall population decline in the western range, has been recently proposed (Morales and Bretagnolle 2021), while hunting pressure in the eastern populations may be also a reason for concern (BirdLife International 2015; Collar et al. 2017). Indeed, the little bustard seems to meet the requirements for listing as globally threatened under IUCN criteria A2abcd + 3bcd + 4abcd. A thorough review of population size and trends at national levels is presented in chapter "The little bustard around the world: distribution, global conservation status, threats and population trends".

In countries in which the little bustard is included in red data books as *Vulnerable* or *Endangered*, which are the commonest situations (Gauger 2007; Iñigo and Barov 2010; Table 1), periodic assessments of population status should be carried out. However, no general information about periodic national assessments is available, except for three countries. National censuses take place every 4 years in France, 5 years in Portugal and 10 years in Spain (despite national legislation stating that these assessments should be carried out every 6 years), and no periodic monitoring has been done in the remaining countries. But even when these surveys are done, uplisting or downlisting conservation status may not be immediate: in Spain for instance, the species should be upgraded to *Endangered* according to the results of the Second National Survey, since the situation complies with IUCN criteria a2abc + a3b (IUCN 2012; García de la Morena et al. 2018). Then, following upgrading, a National Recovery Plan should be produced, including the delineation

Table 1 Conservation status of little bustard in different European and non-European countries, and at global and European levels

Region	Status	Criteria	Justification	Source	Date	Local name
World	NT	IUCN	A2abcd + 3bcd + 4abcd	IUCN Red List	2015	Little bustard
Europe	VU	IUCN	A2abcd + 3bcd + 4abcd	IUCN Red List	2015	Little bustard
Albania	CR EN	IUCN	Not defined	IUCN Red List assessment	2013	Pula e livadheve
Austria	RE EX	IUCN	Modified IUCN	Rote Listen gefährdeter Tiere Österreich	2005	Zwergtrappe
Bulgaria	EX	IUCN	Rare passage migrant and winter visitor. Possibly breeding in the past	Red Data Book of the Republic of Bulgaria	2011	
China	DD	IUCN	Not defined	Biodiversity Science	2015	
Croatia	CR EN	IUCN	D	IUCN Red List assessment	2007	Mala droplja
Croatia	EX	IUCN	Not defined	Croatia Red List: Birds	2007	
France	EN	IUCN	Not defined	La Liste rouge des espèces menacées en France	2016	Outarde canepetière
Germany	EX	Non-IUCN	Not defined	Roten Listen der in Deutschland gefährdeten Tier- und Pflanzenarten.—Natur und Landschaft	1998	Zwergtrappe
Greece	CR EN	IUCN	D	The Red Data Book of Threatened Animals of Greece	2009	
Italy	EN	IUCN	C2a(i)	Lista Rossa dei vertebrati Italiani	2013	Gallina prataiola
Poland	EX	IUCN	Not defined	Red List of Threatened Animals in Poland	2002	
Portugal	VU	IUCN	A2c + 3c + 4c	Livro Vermelho dos Vertebrados de Portugal	2005	Sisão
Spain	VU	IUCN	Modified IUCN	Spanish National List of Endangered Species	2011	Sisón común
Spain	VU	IUCN	A2c + 3c + 4c	Spanish Red List	2011	Sisón común

Criteria for listing are provided when known, as well as the source

EX Extinct, *RE EX* Regionally Extinct, *CR* Critically Endangered, *EN* Endangered, *VU* Vulnerable, *NT* Near Threatened, *DD* Data Deficient

of critical areas as well as dedicated conservation funding. Finally, beyond the legal consideration of the species, insufficient law enforcement seems to be a problem in several countries (e.g. Turkey, Ukraine and Italy) (Íñigo and Barov 2010).

Census and Monitoring

Monitoring populations from large to local scales is critical to document general population trends (and possibly detect sudden shifts in population trajectories), as well as to inform management and conservation. Coordinated surveys with a multi-scale approach (local, regional, national and European) are necessary to obtain an up-to-date estimate for the population at any given scale. These must be time continued, to allow for real-time monitoring of population trends. Censuses of males are now carried out, more or less routinely, in several countries (see chapter "The little bustard around the world: distribution, global conservation status, threats and population trends"), less often at regional scales (but see below), and rather regularly in some protected areas of the Natura 2000 network. However, information on females is urgently needed, particularly relating to productivity (number of young/female/year), since any value below 1 young of at least 30 days per female per year is probably below the minimum threshold for guaranteeing the long-term survival of populations (Morales et al. 2005; Inchausti and Bretagnolle 2005).

In the eastern part of the distribution range, there is a need for a nationwide survey in Iran (Sehhatisabet et al. 2012), as well as in Kazakhstan where numbers and habitat use are virtually unknown (see chapter "The little bustard around the world: distribution, global conservation status, threats and population trends"). Regular national censuses are currently being carried out in Portugal, Spain and France, irregularly in Sardinia. In Portugal, three censuses were carried out in 2003–2006, 2010 (partially, only within SPAs) and 2016. In France, little bustards have been counted every year in Poitou-Charentes Region from 2008 (V. Bretagnolle), and at regular intervals in La Crau and southern France in winter. They are also yearly monitored in every SPA in western France, and at regular intervals in southern France SPAs. Post-breeding flocks are counted four times a year in western France, and irregularly in La Crau and southern France. In France, therefore, the little bustard is monitored from local up to national scale on a fairly regular and constant basis.

In Spain, as the species is listed as Vulnerable, legislation requires that national censuses must be carried out every 6 years, documenting changes in distribution (both occupation and numbers), population dynamics and viability, evaluating habitat quality (extent and degree of fragmentation, carrying capacity and main threats), and providing a risk assessment. However, the only information available comes from the two National Censuses carried out with an 11-year interval between them. SEO/Birdlife carried out the second national census in 2016, with the explicit support of relevant regional administrations, and showed a severe decline in populations for the 2005–2016 period, with a mean density reduction of 53% (17–79% depending on the region) (García de la Morena et al. 2018). Besides,

SEO/BirdLife's common bird monitoring programme (SEO/Birdlife 2018) detected a 76% decline in the Spanish little bustard population in the period 1998–2017, which means a 3.5% annual decline (García de la Morena et al. 2018). At a local scale, some regular regional censuses have been carried out every 5–6 years from 2004 onwards in areas like Catalonia (Mañosa et al. 2020). In some relevant SPAs like the Reserve of Villafáfila (north-west Spain), breeding male surveys are carried out on a yearly basis.

European and National Action Plans

Despite being one of the most threatened birds in Europe because of the very steep decline observed during the last three decades, only a single Action Plan at a European scale has been developed so far (first draft in De Juana 2001; later update in Iñigo and Barov 2010). An action plan details conservation problems and threats to the target species, and the actions to be carried out to reverse the decline. But in the little bustard (as in many other species) the time lag between science-based proposals and direct action has been huge in key countries like Spain (more than 10 years), probably due to political and management difficulties in their implementation, and in the meantime the conservation status of the species has deteriorated further.

In the first draft of the Action Plan (De Juana 2001), conservation priorities were focused on general actions and the need for improving knowledge of the species' ecology as well as on the effectiveness of agri-environmental schemes (AES) (Table 2). Ten years later, the revised Action Plan (Iñigo and Barov 2010) considered more specific measures aimed at improving female and juvenile survival as well as habitat quality, both being considered key parameters for the maintenance of little bustard populations (Morales et al. 2005). As the latter authors stated, management measures aimed at increasing little bustard productivity had to be considered as a priority in the conservation strategy. Therefore, measures focusing on agrarian extensification (i.e. de-intensification) and promoting practices that both minimize nest losses and favour insect availability for chicks (e.g. fallow and field border management, reduction of pesticide input) were included in this new Action Plan (Table 2).

Rather few national action plans have been produced, however. None for instance exists for Italy, where the last population survives in Sardinia (Santangeli and Dolman 2011). In Portugal, a national action plan for the conservation of steppe birds was drafted in 2007 (Cardoso et al. 2007), including the little bustard among many others. No great advance since then has occurred, besides the definition of main threats and proposal of conservation measures, but this has not been updated. In Spain, no national action plan or national conservation strategy has been prepared until now, despite the species' national threat category of *Vulnerable*, which makes both documents mandatory. Only an academic initiative has been addressed so far to draft the National Conservation Strategy (García de la Morena et al. 2019) and to request the re-listing of the species as *Endangered*. At a regional scale, Andalucía

Table 2 Priority measures included in the first draft (2001, top) and the final Action Plan (2010, bottom) for the conservation of the little bustard in Europe

2001 (De Juana 2001)	**Priority**
Maintaining large areas of extensive arable/pastoral land and enhance habitat quality within them, through application of agri-environment policies (i.e. by increasing legume crops, unploughed fallows and set-aside fields, and by preventing irrigation, concentration of landholdings, afforestation, tree crops, pesticide application and inappropriate stocking densities)	Essential
Designate and manage protected areas	High
Determine movement patterns and locate main wintering concentrations in Iberia	High
Research to determine factors that may influence breeding success	High
Evaluate the effectiveness of management/conservation measures	High
Inform farmers of the need to protect the Little Bustard and the farming practices which may affect it	High
2010 (Iñigo and Barov 2010)	
Develop and implement a set of suitable mitigation measures to be applied to harvest machinery and harvesting methods to protect nesting females and juveniles. Take into account crop type and calendar in relation to the species phenology	High
Maintain diversity of crops and farmland mosaics in areas of predominantly arable farming	High
Promote crops that require later (e.g. wheat) instead of earlier harvesting (e.g. barley)	High
Maintain fallow land, avoid tilling of fallow between April–July and ensure vegetation does not become too dense on older fallows	High
Promote less pesticide use (e.g. organic cereals), and farm manure instead of mineral fertilizer	High
Leave grassy strips and bare ground uncultivated to maintain diversity of microhabitats	Medium
Improve the impact assessment of irrigation plans at national/regional level (ES) including the analysis of cumulative impacts	High
Improve the impact assessment of large-scale crop conversion schemes (in Rural Development Plans)	High
Discourage the concentration of landholdings to prevent intensification of farming (through, e.g. SPA management plans, incentives to small farms, protection of landscape features and obligatory maintenance of set aside areas)	High
Ensure that steppe habitats are conserved and not damaged by infrastructure development (e.g. power lines, transport, renewable energy)	Medium
Develop management plans for individual SPAs or for groups of SPAs with similar habitats, threats and management objectives	High

adopted a Recovery Plan for Steppe Birds that includes the little bustard, outlining the most important areas where to carry out conservation measures. However, it has been unable to stop the decline (app. −41% between 2006 and 2016; García de la Morena et al. 2018). Other regional plans have been created, although the results are discouraging due to the huge decreases in a number of males, e.g. in Navarra (Astrain and Etxeberria 1997; ca. −85% between 2006 and 2016; García de la Morena et al. 2018) or La Rioja (Astrain and Etxeberria 1998; ca. −57% between

2006 and 2016; García de la Morena et al. 2018), where the populations are very small or almost extinct, respectively. Active habitat management for the conservation of the little bustard has been carried out in Catalonia (Mañosa et al. 2015), where several conservation actions within SPAs have been developed, with focus on fallow management, and results seem encouraging (Mañosa et al. 2020).

In France, where the species faces a critical situation, with dramatic declines in the 1980s and 1990s (Jolivet and Bretagnolle 2002), the species is red-listed and classified as *Endangered*, and three successive National Action Plans plus a specific recovery plan have been in place. A first national action plan covered the period 2002–2006, at the end of which 75% of the French population of little bustard was present within Natura 2000 Special Protection Areas (SPAs), in which management measures could be more easily implemented. As a result, more than 1300 ha of AES were specifically implemented for the little bustard and female productivity significantly increased. AES aimed to impose constraints on mowing dates in order to prevent nest destruction and female mortality, and to increase insect abundance and hence food availability for chicks. In one particular area with intensive monitoring and conservation action (Deux-Sèvres), these measures, coupled with population reinforcement from ex situ programmes, helped reverse the negative trend, increasing the population from six males in 2003 to around 30 in 2009 (Bretagnolle et al. 2011).

Despite the overall positive result after the first national plan, strong geographical disparities persisted, and a special effort had to be made to help the last migratory populations of the west-centre of the country (Bamière et al. 2011; Bretagnolle et al. 2011). A second National Plan was therefore run for the 2011–2015 period, based on rigorous scientific evidence. In addition, a National Recovery Plan was submitted, aimed at reducing the risk of extinction of the migratory population in the Atlantic area of France. One of the objectives was to incorporate new, unprotected areas into the Natura 2000 network, both in the Atlantic and in the Mediterranean regions. Seven key goals were included in this second national action plan, namely (1) to improve the implementation of agri-environmental policy, (2) purchase land and manage it appropriately, (3) work with military and civil aviation to maintain or even improve the conditions for the species at the airfields, airports and military camps, (4) release little bustards bred in captivity, (5) monitor accurately changes in little bustard numbers, (6) intervene in development projects (e.g. urbanization, powerlines, roads and railways, wind farms and solar fields) and (7) disseminate information on these conservation actions to the general public. Currently, a third national action plan for the period 2019–2028 is running, including the population reinforcement program started in 2005 (see chapter "Captive breeding, handling and care, and the impact of releases on wild populations"), along with conservation actions aiming at improving agri-environmental measures, and attempts to acquire private lands for the conservation of steppe birds. Although the dramatic decline of the species in France stopped after 2008, the situation is far from favourable, and decisive urgent actions are still needed to prevent the extirpation of the species from France (Bretagnolle et al. 2018).

Protected Areas and Natura 2000

The distribution of the little bustard throughout lowland agricultural landscapes explains a large number of SPAs ('Special Protection Areas' for safeguarding threatened bird species) declared for its effective conservation in Europe under the Natura 2000 network. Indeed, the European Union has designated no fewer than 249 SPAs (up to July 2019) explicitly based on the presence in them of little bustards (Table 3). However, it has been estimated that the species in fact occurs or occurred in over 750 SPAs, which would include half the European breeding population (Íñigo and Barov 2010). Most of the 249 SPAs explicitly established for little bustards hold breeding populations, either as residents (120; 47.1%) or as spring and summer visitors (60; 23.5%).

Official EU documentation reports around 28,643–30,580 birds within these 249 SPAs (Table 3), but this information seems to be outdated, at least for several countries. In France, 75% of the population is located within SPAs. In the case of Italy, more than 50% of the species' breeding range lies outside SPAs (Barov and Derhé 2011). Recent information from Portugal indicates that only 5008 little bustard males (min. 2701; max. 7790) breed within Portuguese SPAs (Silva et al. 2018). These authors assessed the effectiveness of the Portuguese Natura 2000 network for little bustard conservation, by comparing population trends across time (between 2003–2006 and 2016) in 51 areas, 30 without EU protection and 21 within 12 Special Protection Areas (SPA) mostly designated for farmland bird conservation. Overall the Portuguese little bustard population declined by 49% over the previous 10–14 years, more sharply outside SPAs (64% decline) than within them (25% decline). As a consequence, the estimated proportion of the national population (males) within SPAs increased from ca. 38% to ca. 56%. However, the

Table 3 Status of the little bustard within European network of protected areas Natura 2000

Country	Number of SPAs	Total area (ha)	Minimum number of birds within	Maximum number of birds within	Minimum density (birds/ 100 ha)	Maximum density (birds/ 100 ha)
Spain[a]	158	3,815,271	14,334	15,335	0.38	0.40
France	37	799,117	1096	1270	0.14	0.16
Italy	24	439,646	435	655	0.10	0.15
Latvia	1	51,353	55	150	0.11	0.29
Portugal[a]	28	999,416	12,713	13,140	1.27	1.31
Slovakia	1	7002	10	30	0.14	0.43
Total EU	249	6,111,805	28,643	30,580	0.45	0.48

[a]See text for more information on recent data

We show the number of Special Protection Areas (SPA) per country (and overall for the European Union) that included the occurrence of little bustard as a criterion for their designation. We give the total area (ha) of these SPAs, as well their estimated minimum and maximum number of birds and density (only for resident and breeding birds). Source: European Environment Agency (last consultation 2020/04/03; EEA 2019). See text for more information on recent data

absolute male density decline was significantly larger within SPAs. As in other countries, Natura 2000 alone seemed ineffective in preventing the population decline (Silva et al. 2018), indicating that the designation of steppe or farmland SPAs is not enough to secure little bustard populations, but has to be combined with specific agricultural measures that guarantee both habitat availability and quality.

In Spain, considering an average estimation of around 38,900 males (app. 27,000–59,000, 95% CI) for the whole Spanish population (García de la Morena et al. 2018), between 36.8 and 39.4% (14,334–15,335 males) would be within SPAs according to Table 3. However, some of these data could be outdated. We have intersected data from the European Breeding Bird Atlas (centroids of 50 × 50 km UTM grids, later assigning them a 5000 m buffer) for the little bustard and the location of all Natura 2000 sites, and found that ca. 30% of little bustard pixels were outside any Natura 2000 site, considering both SPAs and Special Areas of Conservation (SACs). A complementary analysis was carried out to evaluate the effectiveness of SPAs in the protection of little bustards in Spain. Using unpublished data from the second National Census (García de la Morena et al. 2018), the breeding population size within each SPA was estimated for each UTM 10 × 10 km cell with little bustard presence. In addition, habitat data from Information System on Land Occupation of Spain (SIOSE) was used to provide information about the area of potential habitat for the species in each UTM cell. Assuming that the entire habitat within SPAs was optimal for the species (which is certainly an overestimation), the difference between total available habitat and SPA area, within each cell, was used as a proxy for optimal habitat outside the SPA. The comparison of little bustard abundance inside and outside SPAs for each cell showed that only ca. 30% of the Spanish little bustard population was indeed protected under the Natura 2000 SPA network (Table 4; Fig. 1). This was already suggested in the 2000s by Traba et al.

Table 4 Population size, and number of males and percentage within SPA in Spain and the Spanish regions with presence of the species (unpublished data and own elaboration based on García de la Morena et al. 2018)

Autonomous region	Number of males	Males within SPA	% males within SPA
Andalucía	3387	527	16
Aragón	1804	625	35
Castilla—La Mancha	25,436	5874	23
Castilla y León	1871	1092	58
Cataluña	1162	923	79
Comunidad Valenciana	44	25	57
Extremadura	3871	2512	65
Galicia	26	6	23
La Rioja	15	0	0
Madrid	909	363	40
Murcia	140	75	53
Navarra	148	39	26
Spain	38,813	12,062	31

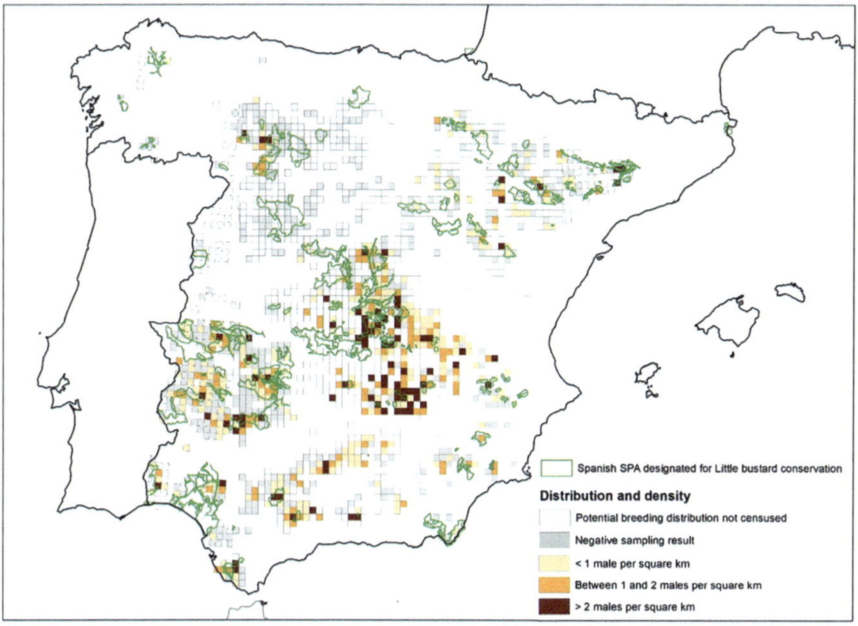

Fig. 1 Abundance of little bustard males during the breeding season in Spain (in UTM 10 × 10 km cells) and distribution of Spanish SPAs designated based on little bustard occurrence. Source: Own elaboration from García de la Morena et al. (2018)

(2007), who detected that less than 20% of the Spanish range of little bustard was included in the SPA network. Using data from García de la Morena et al. (2018) we have estimated that 69% of the little bustard population is outside protected areas, and male density being significantly higher inside (mean ± sd = 0.21 ± 0.01 males within 250 m buffers around census stations) than outside SPAs (0.13 ± 0.01; ANOVA F = 57.22; df = 1; $p < 0.0001$). These figures, however, only refer to males, while a complete assessment of the conservation status would also require estimates of female numbers and distribution, which are unavailable.

In addition, at a more local scale, numbers of little bustards within protected areas may be even lower. Castilla-La Mancha, the administrative region currently holding the largest little bustard breeding population in Spain and hence Europe (it represents 65.6% of the Spanish breeding population), is a good example. Only just over 20% of its breeding population is within SPAs (Table 4; Fig. 2). In Andalusia, even fewer little bustards are within SPAs (16%; Table 4; Fig. 2). However, in Extremadura, which was also historically important for the species at Spanish and European levels, ca. 65% of breeding birds are within SPAs, in spite of which the population has strongly declined in recent years (García de la Morena et al. 2018) (Table 4; Fig. 2).

We compared the two Spanish national breeding surveys to examine extinction rates estimated at 100 km^2 scale, assuming that a UTM 10 × 10 km cell with species presence was in a protected area when >10% of cell surface was within an SPA.

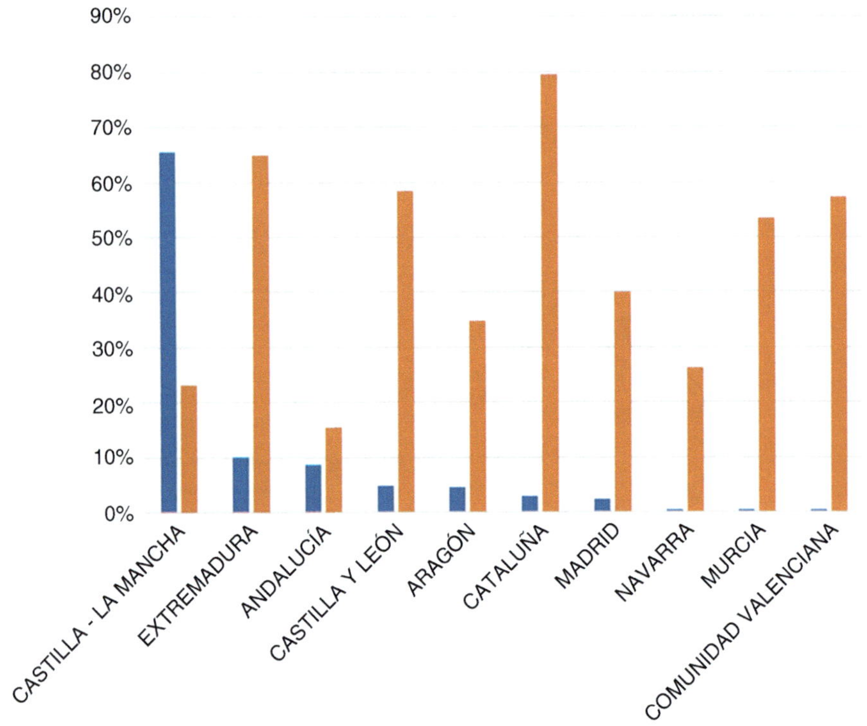

Fig. 2 Percentage of little bustard breeding males for each region in relation to total Spanish population size (blue) and that within SPA (orange). Data presented for the 10 Spanish regions with presence of the species. Source: Own elaboration based on García de la Morena et al. (2018)

Overall, 19% of cells with extinction events between the first (2006) and second (2016) census were within SPAs (Table 5). Finally, we assessed the SPA cover during the non-breeding period and found that 80% of the birds and 74% of the flocks were detected outside SPAs during the last winter national census in Spain (Fig. 3), an even lower proportion compared to that for the breeding period.

SPAs differed between EU countries regarding their habitat quality, and little-bustard conservation efficiency and status: only 20 SPAs (8.0%) showed excellent conservation status (Table 6), according to information provided by member states. In 86 SPAs (34.5%) with little bustards, the conservation status is moderate or unsatisfactory, while in 49 other SPAs (19.7%) information is simply lacking (Table 6). In the case of Spain, comparison of little bustard distribution maps with the SPA network revealed major shortcomings and inconsistencies, since many of the best areas for the little bustard have been simply omitted from the Natura 2000 network (Suárez-Seoane et al. 2002; Traba et al. 2007).

Table 5 Total number of UTM 10 × 10 km cells where the species has become extinct between First (2006) and Second (2016) Spanish National Censuses, and number of extinct cells with recorded local extinctions outside and within SPA, and percentage of extinct cells within SPA in relation with the total of extinct cells

Autonomous region	Extinct total	Extinct no SPA	Extinct SPA	% Extinct SPA
Andalucía	102	91	11	11
Aragón	46	39	7	15
Castilla—La Mancha	4	4	0	0
Castilla y León	149	123	26	17
Extremadura	51	35	16	31
Galicia	9	6	3	33
La Rioja	2	2	0	0
Madrid	8	3	5	63
Navarra	15	8	7	47
Spain	386	311	75	19

Based on data from García de la Morena et al. (2006, 2018)

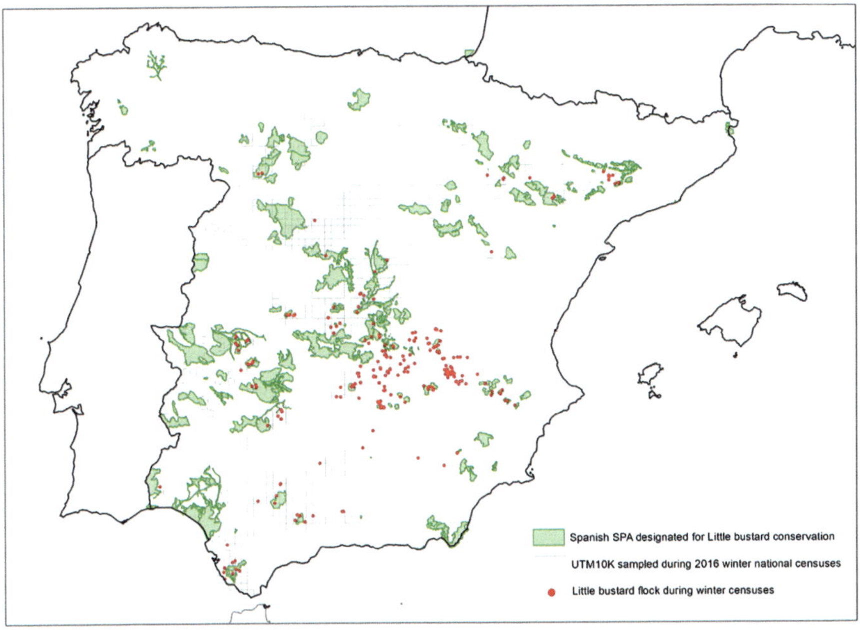

Fig. 3 Abundance of little bustard flocks during winter in Spain (in UTM 10 × 10 km cells) and distribution of Spanish SPAs designated for little bustard presence. Source: Own elaboration based on García de la Morena et al. (2018)

Table 6 Status of little bustard conservation within European network of protected areas Natura 2000. Values in parentheses indicate %

		Conservation status			
Country	Number of SPAs	Excellent	Good	Averaged or reduced	No data
Spain	158	10 (6.3)	63 (39.9)	60 (38.0)	25 (15.8)
France	37	4 (10.8)	7 (18.9)	16 (43.2)	10 (27.0)
Italy	24	3 (12.5)	6 (25.0)	8 (33.3)	7 (29.2)
Latvia	1	–	1 (100)	–	–
Portugal	28	3 (10.7)	17 (60.7)	1 (3.6)	7 (25.0)
Slovakia	1	–	–	1 (100)	–
Total EU	249	20 (8.0)	94 (37.8)	86 (34.5)	49 (19.7)

We show the number of Special Protection Areas (SPAs) per country (and overall for the European Union) that in their designation included the occurrence of little bustard. Only countries with resident or breeding little bustards are included. Conservation status refers to a global assessment of conservation status of the SPA (not specifically of little bustard): Excellent value; Good value; Averaged or reduced value; and No data available. Source: European Environment Agency (last consultation 2019/07/15; EEA 2019)

LIFE Projects

The LIFE programme, created in 1992, is the European Union's funding instrument for the environment, nature conservation and climate action. LIFE projects are the only European-scale funding opportunity for applying conservation measures for this and other threatened species and habitats, and it is principally focused on Natura 2000 sites.

Since 1993, the little bustard has been the main or partial subject of no fewer than 23 LIFE Nature projects (Table 7) in France ($N = 7$), Spain ($N = 6$), Italy ($N = 4$) and Portugal ($N = 6$). The total cost of these projects has been 40.1 million euros, with an EU contribution of more than 23.5 million euros. Regarding little bustard-specific projects (i.e. excluding those considering other target species), seven LIFE projects (four in Italy, two in France, one in Portugal) have addressed the conservation of the species, with more than 9.03 million euros (over 5.6 million euros of EU contribution) (Table 7).

No species-specific LIFE project has yet been carried out in Spain, even though this country is the stronghold of the little bustard in the EU. However, six LIFE projects have included the species as a secondary target, some of them aimed at making crop-farming and livestock-rearing compatible with the little bustard's requirements, and whose eventual success on the species conservation should be fully evaluated. The latter include the promotion of fallows, sowing of small plots to improve feeding resources for steppe birds, modification of harvest and mowing times and methods (e.g. avoiding agricultural work during nesting), and sowing legume crops and improving farm infrastructures (pools, enclosures, etc.) by way of compensations to private owners. These measures are very close to the active habitat management for the little bustard that has been implemented in other projects in central and western France (Bamière et al. 2011; Bretagnolle et al. 2011), and in

Table 7 Summary of LIFE projects carried out in Europe dedicated mainly or partially to the conservation of the little bustard

Funding (€)	EU contribution (€)	Project title	Project number	Initial year	Lead country	Little bustard exclusive
1,765,255	1,059,153	LIFE steppe farming—Sustainable farming in SPAs of Castilla-La Mancha for steppe birds conservation	LIFE15 NAT/ES/000734	2015	ES	No
2,388,011	1,791,008	LIFE des Alpilles—Gestion intégrée des habitats pour les oiseaux d'intérêt communautaire des Alpilles	LIFE12 NAT/FR/000107	2012	FR	No
1,865,964	1,399,473	LIFE + Tetrax—Conservation of the last Italian peninsular population of *Tetrax tetrax*	LIFE12 NAT/IT/001052	2012	IT	Yes
8,636,466	4,722,420	ZEPA ESTEPARIAS ANDALUCIA—Conservation and management of special protection areas for steppe birds in Andalusia	LIFE08 NAT/E/000068	2008	ES	No
1,455,000	970,048	M.As.Co.T.T.S.S.—Management Actions for Conservation of *Tetrax tetrax* in Steppic Sardinia	LIFE07 NAT/IT/000426	2007	IT	Yes
1,604,021	1,203,016	EstepÁrias—Conservation of Great Bustard, Little Bustard and Lesser Kestrel in the Baixo Alentejo cereal steppes	LIFE07 NAT/P/000654	2007	PT	No
1,583,220	870,771	RENF TETRAX—Reinforcement of the migratory breeding populations of the Little Bustard, *Tetrax tetrax* in France	LIFE04 NAT/FR/000091	2004	FR	Yes
553,749	415,312	Albuera Extremadura—Conservation and management of the SPA for Birds site of Community interest wetland "La Albuera" in Extremadura	LIFE03 NAT/E/000052	2003	ES	No
106,762	100,084	Evaluation of bustard conservation best practice in Western Europe	LIFE03 AT/CP/P/00008	2003	PT	No
967,442.1	720,873	Tetrax—Project Tetrax—the conservation of Little Bustard in Alentejo	LIFE02 NAT/P/008476	2002	PT	Yes
736,551	552,413	Serena—Tiros—Habitat management model of the SCI La Serena—Sierra de Tiros (Extremadura, Spain)	LIFE00 NAT/E/007327	2000	ES	No
1,853,176	1,297,223	ZEPA La Serena—Management of the PSA-SCI 'La Serena y Sierras periféricas'	LIFE00 NAT/E/007348	2000	ES	No

(continued)

Table 7 (continued)

Funding (€)	EU contribution (€)	Project title	Project number	Initial year	Lead country	Little bustard exclusive
739,549.51	369,774.75	Gargano—Urgent actions for the NATURA 2000 sites protection in the Gargano N. P.	LIFE98 NAT/IT/005121	1998	IT	Yes
3,982,977.14	2,788,084	Setúbal/Sado—NATURA 2000 network of the Setúbal Peninsula/Sado	LIFE98 NAT/P/005235	1998	PT	No
428,577.92	321,433.44	ZEPAS/Extremadura—Preservation of the Little Bustard, Great Bustard and Lesser Kestrel in Extremadura	LIFE96 NAT/E/003102	1996	ES	No
1,840,676.82	920,338.57	Experimental programme on the conservation of Little Bustard *Tetrax tetrax* and associated fauna in France	LIFE96 NAT/F/003207	1996	FR	Yes
579,638.23	434,728.67	Steppe—Urgent actions for the protection of the Sardinian steppic habitats	LIFE96 NAT/IT/003165	1996	IT	Yes
1,787,500	446,875	Protection and management of the Crau Sèche—second phase—part A	LIFE95 NAT/F/000524	1995	FR	No
49,650	24,825	Protection and management of the Crau sèche (second phase)—Part B	LIFE95 NAT/F/001215	1995	FR	No
819,867	614,900	Second phase of the project for the conservation of steppe birds in Castro Verde	LIFE95 NAT/P/000178	1995	PT	No
1,200,000	600,000	Grands Causses and Causses du Quercy	LIFE93 NAT/F/010300	1993	FR	No
3,850,000	950,000	Protection and management of the Crau sèche—second phase (volet A)	LIFE93 NAT/F/010400	1993	FR	No
1,333,333	1,000,000	First phase of the conservation of steppic birds in Castro Verde	LIFE92 NAT/P/013900	1992	PT	No

Data extracted from LIFE project database (https://ec.europa.eu/environment/archives/life/index.htm)

Catalonia (Mañosa et al. 2020), where focused renting or acquisition and management of fallow land for little bustard coupled with more generic development of agri-environmental schemes for steppe birds is halting the population decline.

In Portugal, six LIFE projects have been implemented in key ecological areas for the little bustard and other threatened birds, including land purchase, as well as the implementation of pilot projects in order to design specific measures to strengthen the preservation of the species and/or its pseudo-steppe habitats. As in Spain, post-LIFE evaluation is needed in order to know the level of success in the implementation of such measures.

In France, two LIFE projects were specifically dedicated to the little bustard: an "Experimental programme on the conservation of the little bustard and associated fauna in France (1996–2001)" and a project for the "Reinforcement of migratory populations (2004–2009)". Both increased scientific knowledge of the species and tested conservation measures. The latter project successfully reinforced little bustard populations in target areas (see chapter "Captive breeding, handling and care, and the impact of releases on wild populations"), although it showed that care had to be taken in regard to source populations, since differences in migratory behaviour were shown in relation to egg origin: chicks hatched from Spanish eggs did not migrate during winter, suggesting a genetic component to migratory behaviour (Villers et al. 2010). In addition, these projects allowed two breeding centres to be established—one for rearing young chicks and another for breeding adults (chapter "Captive breeding, handling and care, and the impact of releases on wild populations"). Monitoring confirmed that released birds of local origin integrated with the wild populations and migrated over the winter (Villers et al. 2010).

Agrarian Policy and the Little Bustard in Europe

Reversing the habitat and land use changes over recent decades in European agrarian landscapes is an unavoidable requisite for improving the little bustard's conservation status, and this implies a serious reform of the CAP after 2020. Unlike other wildlife species of conservation concern, whose preferred habitats can be located in more or less pristine areas, the European populations of little bustard, as well as other European pseudo-steppe birds, depend on anthropogenic ecosystems, i.e. traditional farming systems, including both for agriculture and livestock (chapter "Habitat selection and space use"). The crucial importance for the little bustard of traditional farming systems, which provide suitable cover of pastures and fallows, has already been mentioned in this volume (see chapter "Habitat selection and space use"). Hence, the relationship between the little bustard (and other coexisting bird species) and humans may be seen as a tense trade-off between food production and wildlife conservation. This has been the case since the mid-1960s, when agricultural intensification steeply increased in central and northern Europe during the 'Green Revolution' (Pain and Dixon 1997). In Spain and Portugal, these changes occurred later, from the mid-1980s, after the entry of these countries into the European Union.

There is strong consensus that agricultural intensification is the main reason for the decline of little bustard (and other steppe birds) in Europe (Goriup 1994; Morales et al. 2005; Osborne and Suárez-Seoane 2007; Delgado and Moreira 2010; Lapiedra et al. 2011; Bretagnolle et al. 2018; Traba and Morales 2019).

In this context, it was expected that the implementation of agri-environmental measures (AESs, see below) might have helped reduce threatening factors. However, although some specific measures seem to have been successful in France (Wolff et al. 2001; Bretagnolle et al. 2011) and Portugal (Ribeiro et al. 2014), in general, the scale of AES application has been insufficient to ensure a positive effect (Barov and Derhé 2011; Bretagnolle et al. 2018). In Spain, AESs have failed to stop the dramatic decline of the species, despite initial hopes that economic incentives to farmers would maintain extensive agriculture (see Suárez et al. 1997).

Agri-environmental Schemes and Little Bustard Conservation

Overall the effectiveness of AES in preserving biodiversity in agrarian landscapes has scarcely been explored (Moreno et al. 2010; Concepción et al. 2008). Until 2020 CAP had two main tools to promote agrarian measures compatible with biodiversity conservation: greening, a set of minimum practices compulsory for farmers to qualify for subsidies of Pillar 1; and voluntary agri-environmental schemes (AESs), locally focused, whose target was the promotion of agrarian practices compatible with biodiversity conservation. Greening included setting ecological focus areas (including fallow land), increasing crop diversity and the maintenance of permanent grasslands, another measure with important implications for little bustards. AESs have been changing ever since 1992, as they were implemented in all successive CAP reforms, but in all cases they showed a common factor: a relatively low uptake by farmers. This has been shown in particular when AESs competed with other subsidies and/or could reduce the productivity of farms (Llusia and Oñate 2005; Santana et al. 2013). Other problems lie in the spatial dispersion of farms under AESs, which undermines their overall efficiency: several studies on different animal and plant taxa have shown that effectiveness of AESs for biodiversity increased with their spatial extent and the proportion of fields at regional scale (e.g. Henckel et al. 2015; Caro et al. 2016). Altogether, these problems call for a re-thinking of AES in the face of the new CAP after 2020 (Pe'er et al. 2019; Tarjuelo et al. 2020).

In the case of the little bustard, only a few local studies have quantified the effects of both greening and AES on the conservation of the species (see for instance Concepción et al. 2020). Indeed, the negative trends in the whole European range suggest little or no success. However, the continuous decline of fallow land in Europe over the last 20 years may have been the main problem, masking any positive effects of AESs, the latter being insufficient to counteract the decrease in the main preferred habitats (Traba and Morales 2019), besides the above-commented effects of spatial dispersion of AES, the level of intensification and landscape structure and the ecological contrast created between under-AES fields and the matrix with conventional management (Caro et al. 2016; Tscharntke et al. 2005). In addition, it is possible if not likely that some AESs, designed and applied for other flagship species such as the great bustard, may have hindered or prevented the success of little bustard conservation measures (Moreno et al. 2010; Tarjuelo et al. 2014). In France, dedicated AESs have been implemented since 2004, including delayed mowing of alfalfa and grasslands, restoration of grassland to replace arable crops, and a ban on insecticides and herbicides. Little bustards responded positively and at first increased (Wolff et al. 2001; Bamière et al. 2011; Bretagnolle et al. 2011); however, successive CAP reforms, with strong changes between them, resulted at least in some areas in a decrease of land area under contract, and a concomitant decrease in little bustard populations (Gendre et al. 2018). In addition, Bretagnolle et al. (2018) showed how grasslands in some contexts could become ecological traps if the ratio of grasslands in AESs to uncontracted land is too low.

Nevertheless, in Spain there is a wide consensus between scientists and managers that AESs represent the main management tool for the long-term conservation of the species (Tarjuelo et al. 2020). Unfortunately, however, AESs in Spain show a very poor uptake by farmers, who generally choose to increase profitability through agrarian intensification (Llusia and Oñate 2005). After nearly 25 years of AES application in Spain, implementation has been insufficient and geographically too scattered, so that the positive effects are weak and dispersed (Llusia and Oñate 2005; Concepción et al. 2008, 2012; see Traba and Morales 2019).

Mixed effects of specific long-term conservation measures on little bustards have been found in Portugal, including those promoting agrarian extensification. Santana et al. (2014) detected positive local effects, which supports the view that targeted efforts plus legal regulations, intended to prevent afforestation, the conversion to permanent crops, and the expansion of irrigated agriculture, might be beneficial to the species. However, Silva et al. (2018) reported a major national population crash of ca. 50% in the period 2006–2016. This decline was likely more associated with a decrease in habitat quality than with overall habitat loss inside SPAs, while the expansion of irrigated and permanent crops resulted in a dramatic loss of habitat outside SPAs (Silva et al. 2018). Thus, general measures may no longer be enough to halt the decline of the species in Portugal, if no specific measures are taken to control the expansion of olive and almond crops. In the Évora region, the area converted to irrigation inside the study area was around 15% (with no significant changes in land uses in the remaining 85%), while the area converted to irrigation around the study area was 40–50%, resulting in an overall little bustard population decline of 80–90% (own unpublished data). In permanent grasslands, changes in livestock management, with increased stocking rates and the replacement of sheep by cattle, seem to have contributed to the population decline (Silva et al. 2018; Marques et al., unpublished). Such generalized habitat degradation outside Portuguese SPAs and nearby areas may have favoured longer post-breeding migration movements, which may expose little bustards to a greater mortality risk (Silva et al. 2018).

Promotion and Conservation of Fallows, Set-Asides and Appropriate Pasture Management

Existing evidence clearly suggests that the conservation of the little bustard and other steppe birds requires management measures aimed at promoting agricultural extensive practices and particularly fallows and other fields with spontaneous herbaceous cover and high floristic diversity, like set-asides and extensively grazed pastures (Giralt et al. 2018). Fallow land may be defined as cultivated land not seeded for one or several growing seasons (Traba and Morales 2019). Thus, fallows include different semi-natural grasslands and pastures that will eventually be ploughed for a new crop cycle, and with temporal or occasional extensive grazing (Peco et al. 1999). Maintenance and promotion of fallows, set-asides and extensively grazed pastures are one of the most important habitat management measures for the conservation of the little bustard and other steppe birds, given the amount of resources they offer to birds (feeding, refuge, nesting substrate) (Delgado and Moreira 2010; McMahon et al. 2010; Morales et al. 2013; Giralt et al. 2018).

The decrease in the area of fallow land is a major factor explaining the decrease in Spanish little bustard populations (Traba and Morales 2019), while changes in pasture management are likely a main driver of the decline in Portugal (Faria et al. 2016; Silva et al. 2018) and France (Bretagnolle et al. 2011). Therefore, the promotion of fallows, set-asides and adequately grazed pastures should be a priority to reverse this trend. We argue that the upcoming revision of CAP after 2020 should encourage the maintenance of fallows, set-asides and adequately grazed pastures. To promote fallow lands in the European agrarian landscape, Traba and Morales (2019) suggested: (1) requiring farmers in receipt of subsidies to leave fallow a minimum area of their land; (2) returning to the obligation to keep at least 10% of the agrarian land as fallow; and (3) reducing field-level intensification, specifically in the use of agrochemical inputs. This last point is of extreme importance, as not only is the presence of fallows critical but also their appropriate management in order to meet little bustard requirements. This includes the reduction in the application of agrochemicals, which could negatively affect food resources for adults (weeds) and chicks (insects), and the use of extensive grazing during the breeding period. Indeed, grazing may allow the coexistence of both high vegetation cover in some areas, which is suitable for nesting, with a relatively open structure important for anti-predator vigilance, the performance of the males' sexual display (Morales et al. 2008) and the capture of insects when females are feeding their offspring (Bretagnolle et al. 2011).

Other fallow management practices have been shown to benefit little bustards, including shredding and early (February) herbicide application in small doses, as this may maintain an appropriate herbaceous cover (Robleño et al. 2017; Sanz et al. 2019), although herbicide application is controversial and questioned by technicians and scientists, and should be avoided until more evidence is made available.

In terms of livestock management, it is important to implement a total decoupling of livestock, particularly cattle, in the next CAP reform, so that farmers are to be paid

Table 8 Summary of general recommendations related to the management of fallows, set-asides and pastures

Measure/action	Minimum value/ viable threshold	Effects on little bustard	References
Spatial distribution of fallows, set-asides and similar	10% of total agrarian area should be dedicated to fallow	Reducing dependency of landscape-scale processes and negative effects of agrarian matrix	Traba and Morales (2019)
Minimum field size	>2–3 ha	Species' specific habitat requirements	Chapter "Habitat selection and space use" (this volume)
Diversity at landscape scale	Network of fallow plots of 5–10 hectares	Easing connectivity	Lapiedra et al. (2011)
Diversity at field scale	Diversify agricultural treatments in fallows	Increase intra-field heterogeneity	Robleño et al. (2017) and Sanz et al. (2019)
Reducing machinery	Single tillage early in the year, two maximum. No herbicide	Avoid nest losses and direct mortality	Robleño et al. (2017) and Sanz et al. (2019)
Increase fallow age	10% of the fallows >2–4 years	Increase habitat diversity	Giralt et al. (2018)
Grazing fallows extensively	Light to moderate stocking rates (0.2–0.6 LU/ha)	Species' specific habitat requirements	Faria et al. (2012)
Excluding grazing and ploughing in stubbles during winter	Only during winter	Promotion of natural or seeded fallows to avoid high-mortality post-breeding movements	Cuscó et al. (2018) and Giralt et al. (2018)

Based on Giralt et al. (2018)

per unit area with a fixed maximum livestock density rather than per head, which just encourages the maximization of livestock numbers.

In relation to the management of fallows, set-asides and pastures, Giralt et al. (2018) made a series of general recommendations applicable to little bustard and other steppe birds, which should be made compulsory in the post-2020 CAP, at least in Natura 2000 areas, including (see a summary in Table 8):

- Spatial distribution. Due to the dependence of little bustards and other species on processes occurring at the landscape scale (Brotons et al. 2004; Traba et al. 2013; Cardador et al. 2015), fallows, set-asides and grazed pastures should be spatially concentrated so that their effects are detectable at a large scale. The minimum total agrarian area dedicated to fallow should be 10% per individual property (Traba and Morales 2019), as was compulsory under CAP until 2008.
- Minimum field size. Each particular fallow field should be above 2–3 ha, to be able to meet the habitat requirements of both males and females (see chapter

"Habitat selection and space use"), avoiding the use of perimeter fences when the field size is below 4–5 ha.

- Diversity at landscape scale. Lapiedra et al. (2011) proposed the creation of a network of fallow plots of 5–10 ha, separated by a maximum of 500 m from each other and connected by grassy margins with herbaceous cover. The promotion of leguminous dry crops, with prescribed harvesting/mowing dates, as well as the maintenance of medium and long-term fallows and set-asides (Traba and Morales 2019), is also advised. However, this is dependent on geographical context, as in several parts of Portugal the landscape is already dominated by extensive grasslands. In these areas, management should focus on maintaining habitat quality through extensive field management measures.
- Diversity at field scale. In order to increase heterogeneity within fields it is recommended to diversify agricultural treatments in fallows, or apply the same treatment but at different times (Robleño et al. 2017; Sanz et al. 2019).
- Reducing machinery. Nest losses and direct mortality due to machinery during breeding are two of the main threats to little bustard and other species (Faria et al. 2016; Bretagnolle et al. 2018; Ponce et al. 2018; chapter "Threats affecting little bustards: human impacts"). Therefore, tillage, application of herbicides and/or fertilizers, or reaping should be avoided during the breeding season.

  To avoid plant overcover in fallows, a single tillage early in the year should be applied, two maximum (Sanz et al. 2019), and only when light to moderate grazing is unavailable (see below). Tillage may also benefit invertebrates, which can be crucial both for feeding fledglings (and adults) and for pest control in the following year's crop. Repeated tilling should be avoided (Robleño et al. 2017; Sanz et al. 2019). Herbicide application, even though it may generate plant structure somehow suitable for the species (Robleño et al. 2017), is not recommended. Little bustard males eat mainly plant material (leaves, flowers, stems) (Jiguet 2002; Bravo et al. 2017), so the risk of herbicide poisoning may be expected.

- Fallow age. Old (over 1 year) fallows may offer a greater diversity of habitats, plants and invertebrates, offering greater food availability for little bustards (Traba et al. 2008; Tarjuelo et al. 2019). At least 10% of the fallow area should be maintained for extended periods of 2–4 years (Giralt et al. 2018).
- Grazing fallows extensively. In areas where long-lasting fallows, set-asides and pastures exist and grazing is common, it must be low to moderate (Silva et al. 2004; Faria and Morales 2017). For example, Faria et al. (2012), in a Portuguese area with cattle grazing (a livestock management model increasing in the southwest of the Iberian Peninsula), recommended that grazing management aiming to conserve little bustard populations should consider: (1) the maintenance of larger long-term pastures; but (2) using light to moderate stocking rates (0.2–0.6 LU/ha). In Navarra (Spain), Astrain and Etxeberria (1997) recommended 0.5 sheep grazing units per ha.
- Excluding grazing and ploughing in stubbles during winter. Several studies have indicated the importance of stubbles in the post-breeding and winter periods

(Faria and Silva 2010; Silva et al. 2007). The lack of fallows after haying and cereal harvesting is one of the causes (along with temperature) that forces little bustards to undertake post-breeding movements (García de la Morena et al. 2015), thus incurring higher mortality risks such as collision with powerlines (Silva et al. 2007; Marcelino et al. 2017; chapter "Threats affecting little bustards: human impacts"). In wintering grounds, the promotion of natural or seeded fallows is also recommended. In these areas, alfalfa under irrigation was a crop frequently used by the species during winter, at least in central Spain and Catalonia (García de la Morena et al. 2015; Cuscó et al. 2018), and alfalfa is an important component of the species' diet during winter (Bravo et al. 2017). In recent years, alfalfa fields have decreased in the landscape, probably due to CAP recommendations and frequently replaced by irrigated woody crops, and these changes may have had negative effects on the winter survival of the species (see chapter "Migration, movements and non-breeding ecology").

Concluding Remarks

In this chapter, we have reviewed the main conservation actions taken during the last 30 years for the protection of the little bustard, with a special focus on Portugal, France and Spain, the stronghold (>99%, see chapter "The little bustard around the world: distribution, global conservation status, threats and population trends") of the species in Western Europe. Despite the time elapsed since the first draft of the European Action Plan (De Juana 2001) and the scientific knowledge acquired, the situation of the species not only has not improved but has clearly worsened. Despite its critical status in several countries, the little bustard is still listed as Near Threatened and Vulnerable under the Global and European IUCN Red Lists, respectively, and Vulnerable in Spain, requiring an urgent reclassification at all the three levels that realistically reflects the dramatic decline of the species. Greater efforts in census and field studies are needed to maintain a constant monitoring of population trends, and to address female status and productivity.

During the last 20 years, the European Union has declared a relatively high number of Special Protected Areas designated specifically for the species. However, the conservation status of the little bustard is not greatly helped by such intervention, since many populations are located in farmland outside protected sites. Protection of the little bustard by the Natura 2000 network is clearly inadequate, and this is particularly well exemplified in Spain, which holds the main European populations yet where only about 30% of little bustard numbers are within Natura 2000 sites. Furthermore, the populations within SPAs appear not to be feasibly maintained either. There is a wide consensus between both technical officers of regional governments and scientists that SPA cover is insufficient to guarantee the conservation of the species in Spain (Bota et al. 2020). Portugal is in a similar situation (Silva et al. 2018). Time is running out in some regions of the species' western range, and efforts are needed to designate new SPAs, focusing on both breeding and wintering grounds. An important part of the population is migratory (García de la Morena et al. 2015), spending most of the year outside the breeding grounds, so identification and protection of non-breeding areas is critical for the long-term maintenance of the species.

The LIFE programme has dedicated general and specific projects to improving the conservation status of the little bustard. However, perhaps with the exception of France, the situation has not notably changed, probably due to the need for conservation measures in wide areas outside of protected sites, and beyond the time-frame of specific conservation projects. Positive results have been achieved only locally, where targeted funding has been provided in combination with specialized expert support (Barov and Derhé 2011). In most part of the western range, efforts and money invested have not prevented the degradation of the agrarian matrix nor the decline of populations. Notably, CAP continues to be the major driver of agricultural intensification leading to habitat loss and degradation that threaten little bustards (see chapter "Threats affecting little bustards: human impacts"). The application of AESs has not reversed the dramatic decline of the species, owing to difficulties associated with the specific design of measures (Moreno et al. 2010) and irregular implementation (Llusia and Oñate 2005; Oñate et al. 2007).

There is an urgent need for an agricultural policy that takes into account the needs of biodiversity conservation such as fallow maintenance and effective schemes that promote appropriate habitat conservation and management. This entails the maintenance of fields with permanent plant cover on arable land and reducing the use of agrochemicals, achieved by means of well-designed financial and policy instruments. Berthet et al. (2012) and Traba and Morales (2019) have suggested that leaving a minimum of 10% cover of fallow land, as was compulsory under CAP until 2008, could represent the threshold to maintain viable little bustard populations. To achieve this, landowners, farmers and agricultural administrations need to be involved. The non-linear nature of the effects of intensification on biodiversity may require the implementation of more flexible measures than current AESs (Concepción and Díaz 2013). The view that there are win–win strategies for biodiversity conservation as well as for farming profitability if AESs are adequately applied (Oñate et al. 2007; Tarjuelo et al. 2020) should be promulgated among farmers. Their engagement should be pursued on a scale well beyond current levels. Given that it is voluntary, efforts should be made to make farmers prone to uptake AES.

Some views of nature conservation and management advocate for a return to wilderness, which would imply the abandonment of farming in Europe and other regions whose landscapes have been shaped by centuries-long agricultural activity. Wilderness should be understood as ecosystems that develop in the absence of human management (Ceaușu et al. 2015). However, abandoned farmlands in Europe often evolve into structurally complex forested ecosystems (Rey-Benayas and Bullock 2015), which are avoided by the little bustard and other steppe birds. In regions where steppes or natural grasslands are in equilibrium with prevailing geological and climate factors forests do not develop and their rewildering might be an option for these species. A complementary and ongoing debate about the future of farming is the land sharing—land sparing dichotomy. While land sharing can be interpreted as a "wildlife-friendly farming", promoting the improvement of the farmed environment as discussed in this chapter, land sparing advocates for a separation of land designated for farming from that destined for conservation (Phalan et al. 2011). This may confine steppe bird populations to a few designated areas, jeopardizing the natural dynamics of populations and metapopulations. However, the sparing from intensive land use of a large network of farmland areas managed in a wildlife-friendly manner so that biodiversity and associated ecological functions are maintained (a mixed model between land sharing and sparing), could offer a chance for the conservation of these species.

Finally, it is critical to consider the effects of climate change in any agricultural policy aimed at conservation. Due to dispersal constraints and site fidelity, current geographical distribution likely restricts future changes in little bustard (and other steppe birds') distribution tracking climate changes (Estrada et al. 2016). Such constraints should be considered in any long-term conservation policy and local climate mitigation measures implemented.

Of course, other aspects need also to be addressed, such as working with landowners, farmers and other stakeholders to reduce hunting pressure in some countries (Yousefi et al. 2017), as well as other causes of non-natural mortality such as collision with powerlines or poaching (see chapter "Threats affecting little bustards: human impacts"). The current and future development of energy infrastructures (solar and wind farms) should be closely monitored. The conservation of the little bustard should be a priority in coming years at European, national and regional levels, with clear social and science-based strategies coordinated among the different key actors at different scales. In synthesis, we encourage a strategy framed under the agroecological transition concept, in which a network of sufficiently large land expanses (important for farmland birds) should be managed with the joint efforts of main stakeholders, citizens and many farmers. Those areas would function as population sources for the little bustard and other species, sustaining metapopulation and thus species' persistence at a large scale.

Scientific research deserves a final mention. The little bustard is now a well-studied species, and much specific research has been published in the last 20 years as reviewed in this volume. Around 47 studies were carried out in the Iberian Peninsula (Spain and Portugal) and another 17 in France in recent years (Morales and Traba 2016; see chapter "Why a monograph on little bustard?"). Most of these studies involved the estimation of population size, analyses of habitat selection, efficiency of habitat management measures and AES associated with the Common Agricultural Policy (CAP), and general biology and ecology (Morales and Traba 2016). As a consequence, our knowledge of the species and its requirements has strongly improved and has contributed to defining and refining conservation measures. However, the translation of this knowledge into practical conservation is still scarce, although ongoing experiences in some regions provide hope in this respect.

References

Astrain C, Etxeberría A (1997) Plan de conservación del Sisón (*Tetrax tetrax*) en Navarra. Unpublished report. Departamento de Medio Ambiente, Ordenación Territorial y Vivienda del Gobierno de Navarra

Astrain C, Etxeberría A (1998) Plan de Recuperación del Sisón común (*Tetrax tetrax*) en La Rioja. Unpublished report. Consejería de Medio Ambiente del Gobierno de La Rioja

Ausden M (2007) Habitat management for conservation: a handbook of techniques, Techniques in ecology & conservation series. Oxford University Press, Oxford

Bamière L, Havlík P, Jacquet F, Lherm M, Millet G, Bretagnolle V (2011) Farming system modelling for agri-environmental policy design: the case of a spatially non-aggregated allocation of conservation measures. Ecol Econ 70(5):891–899

Barov B, Derhé M (2011) Review of the implementation of species action plans of threatened birds in the European Union (2004–2010). Birdlife International

Berthet ETA, Bretagnolle V, Segrestin B (2012) Analyzing the design process of farming practices ensuring little bustard conservation: lessons for collective landscape management. J Sustain Agric 36:319–336

BirdLife International (2015) *Tetrax tetrax*. The IUCN red list of threatened species 2015: e. T22691896A60011308. Downloaded on 28 May 2020

Bota G, Canessa S, García de la Morena EL, Mañosa S, Morales MB, Traba J (2020) Sisón común: punto crítico en la conservación de la especie. Quercus 411:46–47

Bravo C, Cuscó F, Morales MB, Mañosa S (2017) Diet composition of a declining steppe bird the Little Bustard Tetrax tetrax in relation to farming practices. Avian Conserv Ecol 12(1)

Bretagnolle V, Villers A, Denonfoux L, Cornulier T, Inchausti P, Badenhausser I (2011) Rapid recovery of a depleted population of Little Bustards *Tetrax tetrax* following provision of alfalfa through an agri-environment scheme. Ibis 153(1):4–13

Bretagnolle V, Denonfoux L, Villers A (2018) Are farming and birds irreconciliable? A 21–year study of bustard nesting ecology in intensive agrosystems. Biol Conserv 228:27–35

Brotons L, Mañosa S, Estrada J (2004) Modelling the effects of irrigation schemes on the distribution of steppe birds in Mediterranean farmland. Biodivers Conserv 13:1039–1058

Cardador L, De Cáceres M, Giralt D, Bota G, Aquilué N, Arroyo B, Mougeot F, Cantero-Martínez C, Viladomiu L, Rosell J, Casas F, Estrada A, Álvaro-Fuentes J, Brotons L (2015) Tools for exploring habitat suitability for steppe birds under land use change scenarios. Agric Ecosyst Environ 200:119–125

Cardoso AC, Cruz CM, Claro JC, Silva JP, Almeida J, Pinto M, Rocha P (2007) Plano de ação para a conservação das aves dependentes da estepe cerealífera. ICN

Caro G, Marrec R, Gauffre B, Roncoroni M, Augiron S, Bretagnolle V (2016) Multi-scale effects of agri-environment schemes on carabid beetles in intensive farmland. Agric Ecosyst Environ 229: 48–56

Ceausu S, Carver S, Verburg PH, Kuechly HU, Hölker F, Brotons L, Pereira HM (2015) European wilderness in a time of farmland abandonment. In: Pereira HM, Navarro LM (eds) Rewilding European landscapes. Springer, Cham, pp 25–46

Collar N, García EFJ, de Juana E (2016) Little Bustard (Tetrax tetrax). In: Handbook of the birds of the world alive. Lynx Editions, Barcelona

Collar NJ, Baral HS, Batbayar N, Bhardwaj GS, Brahma N, Burnside RJ, Choudhury AU, Combreau O, Dolman PM, Donald PF, Dutta S, Gadhavi D, Gore K, Goroshko OA, Hong C, Jathar GA, Jha RRS, Jhala YV, Koshkin MA, Lahkar BP, Liu G, Mahood SP, Morales MB, Narwade SS, Natsagdorj T, Nefedov AA, Silva JP, Thakuri JJ, Wang M, Zhang Y, Kessler AE (2017) Averting the extinction of bustards in Asia. Forktail 33:1–26

Concepción ED, Díaz M (2013) Medidas agroambientales y conservación de la biodiversidad: Limitaciones y perspectivas de futuro. Ecosistemas 22:44–49

Concepción ED, Díaz M, Baquero RA (2008) Effects of landscape complexity on the ecological effectiveness of agri-environment schemes. Landsc Ecol 23:135–148

Concepción ED, Mario D, David K, András B, Péter B, Yann C, Doreen G, Felix H, Andrea H, Eva K, Marshall EJP, Teja T, Jort V (2012) Interactive effects of landscape context constrain the effectiveness of local agri-environmental management. J Appl Ecol 49(3):695–705. https://doi.org/10.1111/j.1365-2664.2012.02131.x

Concepción ED, Aneva I, Jay M, Lukanov S, Marsden K, Moreno G, Oppermann R, Pardo A, Piskol S, Rolo V, Schraml A, Díaz M (2020) Optimizing biodiversity gain of European agriculture through regional targeting and adaptive management of conservation tools. Biol Conserv 141:108384

Cuscó F, Cardador L, Bota G, Morales MB, Mañosa S (2018) Inter-individual consistency in habitat selection patterns and spatial range constraints of female little bustards during the non-breeding season. BMC Ecol 18:56

De Juana E (2001) European species action plan. Little Bustard *Tetrax tetrax*: draft action plan. BirdLife International, Cambridge

Delgado A, Moreira F (2010) Between-year variations in Little Bustard Tetrax tetrax population densities are influenced by agricultural intensification and rainfall. Ibis 152(3):633–642. https://doi.org/10.1111/j.1474-919X.2010.01026.x

EEA (2019) Natura 2000 data—the European network of protected sites. https://www.eea.europa.eu/data-and-maps/data/natura-10. Last consultation: 2020/04/03

Estrada A, Delgado MP, Arroyo B, Traba J, Morales MB (2016) Forecasting large-scale habitat suitability of European bustards under climate change: the role of environmental and geographic variables. PLoS One 11(13):0149810

Faria N, Morales MB (2017) Population productivity and late breeding habitat selection by the threatened Little Bustard: the importance of grassland management. Bird Conserv Int 28:521–533

Faria N, Silva JP (2010) Habitat selection of the little bustard during the beginning of an agricultural year. Ardeola 57:363–373

Faria N, Rabaça J, Morales MB (2012) The importance of grazing regime in the provision of breeding habitat for grassland birds: The case of the endangered little bustard (*Tetrax tetrax*). J Nat Conserv 20:211–218

Faria N, Morales MB, Rabaça J (2016) Exploring nest destruction and bird mortality in mown Mediterranean dry grasslands: an increasing threat to grassland bird conservation. Eur J Wildl Res 62:663–671

García de la Morena EL, Bota G, Ponjoan A, Morales MB (2006) El sisón común en España. I Censo Nacional (2005). SEO/Birdlife, Madrid

García de la Morena EL, Morales MB, Bota G, Silva JP, Ponjoan A, Suárez F, Mañosa S, De Juana E (2015) Migration patterns of Iberian little bustards *Tetrax tetrax*. Ardeola 62:95–112

García de la Morena EL, Bota G, Mañosa S, Morales MB (2018) El sisón común en España. II Censo Nacional (2016). SEO/Birdlife, Madrid

García de la Morena EL, Bota G, Traba J, Mañosa S, Morales MB (2019) Bases científicas para la elaboración de la Estrategia Nacional de Conservación del sisón común (Tetrax tetrax). Informe técnico, Fundación Biodiversidad

Gauger K (2007) Occurrence, ecology and conservation of wintering Little Bustard *Tetrax tetrax* in Azerbaijan. Arch Nat Landsc 46(2):5–27

Gendre N, Éraud C, Bretagnolle V, Dalloyau S (2018) L'Outarde canepetière *Tetrax tetrax* en France en 2012 et 2016: effectifs et répartition. Ornithos 25-5:290–302

Giralt D, Robleño I, Estrada J, Mañosa S, Morales MB, Sardà-Palomera F, Traba J, Bota G (2018) Manual de gestión de barbechos para la conservación de aves esteparias. Fundación Biodiversidad – Centre de Ciència i Tecnologia Forestal de Cayalunya

Goriup P (1994) Little Bustard Tetrax tetrax. In: Tucker GM, Heath MF (eds) Birds in Europe: their conservation status. BirdLife International, Cambridge, pp 236–237

Henckel L, Borger L, Meiss H, Gaba S, Bretagnolle V (2015) Organic fields sustain weed metacommunity dynamics in farmland landscapes. Proc Biol Sci 282:20150002. https://doi.org/10.1098/rspb.2015.0002

Inchausti P, Bretagnolle V (2005) Predicting short-term extinction risk for the declining Little Bustard (*Tetrax tetrax*) in intensive agricultural habitats. Biol Conserv 122(3):375–384
Iñigo A, Barov B (2010) Action plan for the Little Bustard *Tetrax tetrax* in the European Union. SEO/BirdLife and BirdLife International for the European Commission
IUCN (2012) IUCN Red List categories and criteria: version 3.1, 2nd edn. IUCN, Gland
Jiguet F (2002) Arthropods in diet of Little Bustards Tetrax tetrax during the breeding season in Western France. Bird Study 49:105–109
Jolivet C, Bretagnolle V (2002) L'outarde canepetière en France: évolution, récente des populations, bilan de mesures de sauvegarde et perspectives d'avenir. Alauda 70:93–96
Lapiedra O, Anna P, Anna G, Gerard B, Santi M (2011) Brood ranging behaviour and breeding success of the threatened little bustard in an intensified cereal farmland area. Biol Conserv 144(12):2882–2890. https://doi.org/10.1016/j.biocon.2011.08.005
Llusia D, Oñate JJ (2005) Are the conservation requirements of pseudo-steppe birds adequately covered by Spanish agri-environmental schemes? An ex-ante assessment. Ardeola 52:31–42
Mañosa S, Bota GC, Bonell JE, Martínez FC (2015) Una oportunidad para el Sisón en Cataluña: La población se localiza en la plana de Lleida. Quercus (356):24–35
Mañosa S, Bota G, Giralt D, Estrada J (2020) The changing status of steppe–land birds in the Lleida Plain of Catalonia. In: Hurford C, Wilson PJ, Storkey J (eds) The changing status of arable habitats in Europe: a nature conservation review. Springer, Dordrecht
Marcelino J, Moreira F, Mañosa S, Cuscó F, Morales MB, García de la Morena E, Bota G, Palmeirim JM, Silva JP (2017) Tracking data of the Little Bustard Tetrax tetrax in Iberia shows high anthropogenic mortality. Bird Conserv Int:1–12
McMahon BJ, Giralt D, Raurell M, Brotons L, Bota G (2010) Identifying set-aside features for bird conservation and management in Northeast Iberian pseudo-steppes. Bird Study 57:289–300
Morales MB, Bretagnolle V (2021) An update on the conservation status of the Little Bustard (Tetrax tetrax): global and local population estimates, trends and threats. Bird Conserv Int 00: 000–000
Morales MB, Traba J (2016) Prioritizing research in steppe bird conservation: a literature survey. Ardeola 63(1):5–18
Morales MB, Bretagnolle V, Arroyo B (2005) Viability of the endangered Little bustard *Tetrax tetrax* population of western France. Biodivers Conserv 14:3135–3150
Morales MB, Traba J, Carriles E, Delgado MP, García de la Morena EL (2008) Sexual differences in microhabitat selection of breeding Little Bustards *Tetrax tetrax*: ecological segregation based on vegetation structure. Acta Oecol 34:345–353
Morales MB, Traba J, Delgado MP, Garcia de la Morena EL (2013) The use of fallows by nesting Little Bustard Tetrax tetrax females: implications for conservation in mosaic cereal farmland. Ardeola 60(1):85–97
Morales MB, Traba J, Arroyo B (2015) El declive del sisón en el centro de España. Quercus 356: 36–43
Moreno V, Morales MB, Traba J (2010) Avoiding over-implementation of agri-environmental schemes for steppe bird conservation: a species-focused proposal based on expert criteria. J Environ Manag 91:1802–1809
Oñate JJ, Atance I, Bardají I, Llusia D (2007) Modelling the effects of alternative CAP policies for the Spanish high-nature value cereal-steppe farming systems. Agric Syst 94:247–260
Osborne PE, Suárez-Seoane S (2007) Identifying core areas in a species' range using temporal suitability analysis: an example using Little Bustards Tetrax tetrax L. in Spain. Biodivers Conserv 16(12):3505–3518
Pain DJ, Dixon J (1997) Why farming and birds in Europe? In: Pain D, Pienkowski MW (eds) Farming and birds in Europe: the common agricultural policy and its implications for bird conservation. Academic Press, London
Pe'er G et al (2019) A greener path for the EU Common Agricultural Policy. Science 365:49–451
Peco B, Malo JE, Oñate JJ, Suárez F, Sumpsi J (1999) Agri-environmental indicators for extensive land-use systems in the Iberian Peninsula. In: Brouwer F, Crabtree R (eds) Agriculture and

environment in Europe: The role of indicators in agricultural policy development. CAB International, Wallingford, pp 137–156
Ponce C, Salgado I, Bravo C, Gutiérrez N, Alonso JC (2018) Effects of farming practices on nesting success of steppe birds in dry cereal farmland. Eur J Wildl Res 64(2):13
Phalan B, Onial M, Balmford A, Green RE (2011) Reconciling food production and biodiversity conservation: Land sharing and land sparing compared. Science 333:1289–1291
Rey Benayas JM, Bullock JM (2015) Vegetation restoration and other actions to enhance wildlife in European agricultural landscapes. In: Pereira HM, Navarro LM (eds) Rewilding European landscapes. Springer, Cham, pp 127–142
Ribeiro PF, Santos JL, Bugalho MN, Santana J, Reino L, Beja P, Moreira F (2014) Modelling farming system dynamics in high nature value farmland under policy change. Agric Ecosyst Environ 183:138–144
Robleño I, Bota G, Giralt D, Recasens J (2017) Fallow management for steppe bird conservation: the impact of cultural practices on vegetation structure and food resources. Biodivers Conserv 26:133–150
Santana J, Reino L, Stoate C, Borralho R, Rio Carvalho C, Schindler S, Moreira F, Bugalho MN, Ribeiro PF, Santos JL, Vaz A, Morgado R, Porto M, Beja P (2013) Mixed effects of long-term conservation investment in Natura 2000 farmland. Conserv Lett 7:467–477
Santana J, Reino L, Stoate C, Borralho R, Carvalho CR, Schindler S, Moreira F, Bugalho MN, Ribeiro PF, Santos JL, Vaz A, Morgado R, Porto M, Beja P (2014) Mixed effects of long-term conservation investment in Natura 2000 Farmland. Conserv Lett 7(5):467–477. https://doi.org/10.1111/conl.12077
Santangeli A, Dolman P (2011) Density and habitat preferences of male little bustard across contrasting agro-pastoral landscapes in Sardinia (Italy). Eur J Wildl Res 57:805–815
Sanz-Perez A, Giralt D, Robleño I, Bota G, Milleret C, Mañosa S, Sardà-Palomera F (2019) Fallow management increases habitat suitability for endangered steppe bird species through changes in vegetation structure. J Appl Ecol 56:2166–2175
Sehhatisabet ME, Abdi F, Ashoori A, Khaleghizadeh A, Khani A, Rabiei K, Shakiba M (2012) Preliminary assessment of distribution and population size of wintering Little Bustards Tetrax tetrax in Iran. Bird Conserv Int 22(3):279–287. https://doi.org/10.1017/S0959270911000281
SEO/BirdLife (2018) Programas de seguimiento y grupos de trabajo de SEO/BirdLife 2017. SEP/Birdlife
Silva JP, Pinto M, Palmeirim JM (2004) Managing landscapes for the little bustard *Tetrax tetrax*: lessons from the study of winter habitat selection. Biol Conserv 117:521–528
Silva JP, Faria N, Catry T (2007) Summer habitat selection and abundance of the threatened little bustard in Iberian agricultural landscapes. Biol Conserv 139:186–194
Silva JP, Correia R, Alonso H, Martins RC, D'Amico M, Delgado A, Sampaio H, Godinho C, Moreira F (2018) EU protected area network did not prevent a country wide population decline in a threatened grassland bird. PeerJ 6:e4284. https://doi.org/10.7717/peerj.4284
Suárez F, Naveso MA, De Juana E (1997) Farming in the drylands of Spain: birds of the pseudosteppes. In: Pain DJ, Pienkowski MW (eds) Farming and birds in Europe. The common agricultural policy and its implication for bird conservation. Academic Press, London, pp 297–330
Suárez-Seoane S, Osborne PE, Alonso JC (2002) Large-scale habitat selection by agricultural steppe birds in Spain: identifying species–habitat responses using generalized additive models. J Appl Ecol 39:755–771
Tarjuelo R, Morales MB, Traba J, Delgado MP (2014) Are species coexistence areas a good option for conservation management? Applications from fine scale modeling in two steppe birds. PLoS One 9:e87847
Tarjuelo R, Morales MB, Arribas J, Traba J (2019) Abundance of weeds and seeds but not of arthropods differ between arable habitats in an extensive Mediterranean farming system. Ecol Res 34:624–636

Tarjuelo R, Margalida A, Mougeot F (2020) Changing the fallow paradigm: a win-win strategy for the post-2020 common agricultural policy to halt farmland bird declines. J Appl Ecol. https://doi.org/10.1111/1365-2664.13570

Traba J, Morales MB (2019) The decline of farmland birds in Spain is strongly associated to the loss of fallow-land. Sci Rep 9:9473

Traba J, García de la Morena EL, Morales MB, Suárez F (2007) Determining high value areas for steppe birds in Spain. Hot spots, complementarity and the efficiency of protected areas. Biodivers Conserv 16:3255–3275

Traba J, Morales MB, García de la Morena EL, Delgado MP, Krištín A (2008) Selection of breeding territory by little bustard (*Tetrax tetrax*) males in Central Spain: the role of arthropod availability. Ecol Res 23:615–622

Traba J, Acebes P, Malo JE, Garcia JT, Carriles E, Radi M, Znari M (2013) Habitat selection and partitioning of black-bellied sandgrouse (Pterocles orientalis), stone curlew (Burhinus oedicnemus) and cream-coloured courser (Cursorius cursor) in arid areas of North Africa. J Arid Environ 94:10–17

Tscharntke T, Klein AM, Kruess A, Steffan-Dewenter I, Thies C (2005) Landscape perspectives on agricultural intensification and biodiversity—ecosystem service management. Ecol Lett 8:857–874

Villers A, Millon A, Jiguet F, Lett JM, Attie C, Morales MB, Bretagnlee V (2010) Migration of wild and captive-bred Little Bustards *Tetrax tetrax*: releasing birds from Spain threatens attempts to conserve declining French populations. Ibis 152:254–261

Wolff A, Paul JP, Martin JL, Bretagnolle V (2001) The benefits of extensive agriculture to birds: the case of the little bustard. J Appl Ecol 38:963–975

Wolff A, Dieuleveut T, Martin JL, Bretagnolle V (2002) Landscape context and little bustard abundance in a fragmented steppe: implications for reserve management in mosaic landscapes. Biol Conserv 107:211–220

Yousefi M, Kafash A, Malakoutikhah S, Ashoori A, Khani A, Mehdizade Y, Ataei F, Ilanloo SS, Rezaei HR, Silva JP (2017) Distance to international border shapes the distribution pattern of the growing Little Bustard *Tetrax tetrax* winter population in Northern Iran. Bird Conserv Int 28: 499–508

Conclusions and Perspectives

Manuel B. Morales, Juan Traba, and Vincent Bretagnolle

Compiling and synthesizing all relevant evidence available on a species' ecology is a formidable challenge. Honestly, we do not know if we have adequately accomplished the task. Nevertheless, we have tried and, with the same honesty, believe that the result (i.e., the present monograph) may be a solid ground for those that today and in future years might continue studying and conserving the little bustard.

The Little Bustard, Not a Stranger Anymore

As highlighted by this book, there is a huge amount of information on little bustard ecology, mainly information acquired over the last two decades. For example, no nest description was published from France until 1997, and yet Bretagnolle et al. (2018) reviewed the nesting ecology of the species at a single French study site analyzing 157 nests! This formidable increase in knowledge of the species is mainly attributable to its threatened conservation status, which has motivated research and conservation efforts from both scientists and NGOs in particular over its western distribution range. The information presented in this book spans from, among other issues, phylogenetic relationships, mating systems, parental care, habitat and niche,

M. B. Morales (✉) · J. Traba
Terrestrial Ecology Group, Departamento de Ecología, Universidad Autónoma de Madrid (TEG-UAM), Madrid, Spain

Centro de Investigación en Biodiversidad y Cambio Global (CIBC-UAM), Universidad Autónoma de Madrid, Madrid, Spain
e-mail: manuel.morales@uam.es; juan.traba@uam.es

V. Bretagnolle
Centre d'Etudes Biologiques de Chizé, CNRS & La Rochelle Université, Beauvoir sur Niort, France
e-mail: vincent.bretagnolle@cebc.cnrs.fr

V. Bretagnolle et al. (eds.), *Little Bustard: Ecology and Conservation*, Wildlife Research Monographs 5, https://doi.org/10.1007/978-3-030-84902-3_14

movements, population dynamics, and biotic interactions, to responses to human land use and management, conservation, climate change, and conditions of captive breeding. This has entailed a tremendous advance in scientific knowledge about the little bustard in particular, but also beyond it, contributing to general aspects of ecological theory such as sexual selection and mating systems, species coexistence, or the role of conspecific attraction in habitat selection and species distribution. Such wealth of information required the major synthesis attempted in this volume.

A Unique Family, a Diverse Species

A second lesson arising from the book is the distinctiveness of bustards among birds and, to a lesser extent, of the little bustard among other bustard species. The bustard family has long been known as a strikingly different group. Indeed, its taxonomic affiliations have been long argued, and as described in chapter "The little bustard and its family: an overview of relationships", the latest proposal, arising from DNA analyses of all avian families, defies more intuitive classifications largely based on morphological traits, by grouping bustards with turacos and cuckoos rather than cranes. However, science is often counter-intuitive and the spectacular advance of phylogenomics will surely test the consistency of these relationships. Beyond phylogeny, chapter "Natural history of the little bustard: morphology, biometry, diet, sexual dimorphism, and social and breeding behaviour" highlights many morphological and behavioural traits of the little bustard that are common among the members of its family, but rare or exceptional in other avian groups. These traits make the little and other bustards fantastic (although challenging) models to test different evolutionary hypotheses. Chapter "Natural history of the little bustard: morphology, biometry, diet, sexual dimorphism, and social and breeding behaviour" also highlights biometric variation between different geographical areas within the species' distribution range which, rather than rehabilitate the abandoned taxonomic split between subspecies, may at least substantiate the recognition of two evolutionary—hence conservation—units. This morphological variation indeed calls for further work with genetic evidence and phylogeographic analyses, if evolutionary significant population units (ESUs, Moritz 1994) are to be identified and conserved.

Diverging Fates Between Increasing Eastern and Collapsing Western Populations

As reviewed in chapter "The little bustard around the world: distribution, global conservation status, threats and population trends", the fates of the western and the eastern populations are very different. This chapter's main message is that, if the strong decline of western populations continues in the coming years, the future of the species will rely on the eastern range, whose populations appear to have grown in the last 30 years, mainly due to the abandonment of intensive agriculture after the fall of the Soviet Union. In fact, the eastern range, comprising the populations found from the Volga basin in European Russia through Kazakhstan to eastern China, has become the stronghold of the species at world level. This raises at least two conservation issues. The first is the possible future extinction of the western range ESUs. Regional extinctions have already occurred in recent years in northern Africa and continental Italy, and we have little or no idea of their ecological, genetic, and evolutionary singularities. A clear example of potential ESU extinction that can be hopefully prevented is that of little bustards from western France. This is the only and last remaining fully and long-distance migrant population of the western range and, as such, it can be considered as evolutionarily distinct from others in that range. The second issue is the risk of agricultural re-intensification in the eastern range, particularly in southern Russia and Kazakhstan, which may hamper or even reverse the current recovery of the species in that region.

Population trends ultimately reflect the behaviour of demographic parameters. Chapter "Breeding biology and demographic traits" shows that, although we still lack quantitative information on some demographic parameters of little bustard populations (e.g. survival and natal dispersal) and how they respond to environmental change, important progress has been made in a few intensively studied populations of Western Europe. They reveal low breeding success and male-biased sex ratios, as well as lower female survival than would be expected for a bird of its size and characteristics. The lack of information is evident for populations in the eastern range, where there are virtually no quantitative studies on the species' breeding biology and demography, which would be necessary to prevent the negative population trends observed in the west.

Little Bustard Habitat Selection: Heterogeneity Is the Key

Habitat selection by the little bustard is synthesized at different spatial scales in chapter "Habitat selection and space use". At a large geographical scale, the association of little bustard distribution with temperature-related variables suggests the sensitivity of the species to temperature changes forecasted by future climate models. Moreover, the important role of conspecific attraction and philopatry in the species' spatial distribution, along with its strong dependence on grasslands and extensive farmland, make it vulnerable to climate change. At the landscape scale, chapter "Habitat selection and space use" highlights the role of natural and semi-natural grasslands (long-term fallows, extensively grazed pastures, extensive alfalfa cultures) as a critical habitat for little bustards, and reinforces the role of young fallows and more temporary grasslands as an alternative habitat, especially since they may offer some inter-annual stability. At an even more local scale, spatial variation in vegetation structure seems to drive little bustard microhabitat selection. Little bustards select a particular range of vegetation structure as a result of trade-offs between anti-predator vigilance, food requirements, and sex-related constraints, regardless of landscape composition or type of land use. Vegetation structure variation also seems important in how little bustards share their habitat with other grassland and farmland birds with similar requirements.

One central aspect emerges when analyzing little bustard habitat selection from different perspectives: heterogeneity at every scale. The species depends on a heterogeneous environment that can provide the different and complementary resources required for its vital purposes. At the landscape scale, the mosaic of agricultural habitats that are managed differently, as is typical of extensive farmland, produces such heterogeneity. In more uniform landscapes like those dominated by grassland or natural steppe, heterogeneity is provided at a smaller scale by variation in vegetation structure, often generated by low or moderate grazing. The loss of such heterogeneity through agricultural intensification is identified throughout this volume as the prime cause of the steep population declines experienced by the little bustard in many parts of its distribution range.

Beyond Reproduction

The yearly non-breeding period for the little bustard comprises at least 8 months, from August to March. It is therefore unsurprising that the fate of populations relies strongly on what happens in these two-thirds of the year. Migration is a crucial process occurring in that period. From chapter "Migration, movements and non-breeding ecology", we learn that little bustard seasonal movements are more complex than the traditional classification of the species suggests (i.e. migratory versus resident). Besides the fully migratory populations that have been long known to perform classic directional winter migration (i.e. those of western France, Russia, and central Asia), there are others that are also fully migratory, but perform shorter migrations, such as many from northern Spain, or partially migratory, such as those from central Spain, in both cases wintering in the south of the Iberian Peninsula. Moreover, there are other seasonal movement patterns different from winter migration that also entail a directional round trip movement after breeding (summer migration) or a combination of the latter with a winter directional migration from summer sites to wintering sites and a return to the breeding grounds. Lastly, there are resident populations in southern Iberia and other regions like Sardinia that perform short-distance movements, tracking resource availability.

Because little bustards often move they are exposed to threats, e.g. those involving anthropogenic features (powerlines, wind farms). In addition, habitat selection outside the breeding season is not constrained by breeding requirements and thus little bustards can be found in places where the proximity of linear infrastructures, or irrigated farmland, is more likely because such threats are usually reduced or absent in protected breeding sites. In synthesis, threats faced by little bustards outside their breeding areas (powerline collision, poaching, disturbance from human activity) are driven by land use changes and infrastructure development.

Lek Mating Systems: What the Little Bustard Has Taught Us

Habitat use and habitat selection are two aspects of a species' general behavioural ecology. In lek-like mating systems, the use that females make of habitat in display arenas is key in determining the type of mating strategy used in a particular population. Chapter "Behavioural ecology of the little bustard: sexual selection and mating systems" shows that little bustards are highly flexible in regard to mating strategies, comprising basically all possible mating systems in which males do not participate in parental care, and again this flexibility depends mainly on habitat heterogeneity, landscape structure and/or resource dispersion. To a lesser extent, the mating system found in a given population further depends on bird density (e.g. monogamy could be found in little bustard populations with very low density). These factors in turn may be influenced by the intensity of agricultural management. Knowledge of little bustard mating strategies, sexual display, and secondary sexual traits has also contributed to sexual selection theory. This contribution should be particularly acknowledged, given the great methodological, logistic, and even ethical challenges posed by the study of little bustards in the wild. As a consequence, some other aspects, such as multiple-paternity clutches, the level of sperm competition, sibling interactions or natal dispersal remain very poorly documented and deserve future attention.

Predators, Parasites, and Competitors

A largely unexplored aspect of little bustard ecology is the role of biotic interactions in regulating little bustard population dynamics. Chapter "Interspecific relationships" presents the first complete account of this issue, compiling results from both historical and recent studies. The little bustard has been shown to suffer asymmetrical pre-emptive competition for its preferred habitat with the great bustard, with which it coexists in many parts of its distribution. This may have important conservation consequences, since the extent of preferred habitats has markedly declined in recent years (Traba and Morales 2019). Predation seems to be the main natural cause of mortality in the species, with raptors and opportunistic mammals being the most important predators, while the role of parasites and their transmission from released gamebirds deserves deeper research. Positive interactions are ubiquitous in nature and may occur in little bustards as well, but remain to be confirmed. The best example concerns the frequent formation of mixed flocks with the pin-tailed sandgrouse, a possible form of either mutualism or mere commensalism by the smaller species. Greater research effort possibly involving challenging pseudo-experimental design is needed to understand the role of biotic interactions in little bustard population dynamics, which would in turn help improve conservation strategies.

Intensive Care Conservation

Ex situ conservation and population reinforcement programmes are usually put in place when wild populations show symptoms of an extinction vortex. That was probably the case of the little bustard population of Poitou-Charentes, in western France, where barely 300 males remained in the year 2000. This programme, whose general aims and technical details are described in chapter "Captive breeding, handling and care, and the impact of releases on wild populations", made possible, in combination with habitat management measures, to halt this population's decline, avoid its extinction, and contribute to its present stability (although the population does not show any sign of increase). But it also generated a huge amount of information on many aspects of the species' biology, such as incubation, hatchability, chick growth, or disease. Such new knowledge has provided feedback to improve many aspects of the captive breeding programme. A nice example was the strategic shift from a captive stock based on birds from Spain to one based on locally rescued eggs, after verifying that little bustards hatched from Spanish eggs did not exhibit the migration behaviour of local birds and stayed in the region through the winter. This allowed two important conclusions to be reached: the survival of released birds from sedentary populations would be compromised if hard winter conditions appear, and if migration behaviour is genetically controlled, then genetic configuration of the local population would be significantly modified.

Responsive Population Dynamics

Although little bustard populations have decreased precipitously in many areas of the western Palearctic, they can show very positive dynamics when conditions are favourable, growing exponentially and reaching high densities from very low numbers in a relatively few years, as discussed in chapter "Little bustard population dynamics". Another lesson from the species' population dynamics is that many aspects of its ecology are largely governed by density-dependent processes, which means that, even in very large populations, declines in abundance produce steeply negative exponential growth rates. This is a challenge for managers since declines are usually faster than they expect, particularly in their initial stages. The importance of density dependence is also reflected in lekking dynamics, so that small lek size may result in strong breeding depression and lead to increased risk of population extinction via an Allee effect and to the accumulation of individuals in the largest aggregations. Population viability analyses (PVAs) show that the persistence of little bustard populations requires high adult survival (>60%) and a fecundity rate of at least 0.9 fledglings per female and year. These parameters are particularly sensitive to habitat changes and govern the species' population trends. As regards sex ratio, populations tolerate a moderate shortage of males, but no shortage of females, which is consistent with the species' lek mating system. Therefore, conservation efforts should addressed improving adult survival, particularly female survival, and fecundity. PVAs also illustrate the relevance of connectivity in the dynamics of little bustard metapopulations, and highlight the importance of guaranteeing a secure landscape matrix for dispersing birds.

Old and New Threats

Factors that negatively impact population parameters and compromise the viability of the species have been fully reviewed in chapter "Threats affecting little bustards: human impacts". Above all, agricultural management is the main determinant of habitat quality for the little bustard worldwide. In the EU, the Common Agricultural Policy (CAP) continues to be a major driver of agricultural intensification, leading to loss and degradation of little bustard habitat. Implementation of an agricultural policy that takes into account biodiversity conservation is more urgent than ever. In this context, the maintenance of ecologically functional fallows and other grassland habitats along with effective agri-environmental schemes that promote adequate habitat conservation and management are critical if the little bustard and associated farmland biodiversity are to be conserved. In addition, the development in steppe habitats and more generally open farmland landscapes of energy production facilities is likely to increase in the future. Solar farms occupy large areas in open landscapes, generally on poorer and less productive soils (thus less intensified) where breeding bustard populations persist, and are expected to have the greatest direct impact. In any case, threats cannot be assessed in isolation, as many factors operate synchronously and may have cumulative or even synergistic effects on the species. Therefore, impacts go beyond land-use change, since they also bring more frequent disturbances and other risks that reduce survival and may act as drivers of the extinction vortex for small and isolated populations. In a global perspective, a key question is whether the threats identified here have a significant impact on populations in the eastern distribution range, which is now the world's stronghold for the species. The evolution of agriculture in Russia and Kazakhstan will be determinant, and probably already is, for the fate of eastern populations, as discussed also in chapter "The little bustard around the world: distribution, global conservation status, threats and population trends", although hunting and persecution in the winter quarters and during migration have been identified as additional threats to these populations.

What Has Been and Can Be Done for Little Bustards?

In spite of this clear identification of threats and the assessment of their impacts, there has been little progress in the implementation of successful conservation measures in the last 20 years in Europe, and across the whole distribution range, as concluded in chapter “Little bustard and humans: conservation and management”. Moreover, the main threats have persisted or even intensified during this period. Positive results of conservation action reversing negative population trends have only been achieved at particular sites where targeted funding has been provided (e.g. via LIFE projects) in combination with specialized expert support on the ground. A good example of a successful experience combining these two elements comes from Poitou-Charentes, western France, where the joint effect of targeted agri-environmental schemes and a population reinforcement programme (see chapter “Captive breeding, handling and care, and the impact of releases on wild populations”) under two LIFE projects has halted the decline and favoured some level of recovery of the French migratory population.

A relevant gap in the strategies to conserve the little bustard in Europe is the lack of effective integration of conservation measures with agricultural practices through legislation and policy actions. European agricultural policies should incorporate biodiversity-targeted measures that farmers view also as a way to maintain or increase the profitability of their activities and/or the added value of their products. To achieve this, the involvement of farmers and agriculture administration is required. Win-win strategies for biodiversity conservation and for farm profitability when AES are adequately applied (Oñate et al. 2007; Tarjuelo et al. 2020) should be better advertised to farmers. Recent studies have also shown that such win-win strategies exist even in the absence of AES or compensatory payment (Gaba and Bretagnolle 2020; Catarino et al. 2019a, b). Farmers’ engagement should be achieved at a sufficiently large scale (not reached at the moment) to guarantee efficiency at the landscape scale. To this end, other than farmers, other stakeholders involved in the food production chain should be empowered (Gaba and Bretagnolle 2020; Bretagnolle et al. 2019). A strategy in which a network of sufficiently large territories (important for farmland birds) would be best suited to achieve this. In those territories which would function as population sources for the little bustard and other species, the main stakeholders, citizens, and many farmers should join efforts for an agro-ecological transition.

Perspectives

What are the challenges of little bustard research and conservation for the coming years? Despite huge gains in scientific knowledge of the species biology and ecology over the last two decades, a few unanswered questions remain. In addition, conservation of the species will surely require more scientific research in order to be more effective. We list below areas that, in our view, should be considered as a priority.

A first aspect is relevant for both basic knowledge and a global conservation strategy: a complete and accurate phylogeography of the species is currently missing. This should consider genetic markers sampled in the main populations of the western and eastern subranges. Detailed biogeographic and phylogeographic analyses should then help define ESUs in order to best secure the main genetic lineages and therefore genetic diversity within the species. The combination of this genetic information with the particular ecological characteristics of each population will additionally allow more focused and efficient conservation actions across the entire distribution range.

Second, basic biological knowledge is still missing in several areas, such as sensorial perception or physiological trade-offs related to sexual selection and reproductive investment. Some life history traits also remain poorly known, such as adult survival which is critical data for population viability analyses (Morales et al. 2005; Inchausti and Bretagnolle 2005). Adult survival analyses should be now feasible since a large number of birds have been tracked in Western Europe.

Third, trophic ecology of the little bustard should be better understood. Although we already know much about its diet, at least in some sites, we need to know the extent to which agricultural habitats are providing the amount and quality of food resources to guarantee the good physiological condition of individuals, particularly females since they must afford the costs of laying, incubation and chick rearing. There is some evidence that the survival of re-laying females may be limited by an impoverished physiological condition in a context of reduced food supply due to agriculture management. A more precise knowledge of feeding behaviour and food selection in relation to habitat type and agriculture management, in combination with eco-toxicological studies, should help elucidate the role of pesticides and other agricultural chemical inputs on little bustard populations. In a wider context, knowledge on how little bustards partition their food niche with coexisting ecologically similar species like the great bustard, pin-tailed sandgrouse, and partridges would contribute both to the understanding of interspecific interactions and to more species-targeted management measures in areas of sympatry.

Fourth, there are also a few areas of little bustard behavioural ecology that need further investigation, such as sperm competition, multiple-paternity clutches (which could reveal the existence of polyandry, as found in other bustards, see Lesobre et al. 2010), sibling and mother–offspring conflict, and natal dispersal. We acknowledge they all pose significant methodological challenges, but advances in genetic analysis techniques or remote monitoring, along with current work on captive stocks, may in the future provide the chance to address them.

Fifth, in regard to conservation, we need to better appreciate the status and trends of the eastern population. Our review work for this volume demonstrates that there is a non-negligible amount of literature in Russian about population sizes, trends, and distribution of the species in different regions of the eastern range, and that there are researchers in the region interested in the little bustard. The collaboration between western and eastern scientists and conservationists should thus be encouraged from national and international institutions. It is critical that comparable and repeatable surveys and estimation methods are put into practice to adequately assess population trends and distribution. In addition, the study of the species' ecology and behaviour in its natural steppe landscapes would be of highest interest. For instance, we currently do not know which type of mating system occurs in pristine habitats. Similarly, we know very little about migration ecology and the migration routes of eastern birds. Are there migration routes other than the massive flyway crossing the

Caucasus? What is their relative importance? How much do post- and pre-breeding flyways differ? Are all little bustard populations fully migratory throughout the extant eastern range? Are there summer migrants in most arid areas as found in the Iberian Peninsula? If so, summering sites could be important for those populations' survival and thus should be considered in conservation plans.

The greatest challenge of the future, however, has to do with the current collapse of western populations. In a more general perspective, meeting this challenge will have to answer the critical question posed by Bretagnolle et al. (2018): are farming and birds, such as little bustards, irreconcilable? In line with this question, we may further ask whether managing farmland to conserve the little bustard would contribute also to the conservation of other species, and to what extent. Preliminary evidence suggests that most farmland birds relying on extensive grasslands could also benefit from conservation measures for the little bustard, although the species' role as an umbrella needs to be further investigated.

As discussed above, the involvement of farmers is needed in win-win agricultural practices incorporated into CAP prescriptions, so that entire regions important for little bustards and other farmland birds could be managed under an agro-ecological framework. Unfortunately, this is a difficult task requiring social and political engagement, and the time required to achieve this may be too long, particularly in countries like Spain and Portugal where agricultural intensification is still underway. Our synthesis suggests that the Western Europe population and the eastern range are two separate conservation units, given their strongly diverging population trends, ecological differences, and possibly some genetic divergence. This would allow classifying the rapidly vanishing western population in the "Globally Endangered" IUCN category, thus favouring a more coordinated and better funded (and therefore more ambitious) conservation action. In such countries, local and national administrations will probably implement specific national conservation strategies, thereby allowing a more targeted application of favourable agricultural management measures in areas important for the species. It should also involve the enforcement of limitations on the development of rapidly growing renewable energy infrastructures, particularly solar farms. As regards the eastern population, although this conservation unit would probably maintain the current status of near threatened, a more efficient conservation strategy focused on its particular ecology and requirements, would be required.

References

Bretagnolle V, Denonfoux L, Villers A (2018) Are farming and birds irreconcilable? A 21–year study of bustard nesting ecology in intensive agrosystems. Biol Conserv 228:27–35

Bretagnolle V, Benoit M, Bonnefond M, Breton V, Church JM, Gaba S, Gilbert D, Gillet F, Glatron S, Guerbois C, Lamouroux N, Lebouvier M, Maze C, Mouchel JM, Ouin A, Pays O, Piscart C, Ragueneau O, Servain S, Spiegelberger T, Fritz H (2019) Action-orientated research and framework: insights from the French long-term social-ecological research network. Ecol Soc 24(3):10

Catarino R, Gaba S, Bretagnolle V (2019a) Experimental and empirical evidence shows that reducing weed control in winter cereal fields is a viable strategy for farmers. Sci Rep 9:9004
Catarino R, Bretagnolle V, Perrot T, Vialloux F, Gaba S (2019b) Bee pollination outperforms pesticides for oilseed crop production and profitability. Proc R Soc B 286:20191550
Gaba S, Bretagnolle V (2020) Social–ecological experiments to foster agroecological transition. People Nat 2:317–327
Inchausti P, Bretagnolle V (2005) Predicting short-term extinction risk for the declining little bustard (*Tetrax tetrax*) in intensive agricultural habitats. Biol Conserv 122:375–384
Lesobre L, Lacroix F, Le Nuz E, Hingrat Y, Chalah T, Saint Jaime M (2010) Absence of male reproductive skew, along with high frequency of polyandry and conspecific brood parasitism in the lekking Houbara bustard *Chlamydotis undulata undulata*. J Avian Biol 41:117–127
Morales MB, Bretagnolle V, Arroyo B (2005) Viability of the endangered little bustard *Tetrax tetrax* population of western France. Biodivers Conserv 14:3135–3150
Moritz C (1994) Defining 'Evolutionary Significant Units' for conservation. Trends Ecol Evol 9: 373–375
Oñate JJ, Atance I, Bardají I, Llusia D (2007) Modelling the effects of alternative CAP policies for the Spanish high-nature value cereal-steppe farming systems. Agric Syst 94:247–260
Tarjuelo R, Margalida A, Mougeot F (2020) Changing the fallow paradigm: a win-win strategy for the post-2020 common agricultural policy to halt farmland bird declines. J Appl Ecol. https://doi.org/10.1111/1365-2664.13570
Traba J, Morales MB (2019) The decline of farmland birds in Spain is strongly associated to the loss of fallowland. Sci Rep 9:9473

GPSR Compliance

The European Union's (EU) General Product Safety Regulation (GPSR) is a set of rules that requires consumer products to be safe and our obligations to ensure this.

If you have any concerns about our products, you can contact us on ProductSafety@springernature.com

In case Publisher is established outside the EU, the EU authorized representative is:

Springer Nature Customer Service Center GmbH
Europaplatz 3
69115 Heidelberg, Germany

Batch number: 10732044

Printed by Printforce, the Netherlands